Polymer Matrix Wave-Transparent Composites

Polymer Matrix Wave-Transparent Composites

Materials, Properties, and Applications

Junwei Gu
Yusheng Tang
Jie Kong
Jing Dang

WILEY VCH

Authors

Prof. Junwei Gu
Northwestern Polytechnical University
127 West Youyi Road
Xi' an
CH, 710072

Prof. Yusheng Tang
Northwestern Polytechnical University
127 West Youyi Road
Xi'an
CH, 710072

Prof. Jie Kong
Northwestern Polytechnical University
Xi'an
CH, 710072

Dr. Jing Dang
AVIC the First Aircraft Institute
Yanliang, Shaanxi, 710089, PR China

Cover: © AndSus/Adobe Stock Photos

Library of Congress Card No.: applied for

British Library Cataloguing-in-Publication Data
A catalogue record for this book is available from the British Library.

Bibliographic information published by the Deutsche Nationalbibliothek
The Deutsche Nationalbibliothek lists this publication in the Deutsche Nationalbibliografie; detailed bibliographic data are available on the Internet at <http://dnb.d-nb.de>.

Print ISBN: 978-3-527-35099-5
ePDF ISBN: 978-3-527-83960-5
ePub ISBN: 978-3-527-83961-2
oBook ISBN: 978-3-527-83962-9

Typesetting Straive, Chennai, India
Printing and Binding CPI Group (UK) Ltd, Croydon, CR0 4YY

C9783527350995_310124

Contents

Preface

With the rapid development of information technologies, the increasing frequency of electromagnetic waves transmission and reception has placed higher demand on the wave-transparent performances, mechanical properties, and environmental resistances of the composites in service. Polymer matrix wave-transparent composites present lightweight, high strength, low dielectric constant (ε) and dielectric loss ($\tan\delta$), and materials/structure/function integration, which have promising applications in the fields of aviation/aerospace, transportation, and 5G communications. In recent years, polymer matrix wave-transparent composites have become one of the key materials in many major engineering fields.

At present, the research progress and achievements in the field of polymer matrix wave-transparent composites are complex, and it is necessary to summarize the basic concept and research progress in this field for the reference of researchers. Currently, there are few books on polymer matrix wave-transparent composites. The authors wrote this monograph to summarize the latest scientific research achievements worldwide.

This book is divided into seven chapters. After the brief introduction to polymer matrix wave-transparent composites, it first introduces the measurement method of the dielectric properties as well as the mechanism and models of the wave transmission. Then, starting from reinforced fibers and polymer matrix widely utilized in recent years, it elaborates on the structures and properties of the common reinforced fibers and polymer matrix. Furthermore, the interface between the reinforced fibers and polymer matrix is also introduced. Finally, the practical applications of the polymer matrix wave-transparent composites are introduced in detail. Chapter 1 is written by Prof. Junwei Gu, where the introduction of polymer matrix wave-transparent composites is given from the basic concept, the composition of polymer matrix and reinforced fibers and their advantages and disadvantages, the factors influencing the wave-transparent performances, and the performance requirements under actual service condition. Chapter 2 is written by Dr. Jing Dang, Dr. Lin Tang, and Prof. Junwei Gu, where the relationship between the wave-transparent performances and ε and $\tan\delta$ of wave-transparent materials is discussed, and the equation for dielectric parameters, test methods for dielectric properties, and wave-transparent models are introduced. Chapter 3 is written by

Prof. Jie Kong, Dr. Zheng Liu, Dr. Jing Dang, and Prof. Junwei Gu, where the basic structure, physical and chemical properties, and modification methods of the polymer matrix are described in detail. Chapter 4 is written by Dr. Lin Tang, Prof. Yusheng Tang, Jiani Zhang, and Prof. Junwei Gu and mainly covers the structures, composition, preparation methods, and physical and chemical properties of the inorganic fibers (glass fibers and quartz fibers) and organic fibers (aramid fibers, ultrahigh molecular weight polyethylene fibers, and poly-*p*-phenylene benzobisoxazole fibers). Chapter 5 is written by Dr. Zheng Liu, Prof. Jie Kong, Dr. Jing Dang, and Prof. Junwei Gu, which describes the basic concepts of interfaces for polymer matrix wave-transparent composites, the formation of the interfaces, the mechanism of interfacial action, the interfacial characterization methods, and so on. This chapter also introduces the research progress of surface functionalization for reinforced fibers and the design and synthesis of interfacial compatibilizers. Chapter 6 is written by Prof. Yusheng Tang, Dr. Lin Tang, Yuhan Lin, and Prof. Junwei Gu, which introduces the structural design and molding principle of the polymer matrix wave-transparent composites as well as the processes of hand paste molding, pressure bag molding, laminated molding, resin transfer molding (RTM), winding molding, and so on. Chapter 7 is written by Dr. Jing Dang and Prof. Junwei Gu and introduces the application of the polymer matrix wave-transparent composites in radomes for aircraft, airborne, shipborne, ground vehicle radar,5G communication, printed circuit boards, and so on.

Part of the research in this book is supported by the following grants: National Scientific Research Project (Basis Strengthening Plan); State Key Laboratory of Solidification Processing in NWPU (SKLSP202103); Lin Tang and Zheng Liu would like to thank the Innovation Foundation for Doctor Dissertation of Northwestern Polytechnical University (CX2021036 and CX2023026). Yuhan Lin thanks the Practice and Innovation Funds for Graduate Students of NPU (PF2023034). This work is also financially supported by the Polymer Electromagnetic Functional Materials Innovation Team of Shaanxi Sanqin Scholars. At the time of publication of this book, the authors would like to express their sincere gratitude to the above-mentioned funding projects.

On the occasion of the publication of this book, I would like to express my sincere thanks to the family, friends, colleagues, and editors of Wiley who have worked so hard to make the book go smoothly. We would like to thank the “Structure/Function Polymer Composites” (SFPC) research group of Northwestern Polytechnical University for their help in preparing this book. This book is mainly aimed at scholars, scientific researchers in universities and research institutes, as well as related technology developers of companies engaged in the research and development of polymer matrix wave-transparent composites.

Prof. Junwei Gu
Northwestern Polytechnical University
Xi’an, Shaanxi, China

1 Introduction

1.1 Outline on Wave-Transparent Composites

Wave-transparent composites are a class of functional composites that can pass through electromagnetic waves. On the one hand, wave-transparent composites can provide electromagnetic windows for the transmission and reception of electromagnetic waves to ensure their efficient operation [1]. On the other hand, they can protect the radar antennas, communication, and microwave systems from the harsh external environment such as heavy rain, strong winds, snow, sand, solar radiation, and salt spray [2], ensuring the stability and reliability of electromagnetic wave transmission. With the rapid development of modern electronic information technology as well as the aviation and aerospace industries, the requirements for comprehensive performance of wave-transparent composites are becoming more and more demanding [3].

As far as matrix classification, wave-transparent composites can be divided into ceramic-based and polymer matrix wave-transparent composites [4]. Ceramic-based wave-transparent composites can meet the electrical performance requirements of radar radomes in the centimeter-band electromagnetic wave range. However, for millimeter-band electromagnetic waves (wavelength in the range of 1–10 mm and frequency in the range of 30–300 GHz), ceramic-based wave-transparent composites have disadvantages such as low strength, thick cover walls, and poor wave-transparent performances, which make it difficult to meet the performance requirements of radar radomes for millimeter wave [5, 6].

Polymer matrix wave-transparent composites have the advantages of lightweight, high strength, low dielectric constant (ε) and dielectric loss ($\tan\delta$), and materials/structure/function integration, which have a wide range of promising applications in satellite antennas, aircraft, missiles, 5G ground communication base stations, printed circuit boards, and so on. (Figure 1.1) [7].

This book will describe the wave-transparent mechanism, polymer matrix and reinforced fibers, their two-phase interfaces, molding process, and application prospects of the polymer matrix wave-transparent composites.

Polymer Matrix Wave-Transparent Composites: Materials, Properties, and Applications, First Edition.
Junwei Gu, Yusheng Tang, Jie Kong, and Jing Dang.

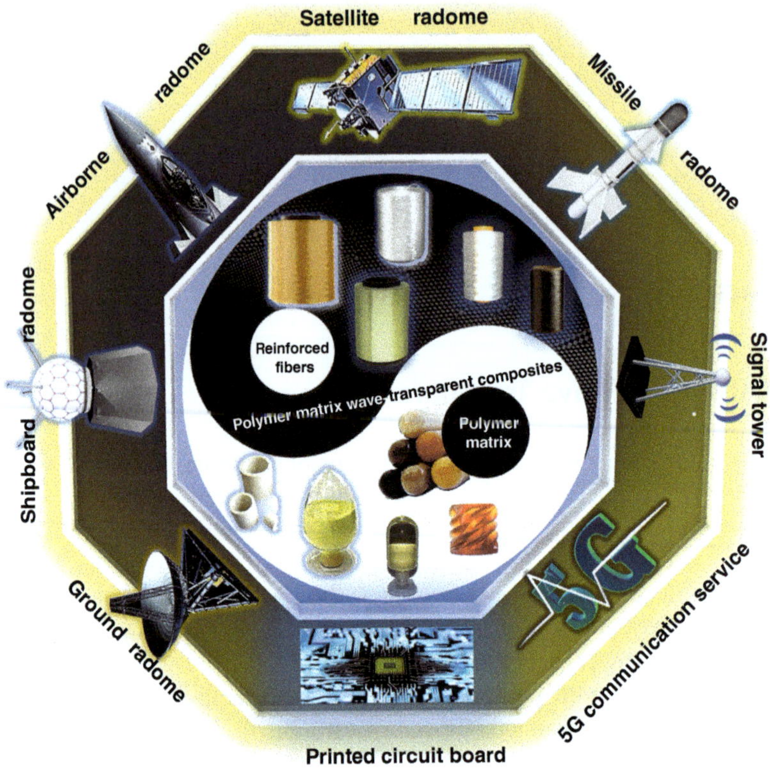

Figure 1.1 Application examples of polymer matrix wave-transparent composites. Source: Polymer matrix wave-transparent composites: A review. Journal of Materials Science & Technology, 2021, 75: 225–251 (Figure 1).

1.2 Composition of Polymer Matrix Wave-Transparent Composites

Polymer matrix wave-transparent composites consist of polymer matrix, reinforced fibers, and two-phase interfaces [8]. Polymers with low ε and $\tan\delta$ values as the matrix fibers with high strength and modulus as reinforced fibers produce advanced polymer-based composites (Figure 1.2) with both mechanical properties and wave-transparent performances *via* hot pressing, vacuum bagging, or resin transfer molding [9].

The heat resistance of polymer matrix determines the thermal stability of the composites in this case, and the fibers mainly serve as reinforcement [10]. Because the dielectric properties of different polymer matrices differ substantially. However, the ε value of reinforced fibers is generally larger than that of polymer matrix. Therefore, the selectively reinforced fibers possess excellent mechanical and thermal properties but also wonderful dielectric properties [11].

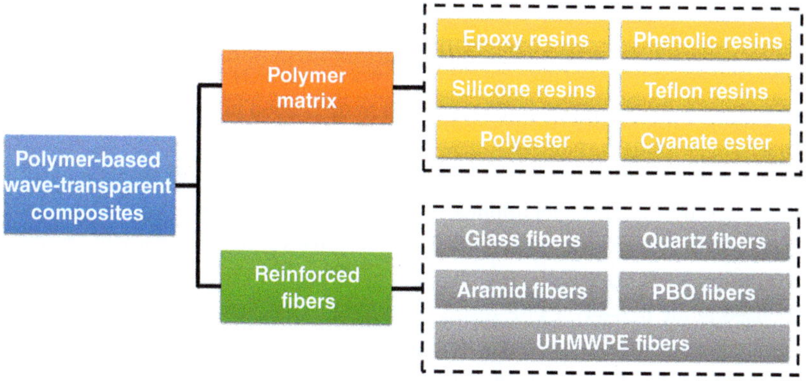

Figure 1.2 Composition of polymer matrix wave-transparent composites (commonly used polymer matrix and reinforced fibers).

1.2.1 Polymer Matrix

Polymers commonly used in wave-transparent composites mainly include epoxy resins [12], phenolic (PF) resins [13], polyimide (PI) resins [14], bismaleimide (BMI) resins [15], silicone resins, polytetrafluoroethylene (PTFE) resins [16], unsaturated polyester (UP) resins [17–19], and cyanate (CE) resins [20]. Table 1.1 shows the main physical and chemical properties of the common polymer matrix.

Epoxy resins have good flowability, low curing shrinkage, and high thermal decomposition temperatures (300–350 °C), but their high ε and $\tan\delta$ values limit their application in high-performance polymer matrix wave-transparent composites [21–23]. PF resins have good heat resistance (long-term service temperature at 250 °C), mechanical properties, and weatherability [24]. However, the ε values of PF resins increase significantly with increasing temperatures [25–27]. PI resins have high heat resistance ($T_g \geq 250$ °C), ε, and $\tan\delta$ values that remain stable over a wide

Table 1.1 Main physical and chemical properties of the common polymer matrix.

Types	Density (g/cm^3)	Flexural strength (MPa)	Flexural modulus (GPa)	ε (10^6 Hz)	$\tan\delta$ (10^6 Hz)
Epoxy	1.30	97	3.8	3.0	0.020
PF	1.30	92	3.5	3.2	0.020
PI	1.36–1.43	170	3.8	3.2	0.007
BMI	1.30	150	3.7	3.0	0.014
Organic silicon	—	85	—	3.0–5.0	0.003–0.050
PTFE	2.20	90	—	2.1–2.3	0.0003–0.0004
UP	1.29	85	3.2	3.0	0.018
CE	1.29	80	2.8	2.8–3.2	0.002–0.008

range of temperatures and frequencies [28]. At the same time, PI resins have excellent mechanical properties, chemical resistance, and dimensional stability [29–31]. However, PI resins are costly and difficult to process [32, 33]. BMI resins are an ideal polymer matrix for advanced composites due to their good heat resistance, excellent mechanical properties, relatively low ε value, resistance to humidity, chemical reagents, and good processability [34, 35]. However, the relatively high $\tan\delta$ values of BMI resins limit their wider application to a certain extent [36–38]. Silicone resins have excellent heat resistance and stable ε and $\tan\delta$ values under a wide range of environmental conditions [39–41], but their poor mechanical strength makes them rarely used alone [42–44]. PTFE resins have the lowest ε and $\tan\delta$ [45, 46] but are not easy to process and have low bonding properties between PTFE matrix and reinforcements [47–49]. UP resins have better mechanical properties than PF resins and have low ε and $\tan\delta$ values [50–52], which can be cured at room temperature. UP resins have a simple molding process, making them suitable for large-scale or large radome production [53–55]. However, UP resins have a short storage period, relatively low heat deflection temperature, and large curing shrinkage, which makes them unsuitable for the preparation of polymer matrix wave-transparent composites with high dimensional accuracy requirements [56–58].

In comparison, CE resins combine the high-temperature resistance of BMI and PI resins with the good processing properties of epoxy resins [59–61]. The highly symmetrical triazine ring structure and low polarity of the cured CE resins also make them low ε (2.8–3.2) [62–64], good heat resistance, and dimensional stability over a wide temperature and frequency range [65]. The structure and properties of commonly used polymer matrix are described in detail in Chapter 3.

1.2.2 Reinforced Fibers

Reinforced fibers for polymer matrix wave-transparent composites mainly include glass fibers [66, 67], quartz fibers [68], Kevlar fibers [69, 70], ultra-high-molecular-weight polyethylene (UHMWPE) fibers [71, 72], and poly(*p*-phenylene-2,6-benzobisoxazole) (PBO) fibers [73, 74]. Their main physical and chemical properties are shown in Table 1.2.

Table 1.2 Main physical and chemical properties of common reinforced fibers.

Properties Types	Density ($g\,cm^{-3}$)	Tensile strength (GPa)	Modulus (GPa)	ε (10^6 Hz)	$\tan\delta$ (10^6 Hz)
E-glass fibers	2.54	3.75	72	6.13	0.0038
S-glass fibers	2.49	4.00	85	5.21	0.0068
D-glass fibers	2.6	2.40	52	4.00	0.0025
Quartz fibers	2.20	1.70	72	3.78	0.0002
Kevlar49 fibers	1.45	3.45	137	3.85	0.0010
UHMWPE fibers	0.97	5.01	193	2.25	0.0002
PBO fibers	1.56	5.80	280	3.00	0.0010

Figure 1.3 Application examples of D-glass fibers reinforced polymer matrix wave-transparent composites: MIRAGE 2000 (France, a); GRIPEN JAS 39 – Credit: Thierry ducros/Airliners.net (Sweden, b); HAWK 200 – Credit: Ben Stacey/Flickr (United Kingdom, c); HARRIER – Credit: Weimeng/Air Team Images (United Kingdom, d). HARRIER GR.9 – Credit: Titan Miller/Airliners.net. Source: (b) Ben Stacey/Flickr.

Glass fibers are the most commonly used inorganic reinforced fibers for wave-transparent composites. The earliest glass fibers used were E-glass fibers [75, 76]. Then, high-strength glass fibers (S-glass fibers) [77–79] and high-silica glass fibers (D-glass fibers) [80, 81] were developed to meet the special needs of aviation, aerospace, military, and other high-tech fields. Compared to E-glass and S-glass fibers, D-glass fibers have relatively lower ε and $\tan\delta$, which has been used in the radomes of MIRAGE 2000 (France), GRIPEN JAS 39 (Sweden), HAWK 200 (United Kingdom), and HARRIER (United Kingdom) (Figure 1.3) [82, 83].

However, with the rapid development of information technology, electronic components receive and transmit electromagnetic waves at increasingly high frequencies [84]. The high content of alkali metal oxides in glass fibers and the strong signal hysteresis and attenuation produced during electromagnetic wave transmission limit their application in high-frequency and high-precision wave-transparent composites [85]. Quartz fibers contain only a single component of silicon dioxide (SiO_2) with purity of over 99.9% and have excellent high-temperature resistance, electrical insulation properties and ablation resistance, low ε and $\tan\delta$ values, and so on [86, 87], which have been one of the most commonly used reinforced fibers in wave-transparent composites in the military and civilian sectors. However, quartz fibers have disadvantages such as high density, poor mechanical properties, and large ε values [88].

With the increasing demand for comprehensive performances of polymer matrix wave-transparent composites in terms of weight reduction, wave-transparency, and loading, researchers have carried out relevant research on organic reinforced fibers such as Kevlar fibers [89], UHMWPE fibers [90] and PBO fibers [91, 92]. Kevlar fibers, with low density, high specific strength, and specific modulus, are one of the most commonly used organic reinforced fibers in polymer matrix wave-transparent composites [93–95]. However, the high moisture absorption of Kevlar fibers is susceptible to moisture swelling and cracking, resulting in the degradation of wave-transparent performances and mechanical properties [96]. UHMWPE fibers, also known as high-strength, high-modulus polyethylene fibers, have a relative molecular mass of over 1 million, which is beneficial to outstanding impact resistance, cut resistance, chemical resistance and UV resistance, excellent low-temperature resistance, and low ε and $\tan\delta$ values [97, 98]. However, as the macromolecular chains of UHMWPE fibers are connected by a highly symmetrical methylene structure, the intermolecular Van der Waals forces are weak, making their T_g and melting point low, resulting in their high-temperature resistance and poor creep resistance [99].

Moreover, the surface of UHMWPE fibers does not contain polar groups, resulting in low surface energy, which creates poor bond strength between the UHMWPE fibers and polymer matrix [100, 101]. As a super fiber of the twenty-first century, the large number of rigid aromatic and oxazole rings in the PBO fiber molecular chain and a highly ordered crystal structure give PBO fibers excellent mechanical properties, heat resistance, chemical stability, and low ε (3.0) and $\tan\delta$ (0.001) values, which are of wide interest in the field of airborne/starborne radar radomes. Furthermore, PBO fibers have higher tensile strength, lower density, and ε values than those of inorganic reinforced fibers such as quartz [102]. Compared to those of other organic fibers, PBO fibers have about twice the strength and modulus of para-Kevlar fibers, and the thermal decomposition temperature of PBO fibers in the air is about 650 °C, which is approximately 100 °C higher than that of Kevlar fibers and much better than that of UHMWPE fibers (300 °C) [103]. As a result, PBO fibers have received a lot of attention as potential reinforcements for light weight/loading/wave-transparent integrated wave-transparent composites [104]. However, PBO fibers still have disadvantages of high cost, smooth and inert

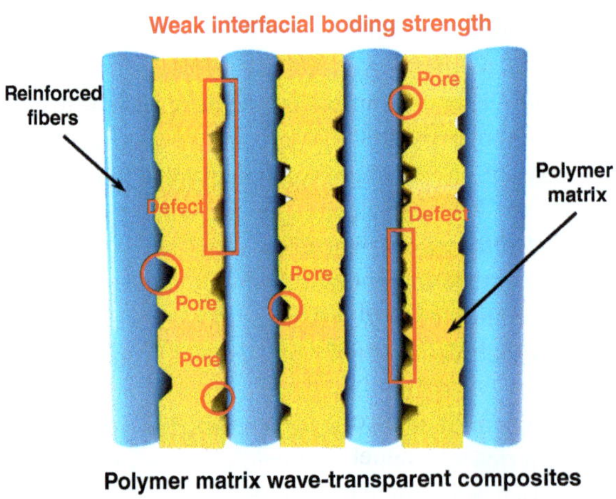

Figure 1.4 Schematic diagram of the two-phase interface for polymer matrix wave-transparent composites.

surfaces, and so on [105–107]. The structure and properties of these commonly used reinforced fibers are described in detail in Chapter 4.

In addition, the microscopic phase interface links the polymer matrix and reinforced fibers [108]. Defects are likely to arise at the poor two-phase interface (Figure 1.4), which would affect the overall performance (especially the interlaminar shear strength, ILSS) of the polymer-matrix wave-transparent composites [109, 110].

Therefore, how to effectively enhance the interfacial compatibility between polymer matrix and reinforced fibers has become a hot and difficult issue in this field [111]. Chapter 5 provides a detailed description of the two-phase interface inner polymer matrix wave-transparent composites and their optimal control strategies.

1.3 Factors Influencing the Wave-Transparent Performances of Polymer Matrix Wave-Transparent Composites

Polymer matrix wave-transparent composites are mainly used for electromagnetic windows and radomes in the fields of aviation/aerospace, 5G communication, and electronic information [112]. In order to ensure that all types of radar and antenna systems remain in stable operating conditions under harsh external environments, polymer matrix wave-transparent composites are required to have excellent wave-transparent performances (low ε and $\tan\delta$ values) [113].

The main factors affecting the wave-transparent performances of polymer matrix composites are divided into internal factors (intrinsic ε and $\tan\delta$) and external factors (thickness and electromagnetic wave frequency) [114, 115]. In general, the lower the molecular polarization rate and the density of polarized molecules of polymer matrix and reinforced fibers, the lower the ε and $\tan\delta$ of polymer matrix wave-transparent composites, the less energy is reflected and lost during the transmission of electromagnetic waves, and the correspondingly higher the wave-transparent rate [116]. In addition, polymer matrix wave-transparent composites are typically multiphase systems, and the interface between polymer matrix and reinforced fibers is prone to interfacial polarization, increasing the ε and $\tan\delta$ values, which is not conducive to improving the wave-transparent performance [117].

In addition, the thickness of polymer-based wave-transparent composites also affects their wave-transparent performances [118]. When the frequency of the electromagnetic wave is constant, the thickness of the wave-transparent composites increases, resulting in a tendency for the wave-transparent rate to decrease and then increase (Figure 1.5) [119]. This is mainly due to the reflection and loss (both absorption and interference shifts) that occur on the surface and inside the wave-transparent composites as the electromagnetic waves pass through [120, 121]. When the thickness approaches an odd multiple ($d = n\lambda/4$, $n = 1, 3, 5$, etc.) of its quarter wavelength ($\lambda/4$, Eq. 1.1), electromagnetic waves cause strong interference cancellation in the wave-transparent composites. This leads to an attenuation of the

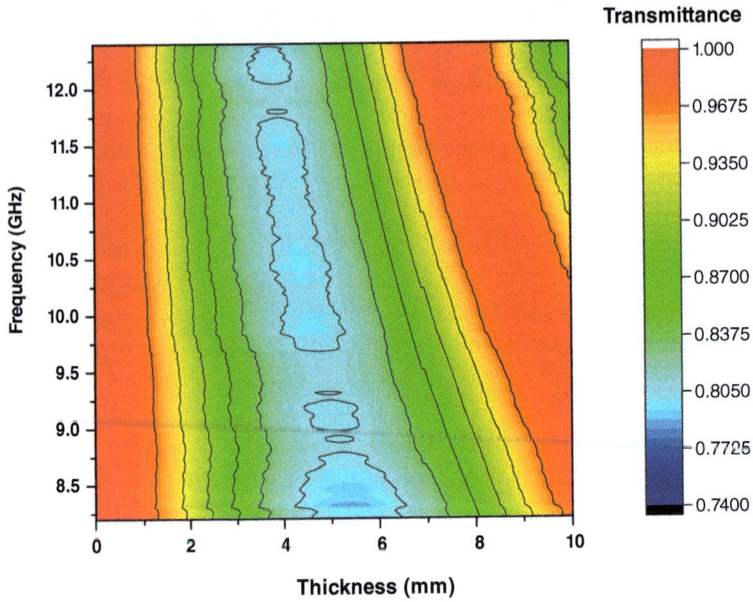

Figure 1.5 Schematic representation of wave transmission versus material thickness for polymer matrix wave-transparent composites.

electromagnetic wave energy and a significant reduction of the transmitted waves, resulting in the reduction of the wave transmission [122–124]. When the thickness is close to an even multiple ($d = n\lambda/4$, $n = 2, 4, 6$, etc.) of $\lambda/4$, the electromagnetic waves reflect less at the incident interface and can enter the interior almost unharmed, with the high wave transmission rate [125, 126].

$$n\lambda/4 = nc/4f_m * (u_r * \varepsilon_r)^{1/2} \tag{1.1}$$

where λ represents the wavelength of the incident waves; c represents the speed of light; f_m represents the frequency of the incident waves; u_r represents the magnetic permeability of the medium; and ε_r represents the dielectric constant of the medium.

1.4 Property Requirements for Polymer Matrix Wave-Transparent Composites

1.4.1 Wave-Transparent Performances

The ε and $\tan\delta$ values of polymer matrix wave-transparent composites are among the most important parameters affecting the wave-transparent performances [127]. In practice, the transmission rate of electromagnetic waves is usually required to exceed 70% in the broad frequency range (0.3–300 GHz), which usually requires the corresponding ε of polymer matrix wave-transparent composites to be stable in the range of 1–4 and $\tan\delta$ in the range of 10^{-2}–10^{-3}. Meanwhile, the ε and $\tan\delta$ are required to remain constant in the broad frequency and temperature range (0–220 °C) [128, 129].

1.4.2 Mechanical Properties

As structural loading materials, polymer matrix wave-transparent composites must have a certain degree of stiffness and strength to ensure the stability and reliability of the antenna system in various complex operating environments [130]. The tensile strength of polymer matrix wave-transparent composites for high-performance radomes is generally not less than 400 MPa. The compressive strength is more than 350 MPa to ensure the integrity of the antenna system under aerodynamic loads and impact of foreign objects, thus ensuring the normal operation of the electronic components inside the radomes [131–133].

1.4.3 Heat Resistant Properties

When the vehicle is flying at ultra-high speed in the atmosphere, the surface temperature of vehicle rises sharply with the increase in Mach number due to the heating of the high-temperature compressed gas between the excitation wave and the body and the strong friction between the surface of the body and the air (usually when the Mach number is 2, the surface temperature of the vehicle is about 150 °C; while when the Mach number increases to 3, the surface temperature rises sharply to about 350 °C, even exceeding the strength limit temperature of aluminum alloy) [134], therefore, when polymer matrix wave-transparent composites are used as radomes for aircraft, they should have excellent heat resistance (pyrolysis temperature greater than 300 °C) to overcome the high thermal stresses of external aerodynamic heating and to avoid deformation or even cracks under rapid temperature change [135, 136].

1.4.4 Environmental Resistance Properties

As protective materials for radar antenna systems, polymer matrix wave-transparent composites are subject to surface aging, polymer matrix degradation, and interfacial debonding between the polymer matrix and reinforced fibers during long-term service, which would seriously affect their service stability and reliability [137, 138]. Therefore, polymer matrix wave-transparent composites are required to have excellent environmental aging resistances. Current research revealed that environmental factors (humidity, heat, high and low-temperature alternation, and light) had a significant effect on the mechanical and dielectric properties of glass fiber-reinforced epoxy resin wave-transparent composites. When the relative humidity increased from 25% to 85%, the ε and $\tan\delta$ increased by 10% and 18.6%, respectively. In addition, the mechanical properties were strongly influenced by the hygrothermal conditions. The retention of tensile and flexural strengths after boiling for 200 hours was about 90%, but the retention of ILSS was only 61% [139].

References

1 Hamid, S., Heberling, D., Junghahnel, M. et al. (2022). Development of a millimeter-wave transparent antenna inside a headlamp for automotive radar application. *International Journal of Microwave and Wireless Technologies* 14 (6): 677–688.

2 Wang, M., Li, J., Shao, B., and Xiang, P. (2022). Progress of low dielectric property modified in 5G. *Plastics* 51 (1): 117–120.
3 Zhao, Y. and Pan, L. (2021). Research progress of aerospace advanced polymer matrix composites and manufacturing technology. *Aerospace Materials & Technology* 51 (4): 29–36.
4 Cao, W., Yang, X., and Zhang, O. (2020). Development and prospect of high performance polymer composites in China. *Engineering Sciences* 22 (5): 112–120.
5 Bin, F., Sun, W., and Liu, Y. (2021). Design technique of millimeter wave transparent window. *Journal of Microwaves* 37 (4): 21–24. 9.
6 Kamutzki, F., Schneider, S., Barowski, J. et al. (2021). Silicate dielectric ceramics for millimetre wave applications. *Joural of the European Ceramic Society* 41 (7): 3879–3894.
7 Gopinathapanicker, J.C., Inamdar, A., Anand, A. et al. (2020). Radar transparent, impact-resistant, and high-temperature capable radome composites using polyetherimide-toughened cyanate ester resins for high-speed aircrafts through resin film infusion. *Industrial & Engineering Chemistry Research* 59 (16): 7502–7511.
8 Li, X., He, R., and Liu, H. (2022). Progress in low dielectric polymer materials. *Journal of Functional Materials* 53 (4): 597–606.
9 Hu, Z.D., Liu, X.Q., Ren, T.L. et al. (2022). Research progress of low dielectric constant polymer materials. *Journal of Polymer Engineering* 42 (8): 677–687.
10 Yang, X.J., Li, B., Li, D. et al. (2019). High-temperature properties and interface evolution of silicon nitride fiber reinforced silica matrix wave-transparent composite materials. *Journal of the European Ceramic Society* 39 (2-3): 240–248.
11 Wu, S.H., Li, C.C., Yu, Z.H. et al. (2020). Nondestructive strategy to effectively enhance the interfacial adhesion of PBO/epoxy composites. *ACS Applied Materials & Interfaces* 12 (40): 45383–45393.
12 Li, P., Du, R., Liu, Y., and Zhao, G. (2021). Preparation of epoxy resin/quartz fiber wave-transparent composites. *Engineering Plastics Application* 49 (2): 29–33. 9.
13 Zhao, Y., Xu, R., Xiao, Y. et al. (2022). Mechanical performances of phenolic modified epoxy resins at room and high temperatures. *Coatings* 12 (5): 643.
14 Wang, S., Jiang, S., Han, X. et al. (2021). Research progress on high-performance polyimide resins and their composites. *Journal of Functional Polymers* 34 (6): 570–585.
15 Zhang, Z.W., Zhou, Y.J., Cai, L.F. et al. (2022). Synthesis of eugenol-functionalized polyhedral oligomer silsesquioxane for low-k bismaleimide resin combined with excellent mechanical and thermal properties as well as its composite reinforced by silicon fiber. *Chemical Engineering Journal* 439: 135740.
16 Zhang, Y., Wang, R., Zhang, X., and Guo, S. (2022). Ingenious sandwich-like adhesive films and controllable introduction of fluorine-containing groups toward strong adhesive strength and low dielectric characteristics. *Industrial & Engineering Chemistry Research* 61 (39): 14494–14507.

17 Gao, Y., Romero, P., Zhang, H.L. et al. (2019). Unsaturated polyester resin concrete: a review. *Construction and Building Materials* 228: 116709.

18 Poorabdollah, M. and Kamran, A. (2019). Optimising cure cycle of unsaturated polyester nanocomposites using directed grid search method. *Polymers and Polymer Composites* 27 (5): 253–261.

19 Zong, D., Zhang, Y., Chen, J., and Liu, Y. (2021). Influencing factors in low temperature curing shrinkage of unsaturated polyester resin. *Journal of Materials Science and Engineering* 39 (1): 77–81. 100.

20 Llevot, A., Grau, E., Carlotti, S. et al. (2016). From lignin-derived aromatic compounds to novel biobased polymers. *Macromolecular Rapid Communications* 37 (1): 9–28.

21 Luo, X., Liu, S., and Zhao, J. (2021). Research progress in low dielectric loss epoxy resin composites. *China Plastics* 35 (3): 130–138.

22 Mi, X.Q., Liang, N., Xu, H.F. et al. (2022). Toughness and its mechanisms in epoxy resins. *Progress in Materials Science* 130: 100977.

23 Muller-Pabel, M., Agudo, J.A.R., and Gude, M. (2022). Measuring and understanding cure-dependent viscoelastic properties of epoxy resin: a review. *Polymer Testing* 114: 107701.

24 Wang, Y.N., Liu, J., and Han, X. (2018). Research progress and application of phenolic resin-based prepreg. *New Chemical Materials* 46 (7): 269–272.

25 Cao, Y., Li, X., Li, Y. et al. (2020). Properties of high silica glass fiber/phenolic composites by soft molding. *Aerospace Materials & Technology* 50 (2): 99–102.

26 Guo, Y., Hu, L., and Zhou, Y. (2018). Research progress on phenolic resin modified with silicon. *New Chemical Materials* 46 (10): 5–9.

27 Tong, Y., Li, L., Liu, H. et al. (2019). Research progress on high temperature resistant resins. *Thermosetting Resin* 34 (3): 51–54. 60.

28 Xue, S., Lei, X., Lian, R. et al. (2021). Progress of high-performance thermosetting polyimide resins. *Polymer Materials Science & Engineering* 37 (5): 149–162.

29 Li, X.T., Liu, T., Jiao, Y.Z. et al. (2019). Novel high-performance poly(benzoxazole-co-imide) resins with low dielectric constants and superior thermal stabilities derived from thermal rearrangement of ortho-hydroxy polyimide oligomers. *Chemical Engineering Journal* 359: 641–651.

30 Sun, G., Zhang, X., Wu, D., and Hou, L. (2021). Research progress in high-performance polyimide composites. *China Plastics* 35 (9): 147–155.

31 Wu, T., Fang, Y.T., Dong, J. et al. (2021). Synthesis and properties of polyimide resin containing acetylene and benzoxazine double crosslinking moieties. *Chemical Journal of Chinese Universities* 42 (6): 1978–1986.

32 Huang, X., Peng, X., Liu, R. et al. (2021). Theprogress of low dielectric constant modified polyimide materials. *Journal of Functional Materials* 52 (5): 5018–5024.

33 Iredale, R.J., Ward, C., and Hamerton, I. (2017). Modern advances in bismaleimide resin technology: a 21st century perspective on the chemistry of addition polyimides. *Progress in Polymer Science* 69: 1–21.

34 Ma, X., Guo, Y., and Zhang, Y. (2019). Research progress of ablative resins for RTM. *Thermosetting Resin* 34 (1): 66–70.

35 Mao, Y.M., Yao, Z.J., Zhou, J.T. et al. (2020). Fabrication and investigations on BMI/OMMT nanocomposites with superior high-temperature wave-transparent performance. *Journal of Materials Science: Materials in Electronics* 31 (18): 16073–16086.

36 Liu, B., Yang, L., Li, F. et al. (2020). Research on properties of bismaleimide/T700 carbon fiber resin composite. *Engineering Plastics Application* 48 (11): 117–120.

37 Wen, Y., Li, B., He, Y. et al. (2018). Study on the curing reaction kinetics of modified bismaleimide resin. *China Plastics Industry* 46 (12): 22–26.

38 Zhang, Z.W., Tian, D., Niu, Z.Q. et al. (2021). Enhanced toughness and lowered dielectric loss of reactive POSS modified bismaleimide resin as well as the silica fiber reinforced composites. *Polymer Composites* 42 (12): 6900–6911.

39 Guan, S.J., Wu, F., Wu, Q. et al. (2021). Modification of silicone resins by Si-N cross-dehydrocoupling with perfect thermal stability and mechanical performance. *New Journal of Chemistry* 45 (35): 15892–15900.

40 Liu, Y., Chen, Z.X., Qin, Y.S. et al. (2020). Silicone resin-based composite materials for high thermal stability and thermal conductivity. *Journal of Electronic Materials* 49 (7): 4379–4384.

41 Zheng, J.Y., Cai, Y.Q., Hu, Y. et al. (2022). Bio-based epoxy functionalized MQ silicone resins: from synthesis to toughened epoxy composites with good mechanical properties, thermal resistance and transparency. *Polymer Chemistry* 13 (37): 5325–5336.

42 Huang, C., Zheng, C., Zheng, G. et al. (2020). Preparation and application of silicone resin based wear-resistant composite fiber material. *New Chemical Materials* 48 (1): 245–249.

43 Ling, Y.Q., Zhang, X.Q., Yan, L.W. et al. (2022). Silicone-grafted epoxy/carbon fiber composites with superior mechanical/ablation performance. *Materials Chemistry and Physics* 275.

44 Zhu, Q.Q., Wang, Z.H., Zeng, H. et al. (2021). Effects of graphene on various properties and applications of silicone rubber and silicone resin. *Composites Part A: Applied Science and Manufacturing* 142: 106240.

45 Pan, C., Kou, K., Jia, Q. et al. (2017). Improved thermal conductivity and dielectric properties of hBN/PTFE composites via surface treatment by silane coupling agent. *Composites Part B: Engineering* 111: 83–90.

46 Zheng, L., Zhou, J., Shen, J. et al. (2019). Effects on the thermal expansion coefficient and dielectric properties of CLST/PTFE filled with modified glass fiber as microwave material. *Chinese Chemical Letters* 30 (5): 1111–1114.

47 Jiang, Z. and Yuan, Y. (2018). Effects of particle size distribution of silica on properties of PTFE/SiO2 composites. *Materials Research Express* 5 (6): 066306.

48 Li, X. and Ge, Z. (2020). A review on PTFE composites filled with inorganic materials. *China Plastics* 34 (1): 92–101.

49 Wang, L., Jin, X., Wu, C., and Zhang, L. (2021). Effect of glass fiber on properties of PTFE/SiO_2 composite substrate. *Engineering Plastics Application* 49 (1): 130–134.

50 Chen, X., Li, Y., Chen, J. et al. (2018). Research progress in unsaturated polyester resin modified by inorganic nanoparticles. *China Plastics* 32 (3): 1–5.

51 Liu, C. and Zhou, Y. (2020). Progress in oil-based unsaturated polyester resins. *Chemistry and Industry of Forest Products* 40 (6): 107–117.

52 Sapuan, S.M., Aulia, H.S., Ilyas, R.A. et al. (2020). Mechanical properties of longitudinal basalt/woven-glass-fiber-reinforced unsaturated polyester-resin hybrid composites. *Polymers* 12 (10): 2211.

53 Farsane, M., Lhasnaoui, S., Anouar, A. et al. (2022). A review of measureing the gelation time in unstrated polyester resins. *Materials and Technologies* 56 (3): 323–329.

54 Moujdin, I.A., Totah, H.S., Abulkhair, H.A. et al. (2022). Development of low shrinkage curing techniques for unsaturated polyester and vinyl ester reinforced composites. *Materials* 15 (9): 2972.

55 Wu, S., Liu, J., Shao, C. et al. (2020). Application of UPR in solid propellant inhibitor. *New Chemical Materials* 48 (3): 6–8.

56 Chen, M.Z., Wang, X.J., Zhang, J., and Zhu, J.J. (2022). Synergistic action of initiator and accelerator on curing thermodynamics and kinetics of chopped carbon fiber-reinforced low-density unsaturated polyester resin with rapid polymerization. *Journal of Thermal Analysis and Calorimetry* 147: 14195–14199.

57 Delaite, C., Bistac, S., Dreyer, E., and Schuller, A.S. (2020). Influence of initiator on the curing of unsaturated polyester resin at 100°C. *Polymer International* 69 (11): 1089–1096.

58 Han, D., Cao, H., Chen, J. et al. (2018). Study on the UV-curing behavior of unsaturated polyester resin by in-situ Fourier transform infrared. *Thermosetting Resin* 33 (5): 9–14.

59 Chen, S., Yuan, L., Wang, Z.H. et al. (2019). Self-constructed nanodomain structure in thermosetting blend based on the dynamic reactions of cyanate ester and epoxy resins and its related property. *Composites Part B: Engineering* 177: 107438.

60 Jayalakshmi, C.G., Salunke, A.D., Joshi, M. et al. (2022). Cyanate ester- epoxy blends toward microwave transparent polymer composites through resin film infusion for wideband electromagnetic applications. *Polymer-Plastics Technology and Materials* 61 (2): 145–160.

61 Zhang, X., Wang, F., Zhu, Y., and Qi, H. (2021). Preparation and property of modified cyanate ester resin. *New Chemical Materials* 49 (6): 164–170.

62 Li, C.P. and Chuang, C.M. (2021). Thermal and dielectric properties of cyanate ester cured main chain rigid-rod epoxy resin. *Polymers* 13 (17): 2917.

63 Qiao, H., Bao, J., Zhong, X. et al. (2019). Modification and thermal analysis for curing properties of cyanate-ester resin. *Journal of Aeronautical Materials* 39 (6): 63–72.

64 Wang, Z.C., Li, R.J., Qin, Q.R. et al. (2022). Curing kinetics and mechanical properties of cyanate ester/hyperbranched benzoxazine copolymers. *Polymers for Advanced Technologies* 33 (5): 1606–1616.

65 Zhou, Y.J., Zhang, Z.W., Wang, P.R., and Ma, X.Y. (2022). High-performance and low-dielectric cyanate ester resin optimized by regulating the structure of linear polyhydroxy ether modifier. *Composites Part A: Applied Science and Manufacturing* 162: 107136.

66 Lu, Y., Dang, Z., and Cao, F. (2021). Preparation and flame retardancy of glass fiber reinforced polypropylene composite. *Plastics* 50 (2): 15–18.

67 Praneeth, A., Kumar, K., Pramod, R., and Kumar, G.B.V. (2019). Investigation of mechanical properties of glass fibre reinforced honeycomb panels. In: *International Conference on advances in Materials and Manufacturing Applications.*

68 Zhang, Y., Zhang, L., Tao, Y. et al. (2018). Effects of laser ablation on electric properties of quartz fiber/cyanate resin wave-transmitting composite. *Acta Materiae Compositae Sinica* 35 (11): 3001–3007.

69 Dai, Y., Meng, C.B., Cheng, Z. et al. (2019). Nondestructive modification of aramid fiber based on selective reaction of external cross-linker to improve interfacial shear strength and compressive strength. *Composites Part A: Applied Science and Manufacturing* 119: 217–224.

70 Zhao, Y., Li, X., Shen, J.N. et al. (2020). The potential of Kevlar aramid nanofiber composite membranes. *Journal of Materials Chemistry A* 8 (16): 7548–7568.

71 Chen, M., Lin, H., Yang, C., and Li, L. (2022). Research progress on modification of ultra high molecular weight polyethylene (UHMWPE) fiber. *Applied Chemical Industry* 51 (2): 520–524.

72 Chhetri, S. and Bougherara, H. (2021). A comprehensive review on surface modification of UHMWPE fiber and interfacial properties. *Composites Part A: Applied Science and Manufacturing* 140: 106146.

73 Chin, J., Forster, A., Clerici, C. et al. (2007). Temperature and humidity aging of poly (p-phenylene-2,6-benzobisoxazole) fibers: chemical and physical characterization. *Polymer Degradation and Stability* 92 (7): 1234–1246.

74 Kuroki, T., Tanaka, Y., Hokudoh, T., and Yabuki, K. (1997). Heat resistance properties of poly(p-phenylene-2,6-benzobisoxazole) fiber. *Journal of Applied Polymer Science* 65 (5): 1031–1036.

75 Al-Furjan, M.S.H., Shan, L., Shen, X. et al. (2022). A review on fabrication techniques and tensile properties of glass, carbon, and Kevlar fiber reinforced polymer composites. *Journal of Materials Research and Technology* 19: 2930–2959.

76 Naito, K. (2022). Static and fatigue tensile properties of carbon/glass hybrid fiber-reinforced epoxy composites. *Scientific Reports* 12 (1): 6298.

77 Cho, K., Wang, G.N., Raju, R. et al. (2019). Influence of surface treatment on the interfacial and mechanical properties of short s-glass fiber reinforced dental composites. *ACS Applied Materials & Interfaces* 11 (35): 32328–32338.

78 Xiong, Q., Liu, Z., Chen, W., and Yan, C. (2021). Study on the performance of glass fiber used for superconducting magnet insulation before and after vacuum heat treatment process. *Chinese Journal of Vacuum Science and Technology* 41 (8): 756–760.

79 Yeon, J., Chowdhury, S.C., and Gillespie, J.W. (2022). Mechanical properties and damage analysis of S-glass: a reactive molecular dynamics study. *Composites Part B: Engineering* 234: 109706.

80 Chen, C., Zhu, Q.O., Wang, H.P. et al. (2022). Excellent mechanical properties of the silicate glasses modified by CeO_2 and TiO_2: a new choice for high-strength and high-modulus glass fibers. *Silicon* 14 (8): 4301–4308.

81 Liu, K., Chen, B., Dai, J., and Zhu, S. (2010). Effect of environment factors on properties of glass fiber reinforced epoxy resin wave-transparent composites. *The Journal of Engineering Plastics Application* 38 (10): 64–67.

82 Feng, Z., Li, J., Zuo, X. et al. (2021). Progress of composite materials for aerospace applications. *Aerospace Materials & Technology* 51 (4): 23–28.

83 Zhang, S., Luo, L., Ding, L. et al. (2021). Effect of acid leaching on structure and properties of high silica glass fibers. *Journal of the Chinese Ceramic Society* 49 (8): 1543–1549.

84 Kong, L.M., Zheng, W., Wang, X.B. et al. (2021). Effects of layering types and fiber off-axis angle on the mechanical properties of s-glass-fiber-reinforced composites. *Mechanics of Composite Materials* 57 (4): 551–562.

85 Zurowski, W., Zepchlo, J., Krzyzak, A. et al. (2022). Effect of the quartz powder on the performance of the two layers glass fiber reinforced polymer composite with emulsion binder. *Composite Structures* 298: 116024.

86 Cui, G., Cheng, Y., Liu, C. et al. (2020). Massive growth of graphene quartz fiber as a multifunctional electrode. *ACS Nano* 14 (5): 5938–5945.

87 Kong, G., Wang, K., Yu, Q. et al. (2022). Properties of aerogel modified quartz fiber reinforced silicone wave-transparent composites. *Acta Armamentarii* 43 (2): 442–450.

88 Wang, X., Yang, S., Li, G. et al. (2021). Research on mechanical properties of carbon/quartz hybrid fiber composite. *China Plastics Industry* 49 (7): 87–90.

89 Liu, Y.L., Ye, L., Shu, Y., and Zhao, X.W. (2018). In situ preparation of intrinsic flame retardant urea formaldehyde/aramid fiber composite foam: structure, property and reinforcing mechanism. *Composites Part A: Applied Science and Manufacturing* 115: 274–282.

90 Li, L., Han, L., Wang, S. et al. (2022). Research progress on surface modification and interfacial properties of UHMWPE fiber. *Journal of Functional Materials* 53 (4): 012016.

91 Liu, Z., Fan, X.L., Zhang, J.L. et al. (2021). Improving the comprehensive properties of PBO fibres/cyanate ester composites using a hyperbranched fluorine and epoxy containing PBO precursor. *Composites Part A: Applied Science and Manufacturing* 150: 106596.

92 Ni, X.Y., Feng, H., Li, L.H. et al. (2021). A novel poly(p-phenylene benzobisoxazole) (PBO)-based three-phase silk-cocoon network structure nanocomposites with enhanced dielectric properties. *Journal of Materials Science: Materials in Electronics* 32 (6): 7574–7586.

93 Lv, J.W., Yin, J.Y., Qin, Y.T. et al. (2021). Post-construction of weaving structure in aramid fiber towards improvements of its transverse properties. *Composites Science and Technology* 208: 108780.

94 Priyanka, P., Dixit, A., and Mali, H.S. (2019). High strength Kevlar fiber reinforced advanced textile composites. *Iranian Polymer Journal* 28 (7): 621–638.

95 Zhang, B., Jia, L.H., Tian, M. et al. (2021). Surface and interface modification of aramid fiber and its reinforcement for polymer composites: a review. *European Polymer Journal* 147: 110352.

96 Sharma, P., Mali, H.S., and Dixit, A. (2021). Mechanical behavior and fracture toughness characterization of high strength fiber reinforced polymer textile composites. *Iranian Polymer Journal* 30 (2): 193–233.

97 Dayyoub, T., Maksimkin, A.V., Senatov, F.S. et al. (2020). Treating UHMWPE surface for enhancing the adhesion properties by cellulose grafting. *International Journal of Adhesion and Adhesives* 98: 102535.

98 Wang, H.Q., Quan, J.Y., Wang, X.W. et al. (2021). The effect of shrinkage on the structure and properties of ultra-high molecular weight polyethylene fibers with different concentration. *Journal of Applied Polymer Science* 138 (15): e50198.

99 Wang, X., Zhang, Y., Sun, Y. et al. (2020). Research progress of ultra high molecular weight polyethylene. *Chemical Industry and Engineering Progress* 39 (9): 3403–3420.

100 Shang, Q., Zhao, H., Zhao, N. et al. (2020). Research progress on surface modification of ultra-high molecular weight polyethylene fiber. *Polymer Bulletin* 4: 22–29.

101 Tang, L., Guo, Z., Lei, S. et al. (2019). Advances in ultra-high molecular weight polyethylene for the production of fiber. *Petrochemical Technology* 48 (8): 864–870.

102 Song, B., Liu, Z.D., Chen, L. et al. (2022). Enhancing interfacial adhesion and mechanical performance of PBO fibers composites through mussel-inspired molecular regulation of interphase structure. *Polymer Composites* 43 (1): 454–466.

103 Naveen, J., Jayakrishna, K., Sultan, M.T.B., and Amir, S.M.M. (2020). Ballistic performance of natural fiber based soft and hard body armour: a mini review. *Frontiers in Materials* 7: 608139.

104 Chen, Y., Zhang, C., and Chen, P. (2021). Surface graft modification of domestic PBO fiber by atmospheric air plasma. *Chinese Journal of Materials Research* 35 (9): 641–650.

105 Chen, F., Zhang, M., Chen, L. et al. (2020). Anti-ultraviolet aging property and functionality enhancement of poly(p-phenylene benzobisoxazole) fiber coated with polydopamine. *Polymer Materials Science & Engineering* 36 (5): 153–160.

106 Liu, Z., Chen, B., and Chen, P. (2020). Treatment of oxygen dielectric barrier discharge plasma on PBO fiber surface and influence on its BMI composites. *Chinese Journal of Materials Research* 34 (2): 109–117.

107 Zeng, J., Kong, H., Du, X. et al. (2021). Surface modification of PBO fibers with 2,2-Bis (3-amino-4-hydroxyphenyl) hexafluoropropane in supercritical carbon dioxide for enhancing interfacial strength. *Materials Today Chemistry* 20: 100426.

108 Zheng, L.H., Yuan, L., Liang, G.Z., and Gu, A.J. (2019). An in situ (K0.5Na0.5)NbO3-doped barium titanate foam framework and its cyanate ester resin composites with temperature-stable dielectric properties and low dielectric loss. *Materials Chemistry Frontiers* 3 (4): 726–736.

109 Wen, Y., Meng, X., Fan, W. et al. (2019). Study on polyimide fiber reinforced resin matrix composites. *New Chemical Materials* 47 (1): 57–61.

110 Yuan, Y., Diao, S., Zhao, C.D. et al. (2020). Preparation of hollow glass microsphere/organic silicone resin composite material with low dielectric constant by in-situ polymerization. *Silicon* 12 (6): 1417–1423.

111 Li, W.C., Huang, W., Kang, Y. et al. (2019). Fabrication and investigations of G-POSS/cyanate ester resin composites reinforced by silane-treated silica fibers. *Composites Science and Technology* 173: 7–14.

112 Meinders, R., Murphy, D., Taylor, G. et al. (2021). Development of fiber-reinforced transparent composites. *Polymers & Polymer Composites* 29 (9): S826–S834.

113 Liao, L., Peng, Z., Zhang, M., and Ruan, W. (2021). Research progress of low dielectric polymers for high-frequency printed circuit boards. *Journal of Functional Polymers* 34 (4): 320–335.

114 Martin, S.J., Godschalx, J.P., Mills, M.E. et al. (2000). Development of a low-dielectric-constant polymer for the fabrication of integrated circuit interconnect. *Advanced Materials* 12 (23): 1769–1778.

115 Wang, L., Yang, J., Cheng, W.H. et al. (2021). Progress on polymer composites with low dielectric constant and low dielectric loss for high-frequency signal transmission. *Frontiers in Materials* 8: 774843.

116 Huan, X., Shi, K., Yan, J. et al. (2020). High performance epoxy composites prepared using recycled short carbon fiber with enhanced dispersibility and interfacial bonding through polydopamine surface-modification. *Composites Part B: Engineering* 193: 107987.

117 Meireman, T., Daelemans, L., Rijckaert, S. et al. (2020). Delamination resistant composites by interleaving bio-based long-chain polyamide nanofibers through optimal control of fiber diameter and fiber morphology. *Composites Science and Technology* 193: 108126.

118 Qi, Q., Zheng, P., Lei, Y., and Liu, X. (2019). Design of bi-modal pore structure polyarylene ether nitrile/SiO2 foams with ultralow-k dielectric and wave transparent properties by supercritical carbon dioxide. *Composites Part B: Engineering* 173: 106915.

119 Yin, X.W., Cheng, L.F., Zhang, L.T. et al. (2017). Fibre-reinforced multifunctional SiC matrix composite materials. *International Materials Review* 62 (3): 117–172.

120 Zhang, C.L., Ye, F., Cheng, L.F. et al. (2021). Electromagnetic wave-transparent porous silicon nitride ceramic prepared by gel-casting combined with in-situ nitridation reaction. *Journal of the European Ceramic Society* 41 (15): 7620–7629.

121 Zhang, N.X., Jiang, H., Bai, M. et al. (2018). An exploration of finite frequency selective surface fringe effect to wave-transparent mechanism. In: *2018 International Applied Computational Electromagnetics Society Symposium in China (ACES-CHINA 2018).*

122 Yin, J., Ma, W., Gao, Z. et al. (2022). An electromagnetic wave-transparent model based on plain fabric structural features. *Textile Research Journal* (8): 00405175221123064.

123 Yuan, H., Meng, Q., Zhang, H. et al. (2021). Preparation and properties of novel silicon-containing polyarylacetylene resin based wave-transparent composite. *Acta Materiae Compositae Sinica* 38 (11): 3629–3639.

124 Zhang, T., Sun, S., Cai, J.Q. et al. (2020). A new precursor to lightweight porous Si-Al-O-B ceramics with enhanced microwave transmissivity. *Ceramics International* 46 (2): 1974–1981.

125 Jiang, W.X., Zhang, X.H., Chen, D. et al. (2021). High performance low-k and wave-transparent cyanate ester resins modified with a novel bismaleimide hollow polymer microsphere. *Composites Part B: Engineering* 222: 109041.

126 Zhang, L., Mao, J.L., Wang, S. et al. (2019). Benzoxazine based high performance materials with low dielectric constant: a review. *Current Organic Chemistry* 23 (7): 809–822.

127 Jianjun, Y., Wensuo, M., Zuobin, G. et al. (2022). Electromagnetic wave-transparent model for 2D woven composites ellipsoid radome. *Mechanics of Advanced Materials and Structures* 77: 1–7.

128 Li, X.T., Zhu, X.M., Dong, J. et al. (2021). Preparation of low-dielectric permittivity polyimide resins with high surface activity from chemically bonded hyperbranched polysiloxane. *Chinese Journal of Polymer Science* 39 (9): 1200–1210.

129 Zu, Y., Zhang, F., Chen, D. et al. (2020). Wave-transparent composites based on phthalonitrile resins with commendable thermal properties and dielectric performance. *Polymer* 198: 122490.

130 Qu, C.Y., Chang, J.Y., Liu, C.W. et al. (2020). Novel allyl and propenyl monomers for modification of the bismaleimide resins, with excellent dielectric properties and high glass transition temperatures. *High Performance Polymers* 32 (1): 116–126.

131 MacDonald, M.E. (2019). An overview of radomes for large ground-based antennas. *IEEE Aerospace and Electronic Systems Magazine* 34 (10): 36–43.

132 Martellosio, A., Pasian, M., Perregrini, L. et al. (2017). High-frequency radomes for polar region ground stations: the state of the art and novel developments of radome technologies. *IEEE Antennas and Propagation Magazine* 59 (6): 88–101.

133 Qamar, Z., Aboserwal, N., and Salazar-Cerreno, J.L. (2020). An accurate method for designing, characterizing, and testing a multi-layer radome for mm-wave applications. *IEEE Access* 8: 23041–23053.

134 Xu, W.Y., Duan, B.Y., Li, P., and Qiu, Y.Y. (2017). Study on the electromagnetic performance of inhomogeneous radomes for airborne applications-part II: the overall comparison with variable thickness radomes. *IEEE Transactions on Antennas and Propagation* 65 (6): 3175–3183.

135 Liu, R.B., Han, Z.W., Li, W.F. et al. (2016). Improvement of the interfacial shear strength of poly(p-phenylene benzobisoxazole) fiber/epoxy resin composite via a novel surface coating agent. *Polymer Composites* 37 (4): 1198–1205.

136 Zimmermann-Ptacek, J., Muggli, M., Wildhack, S. et al. (2018). Thermal, dielectric, and mechanical properties of h-BN-filled PTFE composites. *Journal of Applied Polymer Science* 135 (44): 46859.

137 Wang, Y.Q., Zeng, Z., Gao, M., and Huang, Z.Y. (2021). Hygrothermal aging characteristics of silicone-modified aging-resistant epoxy resin insulating material. *Polymers* 13 (13): 2145.

138 Zou, C.R., Li, B., Liu, K. et al. (2019). Mechanical properties and interfacial characteristics of 2.5D SiNOf/BN wave-transparent composites. *Journal of the European Ceramic Society* 39 (10): 3013–3022.

139 Yang, B., Mao, Y.Y., Zhang, Y.H. et al. (2019). Fast-curing halogen-free flame-retardant epoxy resins and their application in glass fiber-reinforced composites. *Textile Research Journal* 89 (18): 3700–3707.

2

Wave-Transparent Mechanism, Test Methods for Dielectric Properties and Wave-Transparent Models of Wave-Transparent Materials

2.1 Wave-Transparent Mechanism of Wave-Transparent Materials

The ε and $\tan\delta$ of wave-transparent materials are important parameters for evaluating the wave-transparent performances [1–3], usually requiring their $\varepsilon < 4$ and $\tan\delta < 10^{-2}$ [4–6]. The ε is an important physical parameter characterizing the dielectric and polarization performance of wave-transparent materials [7–9], which is the capacitance ratio before and after putting the wave-transparent material in the same capacitors. Tan δ refers to the electromagnetic waves in the transmission process inside the wave-transparent material that consume part of the energy and are converted into heat energy [10–12]. The relationship between the total loss value (L) of the electromagnetic wave transmission in the wave-transparent materials, and ε, $\tan\delta$, and electromagnetic wave frequency (f) is shown in Eq. (2.1) [13–15].

$$L \propto \varepsilon^{1/2} f \cdot \tan\,\delta \tag{2.1}$$

L is proportional to ε and $\tan\delta$. As ε and $\tan\delta$ decrease, the L of electromagnetic waves in the transmission process decreases, and the wave-transparent rate increases. According to the impedance matching theory [16–18], The reflection, energy loss, and transmission of electromagnetic waves occur when they are incident on the wave-transparent materials (Figure 2.1). The larger the ε of the wave-transparent materials is, the more the reflected electromagnetic waves and the smaller the wave-transparent rate are. Electromagnetic wave loss is mainly due to the wave-transparent materials polarization in the electric field, resulting in the electromagnetic waves energy loss (electron polarization, atomic polarization, and orientation polarization are resonance or elastic processes, which would cause energy dissipation).

Polymer Matrix Wave-Transparent Composites: Materials, Properties, and Applications, First Edition.
Junwei Gu, Yusheng Tang, Jie Kong, and Jing Dang.

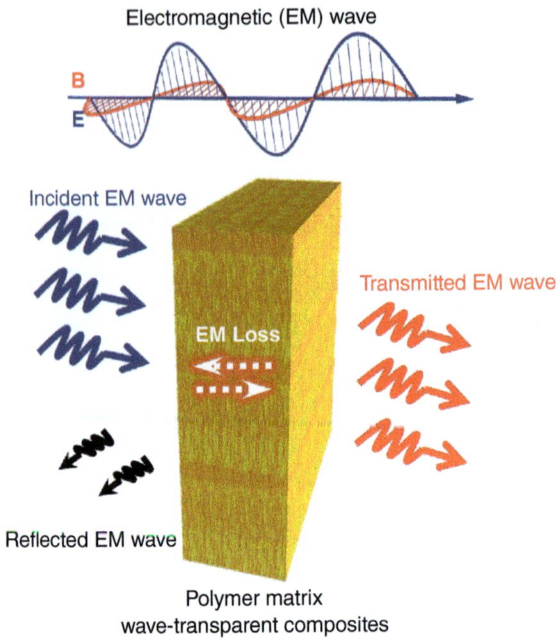

Figure 2.1 Schematic diagram of electromagnetic wave transmission in wave-transparent materials.

The propagation state of electromagnetic waves in wave-transparent materials is usually measured by three physical quantities: energy loss (A), reflection loss ($|\Gamma|^2$), and transmittance ($|T|^2$) [19–21], whose relationships are shown below:

$$A = \frac{2\pi d\varepsilon \tan \delta}{\lambda(\delta - \sin^2 \theta)^{1/2}} \tag{2.2}$$

$$\Gamma = \frac{(\varepsilon - \sin^2 \theta)^{1/2} - \varepsilon \cos \theta}{(\varepsilon - \sin^2 \theta)^{1/2} + \varepsilon \cos \theta} \tag{2.3}$$

$$|T|^2 = \frac{(1 - \Gamma^2)^2}{(1 - \Gamma^2)^2 + 4\Gamma^2 \sin^2 \varphi} \tag{2.4}$$

$$\varphi = \frac{2\pi d}{\lambda}(\varepsilon - \sin^2 \theta)^{1/2} \tag{2.5}$$

$$|T|^2 + A + |\Gamma|^2 = 1 \tag{2.6}$$

where d is the thickness of the wave-transparent materials; λ is the wavelength of electromagnetic waves; θ is the angle of incidence of electromagnetic waves on the surface of the wave-transparent materials.

The wave-transparent performances of the wave-transparent materials are closely related to their ε and $\tan \delta$. When the electromagnetic waves are incident on the surface of the wave-transparent materials, the larger the ε and $\tan \delta$ are, the more reflection and loss occur, resulting in a relatively lower wave-transparent rate. Therefore, the key to preparing wave-transparent materials with excellent wave-transparent performances is to reduce their ε and $\tan \delta$ values.

2.2 Dielectric Parameter Equations for Wave-Transparent Materials

2.2.1 Clausius–Mossotti Equation

The dielectric properties of the wave-transparent materials depend mainly on the polarizability of their molecular structures under the applied electric field [22–24]. The polarizability of the wave-transparent materials is measured by the molecular polarizability (α). In addition, the ε is a macroscopic physical quantity that describes the polarization of mediums. The polarity magnitude of the wave-transparent materials can be discerned from its ε [25, 26]. Generally, wave-transparent materials with $\varepsilon > 3.6$ are polar substances; their ε between 2.8 and 3.6 are weakly polar substances; and their $\varepsilon < 2.8$ are nonpolar substances.

The α is a microphysical quantity that reflects the polarization characteristics of molecules [27–29]. The relationship between macroscopic dielectric properties and microscopic molecular polarizability of the wave-transparent materials can be reflected by the Clausius–Mossotti equation [30–32], as shown in Eq. (2.7).

$$\varepsilon = 1 + 3\frac{N\alpha}{3\varepsilon_0 - N\alpha} \tag{2.7}$$

where N is the number of atoms per unit volume, laterally reflecting the density of polarized molecules; ε_0 is the dielectric constant of the vacuum. It can be deduced that the ε of the wave-transparent materials is proportional to α. Therefore, the most effective way to reduce the ε of wave-transparent materials is to reduce the α. Clausius–Mossotti equation is an empirical formula obtained for single-phase substances at low frequencies, which can provide a theoretical basis for studying the relationship between ε and internal structures of materials. Zhang et al. [30] precisely modulated the ε and $\tan\delta$ values of polyimide resins based on the Clausius–Mossotti equation. The polarization rate of polymer molecular structure and molecular chain stacking would affect the ε and $\tan\delta$, and the introduction of low polarization rate groups (e.g. CF_3) and large free volume molecular structure could be beneficial to reducing the ε and $\tan\delta$ of polymers.

Under the action of external electric field, the wave-transparent materials produce induced dipole moments along the electric field direction and generate polarized charge [33–35]. The corresponding polarization type includes electron, atomic, ion, orientation, and interface polarization [36–38].

2.2.1.1 Electron Polarization

Under the action of external electric field, the peripheral electron cloud in the molecule, atom, or ion constituting the dielectric is elastically displaced relative to the nucleus, and the phenomenon of induced dipole moment is called electron polarization [39, 40] (Figure 2.2a). The corresponding polarization process takes a very short time, about 10^{-15} to 10^{-13} s. When the external electric field is removed, the induced dipole moments are immediately restored without energy loss. Therefore, electron polarization is also called reversible polarization and elastic polarization. Passchier et al. [41] designed and constructed a Compton

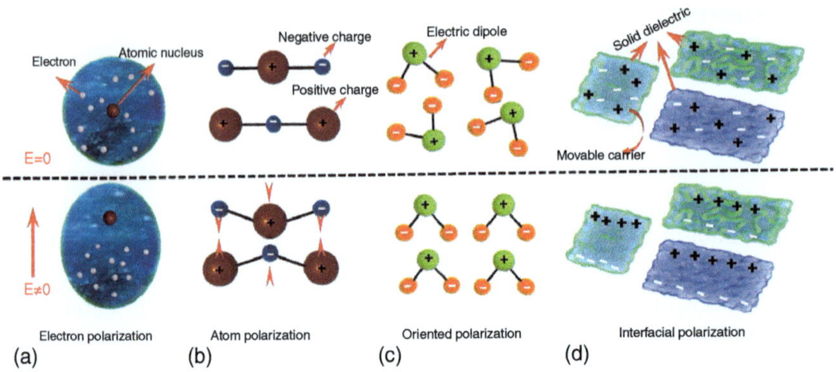

Figure 2.2 Dielectric polarization diagram: Electron polarization (a), atomic polarization (b), orientation polarization (c), and interface polarization (d).

cyclotron to measure the polarization state of longitudinal electron beams. The cyclotron operated normally on the internal target device where the polarized electrons occurred, and the measurement of electron polarization could be performed during the internal electron operation. Jenkins et al. [42] used the nonlocal dielectric theory to describe the electron polarization caused by perturbation field and to characterize the spontaneous quantum mechanical ups and downs in the electron polarization. According to undulation dissipation theorem, the molecules in equilibrium with radiation at a fixed temperature correlated the undulation of spontaneous polarization with the imaginary part of the nonlocal polarizability density, leading to a positive correlation of fluctuating electron polarization with e permittivity function.

2.2.1.2 Atomic Polarization

During atomic polarization, the molecular skeleton of medium is deformed under the action of external electric field, and the relative displacement of the positive and negative charge centers in the molecule occurs, which takes about 10^{-13} s. The deformation process is accompanied by slight energy loss (Figure 2.2b). Li et al. [43] analyzed the process of atomic polarization by using wave function curl, the real and imaginary parts of the wave function, to propose the calculation method of atomic polarization radius. Results showed that the wave function calculation results were generally better than those of Norman's rule.

2.2.1.3 Ion Polarization

The relative elastic displacement occurs between positive and negative ions, resulting in induced dipole moment (ion polarization) under the action of the external electric field. The polarization schematic is similar to that of atomic polarization. Ion polarization generally exists inside the crystalline phase of ionic crystals and ceramic media composed of ionic bonds. Like electron polarization, the formation and disappearance times of the ion polarization are extremely short without energy loss. Wang et al. [44] investigated the ion migration and dielectric behavior of $NaNbO_3$

under the pressure of 29.1 GPa. Results showed that the interaction between Na^+ and NbO_6 octahedra enhanced the vibrational resonance damping of the system. The ion polarization and space charge polarization of the interfacial layer of $NaNbO_3$ jointly affected the dielectric properties at high-pressure state.

2.2.1.4 Orientated Polarization

Orientated polarization, also known as dipole polarization, means that polar molecules with permanent dipole moments are oriented under the action of the external electric field [45] (Figure 2.2c). In the natural state, the distribution of molecules in each direction is equal due to their irregular thermal motion. The turning of polar molecules along the direction of electric field must overcome their own inertia and rotation resistance, resulting in a long time of orientation polarization, generally less than 10^{-9} s. What's more, the orientation polarization of polymer includes the orientation of the side chain groups and the whole molecular chain. Therefore, the time of orientation polarization is similar to that of mechanical relaxation time. Wu et al. [46] studied the oriented polarization of single molecules under the action of external electric field by measuring the changes in the polarization characteristics of single-molecule fluorescence. Results showed that the polar-monomolecular distribution of poly(methyl methacrylate) films was oriented under the external electric field.

2.2.1.5 Interfacial Polarization

Interfacial polarization is the main form of polarization in inhomogeneous mediums. Under the action of the external electric field, carriers of the two-phase medium are trapped by defects or accumulated in the interfaces, which cause the uneven distribution of charge in the mediums, resulting in electric dipole moments [47–49] (Figure 2.2d). Compared with orientated polarization, interfacial polarization takes a relatively longer time, ranging from fraction of second to minutes. Ma et al. [50] studied the dielectric behavior of polymer-derived amorphous silicon carbonitride. Results showed that the interface polarization occurred at the interfaces of silicon carbon nitride and free carbon phases due to the migration and accumulation of charge carriers. In addition, with the increase in temperature, the interface polarization was enhanced, which would increase the ε and $\tan\delta$ of amorphous silicon carbonitride.

Among the above five polarization types, electronic polarization and orientation polarization are collectively referred to as deformation polarization or induced polarization, and their polarizability does not change with changing temperatures. Ion polarization does not normally occur in polymers, and interfacial polarization tends to occur in multiphase mediums. For polymer matrix wave-transparent composites, the interface between the reinforced fibers and polymer matrix is extremely prone to interfacial polarization [51–54]. The carriers of polymer matrix wave-transparent composites tend to accumulate on the weak interface, leading to strong interface polarization, and electromagnetic waves are easily reflected and lost multiple times at the interface (Figure 2.3). In order to prepare polymer matrix wave-transparent composites with a high wave-transparent rate (low ε and $\tan\delta$),

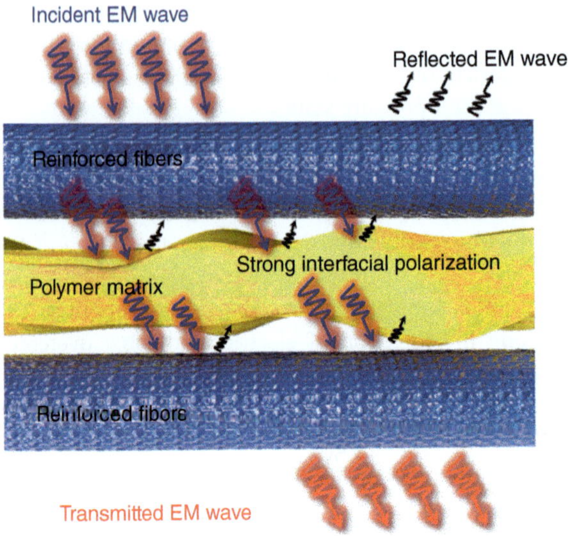

Figure 2.3 Schematic diagram of electromagnetic wave transmission at the interface of polymer matrix wave-transparent composites.

it is necessary to reduce the polarized molecular density and polarizability of the reinforced fibers and polymer matrix [55–57]. On the other hand, it improves the interfacial compatibility between the reinforced fibers and polymer matrix in favor of reducing the interfacial polarization [58, 59]. In our previous work, Gu et al. [60] introduced low polarization and large free volume polysiloxane (HO-POSS) at the interface between the quartz fibers and cyanate ester (CE) resins to prepare quartz fibers/POSS/CE wave-transparent composites. The introduction of HO-POSS enhanced the stability of the interfacial layer and reduced the interfacial polarization of the composites. The ε and $\tan\delta$ of quartz fibers/POSS/CE wave-transparent composites were 2.55 and 0.002, respectively, lower than the ε (3.06) and $\tan\delta$ (0.004) of quartz fibers/CE wave-transparent composites. In addition, Gu et al. [61] introduced epoxy-based polysiloxane (epoxy-POSS) at the interface of PBO fibers and CE resins to prepare modified PBO fibers/CE wave-transparent composites (Figure 2.4). The result showed that the ε and $\tan\delta$ of the modified PBO fibers/CE wave-transparent composites (Sample 3) were 2.85 and 0.0047, respectively, lower than the ε (3.06) and $\tan\delta$ (0.006) of PBO fibers/CE wave-transparent composites (Sample 1).

2.2.2 Maxwell–Garnett Equation

For biphasic dielectric materials, the weakly conductive spherical particles with dielectric constant of ε_i and conductivity of σ_i are uniformly dispersed in the polymer with dielectric constant of ε_m. According to the electromagnetic field theory, the basic relation (Eq. (2.8)) of the effective dielectric constant (ε_eff)

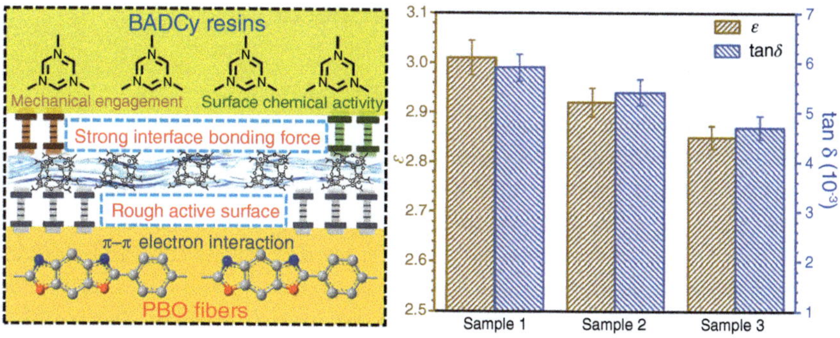

Figure 2.4 Interface diagram of modified PBO fibers/BADCy wave-transparent composites and their ε and $\tan\delta$ [61].

for biphasic dielectric materials is obtained by the method of separation for variables [62–64].

$$\frac{\varepsilon_{\mathrm{eff}} - \varepsilon_{\mathrm{m}}}{\varepsilon_{\mathrm{eff}} + 2\varepsilon_{\mathrm{m}}} = V_1 \frac{\varepsilon_{\mathrm{i}} - \varepsilon_{\mathrm{m}}}{\varepsilon_{\mathrm{i}} + 2\varepsilon_{\mathrm{m}}} \tag{2.8}$$

where V_1 represents the volume fraction of the spherical fillers. When V_1 is greater than 20%, the $\varepsilon_{\mathrm{eff}}$ is closer to the actual ε of the biphasic dielectric materials. Maxwell–Garnett equation is an empirical formula for two-phase insulating substances at low frequencies [65, 66], which can be regarded as the extension equation of Clausius–Mossotti equation. Xue et al. [67] calculated the ε of the two-phase random polystyro-barium titanate composites with interfacial shell based on the Maxwell–Garnett equation and Bruggeman theory, which was in good agreement with the experimental data. In addition, the obtained results were also suitable for studying the dielectric properties of two-phase random composites with spherical interfacial shells. However, Maxwell–Garnett and Clausius–Mossotti equations ignore the influences of temperature on the movement of charged particles and cannot reflect the temperature characteristics of dielectric properties.

2.2.3 Lichtenecker Equation

For multiphase composites, especially polymer matrix wave-transparent composites, the theoretical ε can be calculated using the Lichtenecker equation (Eq. (2.9)) according to the volume fraction and ε of each component [68–70].

$$\log\varepsilon = V_0 \log\varepsilon_0 + V_1 \log\varepsilon_1 + V_2 \log\varepsilon_2 \tag{2.9}$$

where ε_0 is the dielectric constant of air, V_1, V_2, and V_0, respectively, represent the volume fraction of each phase, and $V_0 + V_1 + V_2 = 1$. Lichtenecker equation ties the epsilons of each phase together. According to Eq. (2.9), if the ε of each phase can be obtained, the ε of the multiphase composites would be further predicted. Ghita et al. [71] calculated the ε of glass fibers/PTFE resin composites using Lichtenecker equation, and the calculated values (ε_{c}) were in good agreement with the measured values (ε_{m}).

2.3 Test Methods for Dielectric Properties of Wave-Transparent Materials

Dielectric properties (ε and $\tan\delta$) are the main indexes to evaluate the wave-transparent rate of wave-transparent materials. The corresponding wave-transparent rate can be obtained by direct measurement method or by measuring the ε and $\tan\delta$ and calculating the wave-transparent rate according to Eq. (2.4) (i.e. indirect method). The direct method requires a specific band of transmitting and receiving antennas. It needs high requirements in the measurement environment, so it is generally suitable for measuring the wave-transparent rate of formed products (such as radomes). The indirect method is relatively easy to implement in the laboratory by measuring the ε and $\tan\delta$ of the wave-transparent materials and further calculating the wave-transparent rate. The indirect method can meet the testing requirements in different electromagnetic wave bands. Therefore, this section focuses on the indirect method, emphasizing the testing methods for dielectric properties.

The testing methods for dielectric properties of wave-transparent materials are mainly measured by resonance and nonresonance methods [72–75]. The complex permittivity can be obtained *via* the resonance method by testing the resonant frequency (f) [76] and quality factor (Q) [77, 78] of the cavity before and after loading the samples. The resonant method mainly includes the perturbation method, high Q cavity method, and quasi-optical cavity method, which are suitable for testing dielectric properties of low-loss materials. The nonresonant method measures the reflection and transmission coefficient in the transmission system before and after loading the samples and calculates the complex dielectric constant. Nonresonant methods mainly include transmission line method and free space method [79–81].

2.3.1 Resonance Method

2.3.1.1 Perturbation Method

The perturbation method mainly calculates the ε and $\tan\delta$ of the samples by comparing the changes of the electromagnetic field structures and resonance parameters in the cavity before and after the samples are put into the cavity [82–84]. The perturbation method is generally divided into rectangular and cylindrical resonators, as shown in Figure 2.5 [85, 86].

The f and Q values of the cavity before and after the perturbation can be measured by the vector network analyzer, and then the ε of the samples is calculated according to the perturbation theory formula.

The ε of rectangular samples and round bar samples can be obtained by Eqs. (2.10) and (2.11), respectively:

$$\varepsilon' = 1 + 2K\left[\frac{aL}{\pi d_s^2}\right]\left[\frac{f_{00} - f_{00}}{f_{0s}}\right] \tag{2.10}$$

$$\varepsilon' = 1 + \frac{1}{2}K\left[\frac{aL}{\pi Wt}\right]\left[\frac{f_{00} - f_{00}}{f_{0s}}\right] \tag{2.11}$$

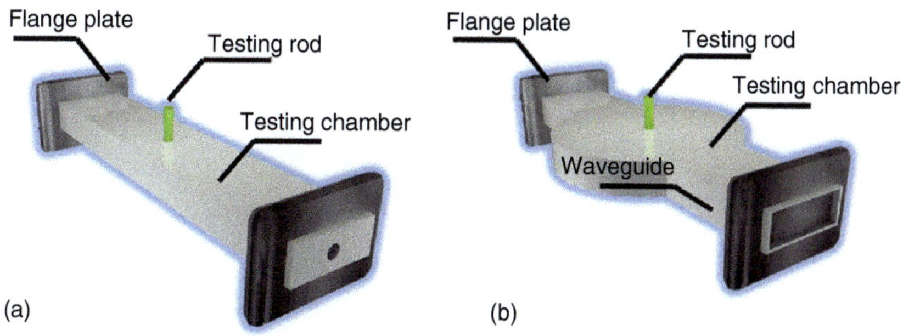

Figure 2.5 Schematic diagram of perturbation test device: Rectangular perturbation resonator (a) and cylindrical perturbation resonator (b).

where f_{00} is the cavity resonance frequency; f_{0s} is the resonant frequency of the test cavity after placing the sample. d_s is the diameter of the round bar sample; a is the width of rectangular cavity; L is the length of rectangular cavity; W is the width of rectangle bar sample; t is the thickness of rectangle bar sample; K is the correction factor of the perturbation formula.

The $\tan\delta$ of rectangular sample and round bar sample can be obtained by Eqs. (2.12) and (2.13), respectively:

$$\tan\,\delta = \frac{aL}{\pi\varepsilon' d_s^2}\left[\frac{1}{Q_{0s}} - \frac{1}{Q_{00}}\right] \tag{2.12}$$

$$\tan\,\delta = \frac{aL}{4\varepsilon' Wt}\left[\frac{1}{Q_{0s}} - \frac{1}{Q_{00}}\right]\eta \tag{2.13}$$

where Q_{00} is the no-load quality factor of the test cavity; Q_{0s} is the no-load quality factor of the cavity after putting the sample into the cavity. The perturbation method based on dielectric perturbation theory is suitable for testing dielectric properties of low-loss materials. However, for high-loss materials, the perturbation method produces too much perturbation, and its test accuracy is greatly reduced.

2.3.1.2 High *Q* Cavity Method

The high Q cavity method is divided into fixed frequency method (fixed f, changed the cavity size) and frequency conversion method (fixed cavity size, changed f) [87–89]. The samples and the wall of high-Q cavity are in noncontact state, and the correction of the slit is generally not required, as shown in Figure 2.6. In addition, the high Q cavity method presents high test accuracy for the samples with low loss. The high Q cavity method has no special limit on the test frequency, but the lower the test frequency, the larger the volume of the high Q resonant cavity system is, and the larger the sample size is, thus increasing the difficulty of sample preparation. When the sample is inserted at the bottom of the high Q cavity, the f value of the cylindrical resonant cavity will be changed due to the difference in ε of the samples and ε_0 of the air, and the Q value of the cylindrical resonator decreases with the increase of $\tan\delta$. Therefore, the ε and $\tan\delta$ can be calculated from the f and Q of the cylindrical resonator before and after loading the samples.

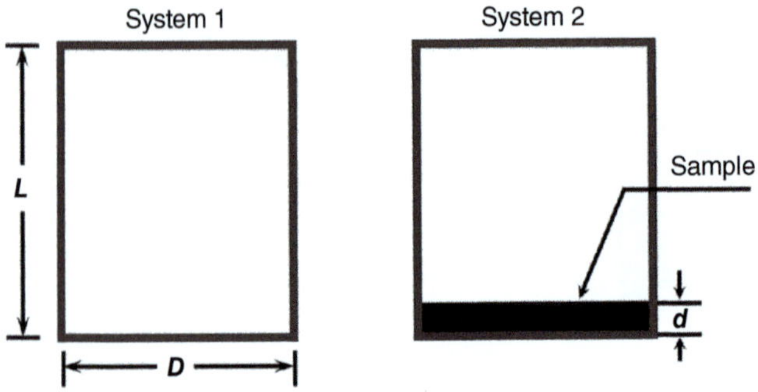

Figure 2.6 Schematic diagram of high Q cavity test principle.

The derivation process of ε for samples is as follows:

For system 1:

$$(f_{00}D)^2 = \left(\frac{c}{\pi}X_{01}\right)^2 + \left(\frac{cn}{2}\right)^2\left(\frac{D}{L}\right)^2 \tag{2.14}$$

For system 2:

$$\beta_\varepsilon^2 = \left(\frac{2\pi f_{0\varepsilon}}{c}\right)\varepsilon' - \left(\frac{2X_{01}}{D}\right)^2 \tag{2.15}$$

$$\beta_0^2 = \left(\frac{2\pi f_{0\varepsilon}}{c}\right) - \left(\frac{2X_{01}}{D}\right)^2 \tag{2.16}$$

$$\frac{\tan\beta_\varepsilon d}{\beta_\varepsilon} - \frac{\tan\beta_0(1-d)}{\beta_0} = 0 \tag{2.17}$$

where, f_{00} is the resonant frequency of the cavity (system 1) of the cylindrical resonant cavity; $f_{0\varepsilon}$ is the resonant frequency of the cylindrical cavity with loading samples (system 2); X_{01} is the root of the Bessel function; D is the diameter of the cylindrical resonant cavity; L is the resonant length of the space-time cavity (system 1) without sample; c is the speed of light; d is the thickness of the sample; β_0 and β_ε are the transmission phase constants of the air part (system 1) and the loading sample part (system 2) of the cylindrical resonator. The D, n, X_{01}, and c are known, L can be obtained by measuring f_{00}, β_0 can be obtained by measuring $f_{0\varepsilon}$, and β_0 can be obtained by Eqs. (2.16) and (2.17), ε' can be calculated.

The relationship between energy storage and energy consumption of system 2, it can be deduced as follows:

$$\tan\delta = \left(1 + \frac{u}{pv\varepsilon'}\right)\cdot\left(\frac{1}{Q_{0\varepsilon}} - \frac{1}{Q_{00}}\right) \tag{2.18}$$

where,

$$\frac{1}{Q_{00}^1} = \frac{1}{Q}\left(\frac{f_{00}}{f_{0\varepsilon}}\right)^{5/2}\cdot\frac{\left(\frac{2X_{01}}{D}\right)^2\cdot(pv+u) + D\left(p\beta_\varepsilon^2 + \beta_0^2\right)}{(pv\varepsilon' + u)\cdot\left[\left(\frac{2X_{01}}{D}\right)^2\cdot\left(1 - \frac{D}{L}\right) + \left(\frac{2\pi f_{00}}{c}\right)^2\cdot\frac{D}{L}\right]} \tag{2.19}$$

$$p = \left[\frac{\sin \beta_0(L-d)}{\sin \beta_0 d}\right]^2 \tag{2.20}$$

$$u = 2(L-d) - \frac{\sin 2\beta_0(L-d)}{\beta_\varepsilon} \tag{2.21}$$

$$v = 2d - \frac{\sin 2\beta_0 d}{\beta_\varepsilon} \tag{2.22}$$

Therefore, the cavity quality factor (Q_{00}) of the cylindrical resonant cavity and the unloaded quality factor ($Q_{0\varepsilon}$) of the test cavity after putting the samples into the test cavity are measured, and the $\tan\delta$ can be obtained by substituting the above equations.

2.3.1.3 Quasi-Optical Cavity Method

Quasi-optical resonator is developed from an optical Fabry–Perot interferometer and quasi-optical cavity mainly contains two forms: the first kind of mirror is composed of two spherical mirrors, the sample loaded in the middle between the two; the second kind of mirror is composed of a plane mirror and a spherical mirror, and the sample is loaded on the plane mirror, as shown in Figure 2.7 [89–91].

where R is the radius of the spherical mirror, D is the distance between the two mirrors (also known as the cavity length), and the part between the two dotted lines is the beam range in the cavity.

The quasi-optical cavity method derives ε and $\tan\delta$ by comparing the changes in resonant cavity frequency (f), cavity length, and quality factor (Q) before and after loading samples. Compared with the closed resonators (the perturbation method and the high Q cavity method), the quasi-optical cavity method presents openness, high Q value, simple working mode, sparse mode, and accurate measurement. It is suitable for the measurement of dielectric properties for mediums in high frequency (millimeter waves, submillimeter waves, and even terahertz).

The measurement accuracy and f of the quasi-optical cavity method are closely related to the thickness of the samples. Therefore, it is necessary to predict the ε and

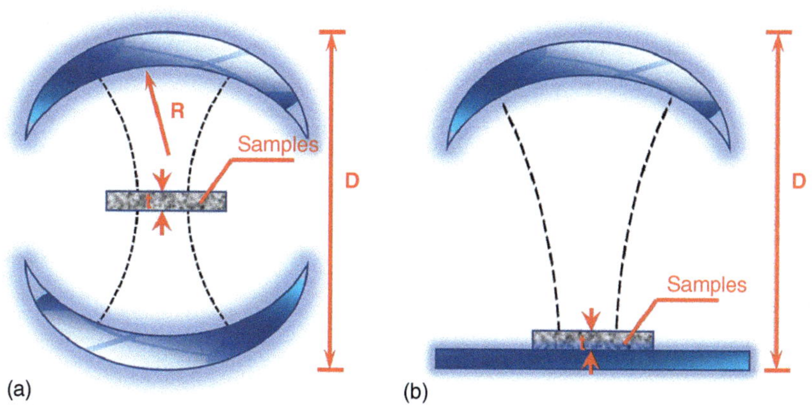

Figure 2.7 Schematic diagram for the testing principle of quasi-optical cavity method: Symmetric structure (a) and asymmetric structure (b).

$\tan\delta$ of the samples before measurement and then design the appropriate thickness to obtain accurate dielectric properties.

The calculation formula of ε can be obtained from the electric field equation:

$$\tan[n\varepsilon t - \phi(t)] = n\tan[-\varepsilon d - \phi_2(t) + \phi_2(t+d)] \tag{2.23}$$

where n is the refractive index of the medium; t is the thickness of the medium; d is the distance between the two mirrors (D) and the thickness difference between the medium (D–t). Make $\phi_{\mathrm{T}} = \phi_1(t)$, $\phi_{\mathrm{D}} = \phi_2(t+d)$–$\phi_2(t)$, giving the formula for calculating the material ε.

$$\frac{1}{n}\tan[n\varepsilon t - \phi_T] = -\tan[\varepsilon d - \phi_D] \tag{2.24}$$

$$\phi_T = \arctan\left(\frac{t}{ns_0}\right) \tag{2.25}$$

$$\phi_{\mathrm{D}} = \arctan\left(\frac{d'}{s_0}\right) - \arctan\left(\frac{t}{ns_0}\right) \tag{2.26}$$

$$w_0^2 = \frac{2}{k}\sqrt{\left(d + \frac{t}{n^2}\right)\cdot\left(R_0 - d - \frac{t}{n}\right)} \tag{2.27}$$

$$s_0 = \sqrt{d'(R_0 - d')} \tag{2.28}$$

$$d = D - t \tag{2.29}$$

$$d' = d + \frac{t}{n} \tag{2.30}$$

$$n = \sqrt{\varepsilon_r} \tag{2.31}$$

where w_0 is the radius of the girth; k is the wave number, $k = 2\pi/\lambda$. The thickness t of the sample is measured by experiment, and the resonant parameter is obtained by testing. The measured ε_{r} of the materials can be obtained by substituting the results into the above equations. In addition, the $\tan\delta$ expression is shown as follows:

$$\tan\delta = \frac{1}{Q_e}\cdot\frac{t\Delta - d}{t\Delta + \sin[2(\varepsilon d - \phi d)]} \tag{2.32}$$

$$\frac{1}{Q_e} = \frac{1}{Q_{0s}} - \frac{1}{Q_1} \tag{2.33}$$

$$\frac{1}{Q_1} = Q_{00}\cdot\frac{2(t\Delta + d)}{D(\Delta + 1)} \tag{2.34}$$

$$\Delta = \frac{n^2}{n^2\cos^2(n\varepsilon t - \phi_t) + \sin^2(n\varepsilon t - \phi_t)} \tag{2.35}$$

where $Q_{0\mathrm{s}}$ is the no-load quality factor after placing the samples; Q_1 represents the no-load quality factor of the ideal loss-free samples; Q_{00} represents the intrinsic quality factor of the resonator.

2.3.2 Nonresonant Method

2.3.2.1 Transmission Line Method

The transmission line method is a classical testing method for dielectric properties [92–94]. It is necessary to put the samples into the transmission line terminal short circuit (SCL) measure the transmission parameters of the transmission line, then calculate the complex dielectric constant (ε_r*) of the samples according to the amplitude and phase. The transmission line method presents simple measurement process and high accuracy and is suitable for liquids or high-loss mediums. At present, the transmission lines are mainly divided into coaxial lines, waveguides, or other types of transmission lines, among which rectangular waveguides are the most widely used. The experimental equipment is shown in Figure 2.8.

Assuming that the single main mode of wave conduction transmission is TE mode, and the characteristic impedance is Z_0, the operating wavelength is λ, the waveguide wavelength is λ_g, and the cut-off wavelength is λ_c. When samples are placed in the waveguide tube, the characteristic impedance (Z_ε) of the filled materials waveguide can be expressed as follows:

$$Z_\varepsilon = \frac{120\pi}{\sqrt{\varepsilon_r u_r - \left(\frac{\lambda}{\lambda_c}\right)^2}} \tag{2.36}$$

For nonconductive dielectric materials, its permeability $u_r = 1$, then:

$$Z_\varepsilon = \frac{120\pi}{\sqrt{\varepsilon_r - \left(\frac{\lambda}{\lambda_c}\right)^2}} \tag{2.37}$$

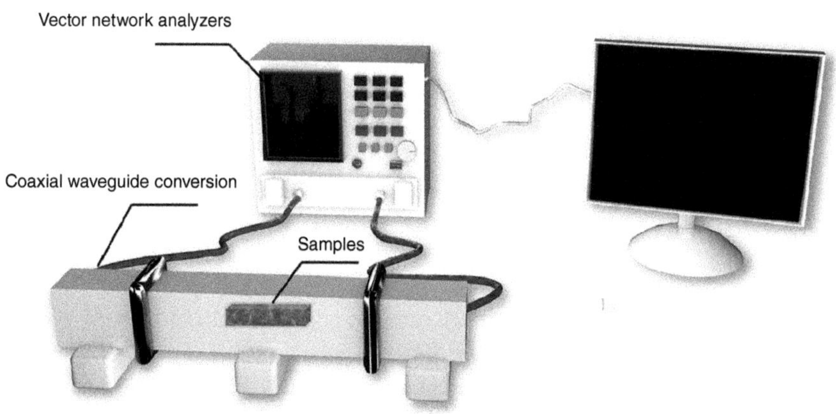

Figure 2.8 Schematic diagram of dielectric properties testing by waveguide transmission method.

Before testing, the relative dielectric constant (ε_r) of the air in the device is 1, then:

$$Z_0 = \frac{120\pi}{\sqrt{1-\left(\frac{\lambda}{\lambda_c}\right)^2}} \tag{2.38}$$

Setting $C_\varepsilon = (Z_0/Z_\varepsilon)^2$, Eq. (2.39) follows from Eqs. (2.37) and (2.38):

$$C_\varepsilon = \frac{\varepsilon_r - \left(\frac{\lambda}{\lambda_c}\right)^2}{1-\left(\frac{\lambda}{\lambda_c}\right)^2} \tag{2.39}$$

Setting $1/\lambda^2 = 1/\lambda_c^2 + 1/\lambda_g^2$, then Eq. (2.39) can be written as:

$$\varepsilon_r = \frac{C_\varepsilon + \left(\frac{\lambda_g}{\lambda_c}\right)^2}{1+\left(\frac{\lambda_g}{\lambda_c}\right)^2} \tag{2.40}$$

When C_ε is a complex number, let $C_\varepsilon = A_\varepsilon + jB_\varepsilon$, then $\varepsilon_r{}^*$ is:

$$\varepsilon_r^* = \varepsilon_r' - j\varepsilon_r'' = \frac{A_\varepsilon + jB_\varepsilon + \left(\frac{\lambda_g}{\lambda_c}\right)^2}{1+\left(\frac{\lambda_g}{\lambda_c}\right)^2} \tag{2.41}$$

Finally, the real and imaginary parts of $\varepsilon_r{}^*$ are:

$$\varepsilon_r' = \frac{A_\varepsilon + \left(\frac{\lambda_g}{\lambda_c}\right)^2}{1+\left(\frac{\lambda_g}{\lambda_c}\right)^2} \quad \varepsilon_r'' = \frac{-B_\varepsilon}{1+\left(\frac{\lambda_g}{\lambda_c}\right)^2} \tag{2.42}$$

It can be deduced from the above equations that the $\varepsilon_r{}^*$ can be calculated by measuring the propagation parameters.

2.3.2.2 Free-Space Method

The free-space method does not damage samples and allows for broadband sweeping and orientation testing. In addition, the free-space method requires large area of the samples to ensure the reception and propagation of electromagnetic waves, and the samples can be tested during wide temperature range [95–97]. The free space method is widely used in microwave remote sensing to measure ε of large area targets. Figure 2.9 shows the testing diagram of the free space method.

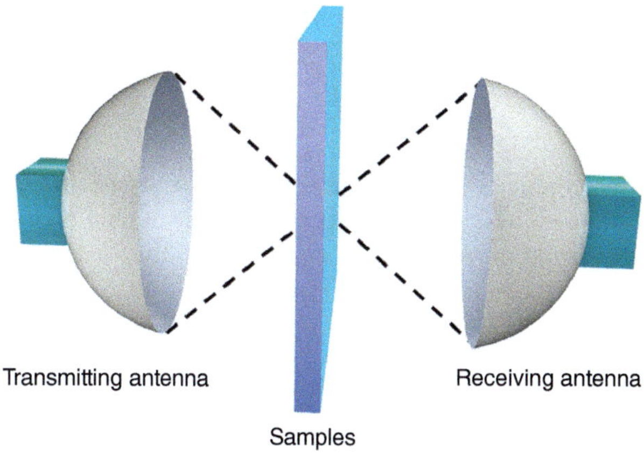

Figure 2.9 Testing diagram of the free space method.

When electromagnetic wave wavelength is much larger than surface roughness for samples, the Fresnel reflection coefficient can be used to describe the relationship between the ε_r of samples and the test angle.

$$R_H = \frac{\cos\theta - \sqrt{\varepsilon_r - \sin^2\theta}}{\cos\theta + \sqrt{\varepsilon_r - \sin^2\theta}} \tag{2.43}$$

$$R_v = \frac{\varepsilon_r\cos\theta - \sqrt{\varepsilon_r - \sin^2\theta}}{\varepsilon_r\cos\theta + \sqrt{\varepsilon_r - \sin^2\theta}} \tag{2.44}$$

where, $\varepsilon_r^* = \varepsilon_r' - j\varepsilon_r''$ is the relative complex dielectric constant of the tested samples, θ is the incident angle of the transmitting antenna, R_H is the horizontal polarization reflection coefficient, and R_v is the vector reflection coefficient in the vertical polarization mode. Usually ε meet $j\varepsilon_r'' \mid (\varepsilon_r' - \sin^2\theta) \mid < 1$, let $P = \sqrt{\varepsilon_r' - \sin^2\theta}$, then Eqs. (2.43) and (2.44) can be expressed as follows:

$$|R_H|^2 = \frac{(\cos\theta - p)^2 + \frac{\varepsilon_r''^2}{4p^2}}{(\cos\theta + p)^2 + \frac{\varepsilon_r''^2}{4p^2}} \tag{2.45}$$

$$|R_v|^2 = \frac{(\varepsilon_r'\cos - p)^2 + \varepsilon_r''^2 \cdot \left(\frac{1}{2}p - \cos\theta\right)^2}{(\varepsilon_r'\cos + p)^2 + \varepsilon_r''^2 \cdot \left(\frac{1}{2}p - \cos\theta\right)^2} \tag{2.46}$$

From the above equations, the ε_r* of the samples is only related to the reflection coefficients of R_H, R_v, and θ. As long as the amplitude and phase of the reflection parameters are measured, the ε_r* of the samples can be obtained through systematic calibration (plate or relative calibration).

2.4 Wave-Transparent Model of Wave-Transparent Materials

2.4.1 Single-Layer Wave-Transparent Model

Electromagnetic waves are oscillating particle waves derived from the electric and magnetic fields, which are in phase and perpendicular to each other [98–101]. When electromagnetic waves incident on a single-layer medium, they will reflect and transmit at the interfaces between the air and the medium [102, 103], as shown in Figure 2.10.

The region of $z < 0$ is air 1, and the region of $z > 0$ is medium 2. The electromagnetic waves of uniform plane are incident vertically from air 1 to the interface of $z = 0$. According to Maxwell's equation [104–106], electromagnetic waves propagating along the z direction are decomposed into the wave equation of electric field intensity E (x direction) and magnetic sensation intensity H (y direction) [10].

The electric and magnetic fields of incident wave in air 1 are, respectively:

$$E_i(z) = e_x E_{\text{im}} e^{-\gamma_1 z} \tag{2.47}$$

$$H_i(z) = e_z \times \frac{1}{\eta_{1c}} E_i(z) = e_y \frac{1}{\eta_{1c}} E_{\text{im}} e^{-\gamma_1 z} \tag{2.48}$$

where, $E_i(z)$ represents the electromagnetic intensity of incident waves; $H_i(z)$ represents the magnetic intensity of incident waves; E_{im} represents the electric field amplitude of incident waves; γ_1 is the propagation constant, $\gamma_1 = j\omega\sqrt{\mu_1\varepsilon_1\left(1 - j\frac{\sigma_1}{2\omega\varepsilon_1}\right)}$; η_1 represents the intrinsic impedance of the medium, $\eta_{1c} = \sqrt{\frac{\mu_1}{\varepsilon_1}}\left(1 - j\frac{\sigma_1}{\omega\varepsilon_1}\right)^{-1/2}$; ω is the change of phase in unit time, and its unit is rad/s;

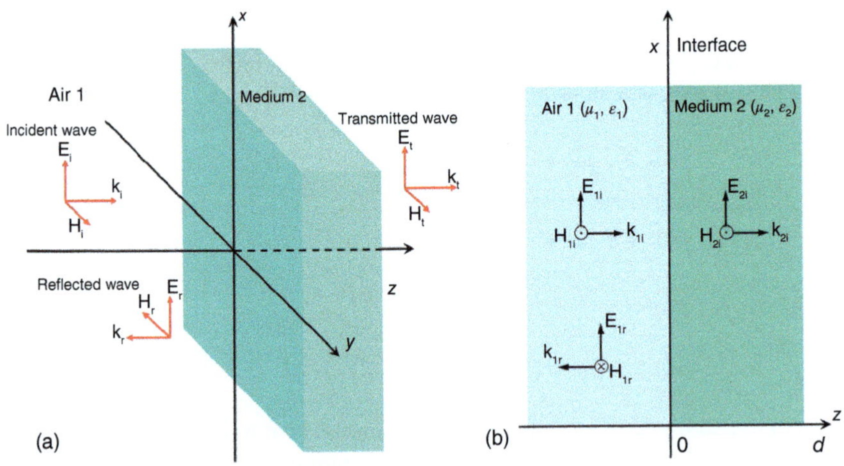

Figure 2.10 Three-dimensional schematic diagram (a) and planar diagram (b) of electromagnetic wave incident on a single-layer medium [10].

μ_1, σ_1, ε_1 are the permeability, electrical conductivity and dielectric constant of air, respectively.

In addition, the electric and magnetic fields of the reflected waves in air 1 are:

$$E_r(z) = e_x E_{rm} e^{\gamma_1 z} \tag{2.49}$$

$$H_r(z) = -e_z \times \frac{1}{\eta_{1c}} E_r(z) = -e_y \frac{1}{\eta_{1c}} E_{rm} e^{\gamma_1 z} \tag{2.50}$$

where $E_r(z)$ represents the electromagnetic intensity of the reflected waves; $H_r(z)$ represents the magnetic intensity of the reflected waves; E_{rm} represents the electric field amplitude of the reflected wave.

Therefore, the electric and magnetic fields of the synthesized waves in air 1 are:

$$E_1(z) = E_i(z) + E_r(z) = e_x \left[E_{im} e^{-\gamma_1 z} + E_{rm} e^{\gamma_1 z}\right] \tag{2.51}$$

$$H_1(z) = H_i(z) + H_r(z) = e_y \frac{1}{\eta_{1c}} \left[E_{im} e^{-\gamma_1 z} - E_{im} e^{\gamma_1 z}\right] \tag{2.52}$$

The electric and magnetic fields of the transmitted waves in medium 2 are:

$$E_2(z) = E_t(z) = e_x E_{tm} e^{-\gamma_2 z} \tag{2.53}$$

$$H_2(z) = H_t(z) = e_z \times \frac{1}{\eta_{2c}} E_t(z) = e_y \frac{1}{\eta_{2c}} E_{tm} e^{-\gamma_2 z} \tag{2.54}$$

where, $E_t(z)$ represents the electromagnetic intensity of transmitted waves; $H_t(z)$ represents the magnetic intensity of transmitted waves; E_{tm} represents electric field amplitude of transmitted waves; γ_2 is the propagation constant, $\gamma_2 = j\omega\sqrt{\mu_2\varepsilon_2\left(1 - j\frac{\sigma_2}{2\omega\varepsilon_2}\right)}$; η_2 represents the intrinsic impedance of the medium, $\eta_{2c} = \sqrt{\frac{\mu_2}{\varepsilon_2}}\left(1 - j\frac{\sigma_2}{\omega\varepsilon_2}\right)^{-1/2}$; ω is the change of phase in unit time, called angular frequency, unit rad/s; μ_2, σ_2, ε_2 are the permeability, electrical conductivity and dielectric constant of the medium, respectively.

According to the boundary condition, there are $E_1(z) = E_2(z)$ and $H_1(z) = H_2(z)$ on the interface of $z = 0$. According to the above formula, the following two equations can be obtained:

$$E_{rm} = \frac{\eta_{2c} - \eta_{1c}}{\eta_{2c} + \eta_{1c}} E_{im} \tag{2.55}$$

$$E_{tm} = \frac{2\eta_{2c}}{\eta_{2c} + \eta_{1c}} E_{im} \tag{2.56}$$

The ratio of E_{rm} to E_{im} is defined as the reflection coefficient at the interface, represented by Γ; the ratio of E_{tm} to E_{im} is the transmission coefficient on the interface, expressed by τ, then:

$$\Gamma = \frac{E_{rm}}{E_{im}} = \frac{\eta_{2c} - \eta_{1c}}{\eta_{2c} + \eta_{1c}} \tag{2.57}$$

$$\tau = \frac{E_{tm}}{E_{im}} = \frac{2\eta_{2c}}{\eta_{2c} + \eta_{1c}} \tag{2.58}$$

In general, Γ and τ are complex numbers, indicating a phase difference between the reflected, transmitted, and incident waves at the interfaces.

Assuming that the wave-transparent materials are ideal mediums, that is, $\sigma_1 = \sigma_2 = 0$, $\mu_1 = \mu_2$, then:

$$\gamma_1 = j\omega\sqrt{\mu_1\varepsilon_1\left(1 - j\frac{\sigma_1}{2\omega\varepsilon_1}\right)} = j\omega\sqrt{\mu_1\varepsilon_1} = j\beta_1 \tag{2.59}$$

$$\gamma_2 = j\omega\sqrt{\mu_2\varepsilon_2\left(1 - j\frac{\sigma_2}{2\omega\varepsilon_2}\right)} = j\omega\sqrt{\mu_2\varepsilon_2} = j\beta_2 \tag{2.60}$$

$$\eta_{1c} = \sqrt{\frac{\mu_1}{\varepsilon_1}}\left(1 - j\frac{\sigma_1}{\omega\varepsilon_1}\right)^{-1/2} = \sqrt{\frac{\mu_1}{\varepsilon_1}} = \eta_1 \tag{2.61}$$

$$\eta_{2c} = \sqrt{\frac{\mu_2}{\varepsilon_2}}\left(1 - j\frac{\sigma_2}{\omega\varepsilon_2}\right)^{-1/2} = \sqrt{\frac{\mu_2}{\varepsilon_2}} = \eta_2 \tag{2.62}$$

where β is called the phase constant, and its unit is rad/m.

Therefore:

$$\Gamma = \frac{\eta_2 - \eta_1}{\eta_2 + \eta_1} \tag{2.63}$$

$$\tau = \frac{2\eta_2}{\eta_2 + \eta_1} \tag{2.64}$$

Equations (2.63) and (2.64) correspond to the wave-transparent model of a single layer medium. η_1 and η_2 are real numbers. As long as the ε of the medium is measured, the wave transmission rate can be obtained. When $\eta_1 > \eta_2$, the reflection coefficient $\Gamma > 0$, indicating that the electric field of the reflected waves at the interface is in the same phase as that of the incident waves; When $\eta_1 < \eta_2$, the reflection coefficient $\Gamma < 0$, which indicates that the phase difference between the electric field of the reflected waves and the electric field of the incident waves at the interface is π, it is called half-wave loss.

2.4.2 Two-Layer Wave-Transparent Model

The single-layer wave-transparent model cannot satisfy the calculation of the wave-transparent rate for multilayer dielectrics in practical engineering. Figure 2.11 shows the corresponding two-layer wave-transparent model. The interface 1 of air 1 and medium 2 is located at $z = 0$, the interface 2 of medium 2 and medium 3 is located at $z = d$. The thickness of medium 2 is d, and interface 1 and interface 2 are parallel to each other. When electromagnetic waves are incident vertically from air 1, they are commonly reflected and transmitted at both interface 1 and interface 2. Therefore, air 1 and medium 2 have incident waves propagating in the $+z$ direction and reflected waves propagating in the $-z$ direction, while medium 3 has only transmitted waves propagating in the $+z$ direction.

The electric and magnetic fields of the incident waves in air 1 are, respectively:

$$E_{1i}(z) = e_x E_{1im} e^{-j\beta_1 z} \tag{2.65}$$

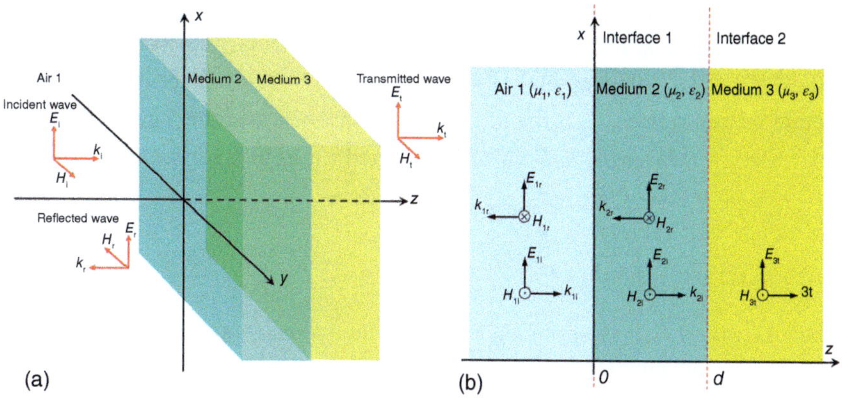

Figure 2.11 Three-dimensional schematic diagram (a) and planar diagram (b) of electromagnetic wave incident on bilayer medium [10].

$$H_{1i}(z) = e_y \frac{1}{\eta_1} E_{1im} e^{-j\beta_1 z} \tag{2.66}$$

The electric and magnetic fields of the reflected waves in air 1 are, respectively:

$$E_{1r}(z) = e_x E_{1rm} e^{j\beta_1 z} = e_x \Gamma_1 E_{1im} e^{j\beta_1 z} \tag{2.67}$$

$$H_{1r}(z) = -e_y \frac{1}{\eta_1} E_{1rm} e^{j\beta_1 z} = -e_y \frac{1}{\eta_1} \Gamma_1 E_{1im} e^{j\beta_1 z} \tag{2.68}$$

The electric and magnetic fields of the synthetic waves in air 1 are, respectively:

$$E_1(z) = e_x E_{1im} \left[e^{-j\beta_1 z} + \Gamma_1 e^{j\beta_1 z}\right] \tag{2.69}$$

$$H_1(z) = e_y \frac{E_{1im}}{\eta_1} [e^{-j\beta z} - \Gamma_1 e^{j\beta z}] \tag{2.70}$$

where $E_{1i}(z)$, $E_{1r}(z)$, and $E_1(z)$ respectively represent the electric field intensity of the incident waves, reflected waves, and synthesized waves in the air. $H_{1i}(z)$, $H_{1r}(z)$, and $H_1(z)$, respectively, represent the magnetic field intensity of the incident waves, reflected waves, and synthesized waves in the air. E_{1im} and E_{1rm} represent the amplitudes of the incident and reflected waves in the air. β_1 is the phase constant of electromagnetic waves in air, $j\beta_1 = j\omega\sqrt{\mu_1\varepsilon_1}$, where μ_1 and ε_1 are the permeability and dielectric constant of air, respectively. $\Gamma_1 = E_{1rm}/E_{1im}$ is the reflection coefficient at interface 1 ($z = 0$). η_1 is the intrinsic impedance of air $\eta_1 = \sqrt{\frac{\mu_1}{\varepsilon_1}}$.

The electric field and magnetic field of the synthesized waves in medium 2 are:

$$\begin{aligned} E_2(z) &= e_x E_{2im} \left[e^{-j\beta_2(z-d)} + \Gamma_2 e^{j\beta_2(z-d)}\right] \\ &= e_x \tau_1 E_{1im} \left[e^{-j\beta_2(z-d)} + \Gamma_2 e^{-j\beta_2(z-d)}\right] \end{aligned} \tag{2.71}$$

$$H_2(z) = e_y \frac{\tau_1 E_{1im}}{\eta_2} \left[e^{-j\beta_2(z-d)} - \Gamma_2 e^{j\beta_2(z-d)}\right] \tag{2.72}$$

where $E_2(z)$ and $H_2(z)$ respectively represent the electric field intensity and magnetic field intensity of the synthesized waves in medium 2. $\tau_1 = E_{2im}/E_{1im}$ is the transmission coefficient at interface 1 ($z = 0$), $\Gamma_2 = E_{2rm}/E_{2im}$ is the reflection coefficient at interface 2 ($z = d$). E_{2im} and E_{2rm} represent the amplitudes of incident and reflected waves in medium 2. β_2 is the phase constant of electromagnetic wave in medium 2 $j\beta_2 = j\omega\sqrt{\mu_2\varepsilon_2}$, where μ_2 and ε_2 are the permeability and dielectric constant of medium 2, respectively. η_2 is the intrinsic impedance of medium 2 $\eta_2 = \sqrt{\frac{\mu_2}{\varepsilon_2}}$.

The electric and magnetic fields of transmitted waves in medium 3 are:

$$E_3(z) = e_x E_{3im} e^{-j\beta_3(z-d)} = e_x \tau_1 \tau_2 E_{1im} e^{-j\beta_3(z-d)} \tag{2.73}$$

$$H_3(z) = e_y \frac{1}{\eta_3} \tau_1 \tau_2 E_{1im} e^{-j\beta_3(z-d)} \tag{2.74}$$

where $E_3(z)$ and $H_3(z)$ respectively represent the intensity of electric and magnetic field for electromagnetic waves in medium 3. $\tau_2 = E_{3tm}/E_{2im}$ is the transmission coefficient at interface 2 ($z = d$). E_{2im} represents the amplitude of the transmitted wave in medium 3. β_3 is the phase constant of electromagnetic wave in medium 3 $j\beta_3 = j\omega\sqrt{\mu_3\varepsilon_3}$, where μ_3 and ε_3 are the permeability and dielectric constant of medium 3, respectively, ω is the phase change in unit time. η_3 is the intrinsic impedance of medium 3 $\eta_3 = \sqrt{\frac{\mu_3}{\varepsilon_3}}$.

In interface 2 ($z = d$), $E_2(d) = E_3(d)$ and $H_2(d) = H_3(d)$, the following two equations can be obtained:

$$1 + \Gamma_2 = \tau_2 \tag{2.75}$$

$$\frac{1}{\eta_2}(1 - \Gamma_2) = \frac{1}{\eta_3}\tau_2 \tag{2.76}$$

It can be deduced that:

$$\Gamma_2 = \frac{\eta_3 - \eta_2}{\eta_3 + \eta_2} \tag{2.77}$$

$$\tau_2 = 1 + \Gamma_2 = \frac{2\eta_3}{\eta_3 + \eta_2} \tag{2.78}$$

At the interface $z = 0$ between air 1 and medium 2, as $E_1(0) = E_2(0)$ and $H_1(0) = H_2(0)$, the following two equations can be obtained:

$$1 + \Gamma_1 = \tau_1 \left[e^{j\beta_2 d} + \Gamma_2 e^{-j\beta_2 d}\right] \tag{2.79}$$

$$\frac{1}{\eta_1}(1 - \Gamma_1) = \frac{\tau_1}{\eta_2} \left[e^{j\beta_2 d} - \Gamma_2 e^{-j\beta_2 d}\right] \tag{2.80}$$

Dividing Eq. (2.79) by Eq. (2.80) yields:

$$\frac{\eta_1(1 + \Gamma_1)}{1 - \Gamma_1} = \frac{\eta_2\left(e^{j\beta_2 d} + \Gamma_2 e^{-j\beta_2 d}\right)}{e^{j\beta_2 d} - \Gamma_2 e^{-j\beta_2 d}} \tag{2.81}$$

Let

$$\eta_{ef} = \frac{\eta_2\left(e^{j\beta_2 d} + \Gamma_2 e^{-j\beta_2 d}\right)}{e^{j\beta_2 d} - \Gamma_2 e^{-j\beta_2 d}} \tag{2.82}$$

The following equations can be obtained:

$$\Gamma_1 = \frac{\eta_{ef} - \eta_1}{\eta_{ef} + \eta_1} \tag{2.83}$$

$$\tau_1 = \frac{1 + \Gamma_1}{e^{j\beta_2 d} + \Gamma_2 e^{-j\beta_2 d}} \tag{2.84}$$

In addition, by substituting Eq. (2.77) into Eq. (2.82), η_{ef} can be further simplified (Eq. (2.85)) by using Euler's formula:

$$\eta_{ef} = \eta_2 \frac{\eta_3 + \eta_2 j \tan(\beta_2 d)}{\eta_2 - \eta_3 j \tan(\beta_2 d)} \tag{2.85}$$

Equations (2.83) and (2.84) are the empirical equations corresponding to the two-layer wave-transparent model. η_{ef} can be calculated by obtaining the μ, ε, and thickness (d) of each dielectric layer in the two-layer dielectrics, and the Γ_1 is further obtained to calculate the corresponding wave-transparent rate. It can be seen that η_{ef} is the ratio of the electric and magnetic fields in medium 2 at $z = 0$, i.e. $\eta_{ef} = E_2(0)/H_2(0)$. It is called the equivalent wave impedance at $z = 0$. In this case, medium 2 and medium 3 are regarded as a whole, and η_{ef} represents the effect of electromagnetic waves on the reflection coefficient of the double-layer mediums at $z = 0$.

For electromagnetic waves incident vertically on multilayer mediums (Figure 2.12), the wave-transparent rate can be obtained by referring to the double-layer wave-transparent model. The uniform plane waves radiate vertically from left to right from the first layer. First, the reflection coefficient Γ_n of electromagnetic waves at the interface n can be obtained according to the nth, and $(n-1)_{th}$ layer of the medium's intrinsic impedance (η_{n+1} and η_n), and then the equivalent wave impedance and reflection coefficient (Γ_{n-1}) of the n_{th} and $(n-1)_{th}$ layer of the medium are further obtained. The reflection coefficient (Γ_{n-2}) of the electromagnetic waves at the interface at n_{th} layer is found from the intrinsic impedance of $(n-2)_{th}$ layer and the equivalent impedance of n_{th} and $(n-1)_{th}$ layer. The process continues until the reflection coefficient of the electromagnetic wave at interface 1 is found, and finally, the wave-transparent rate of the multilayer medium is obtained.

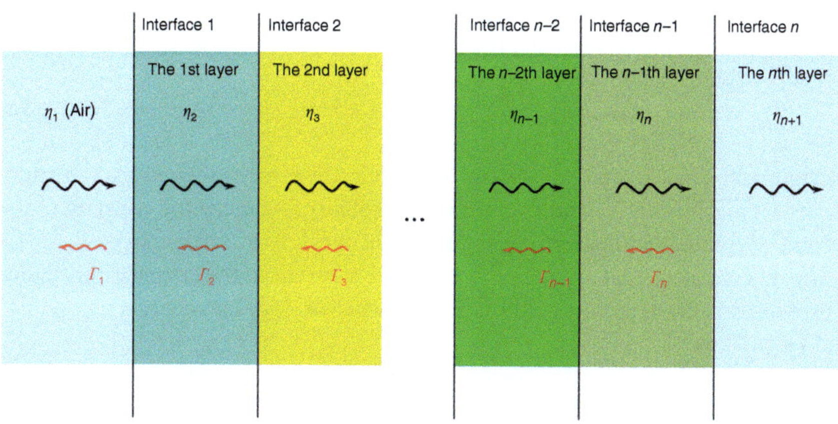

Figure 2.12 Schematic diagram of electromagnetic wave transmission in multilayer media.

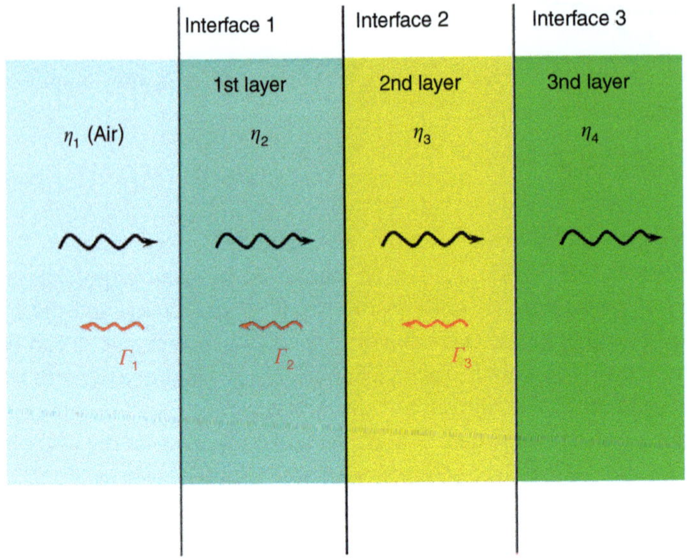

Figure 2.13 Schematic diagram of electromagnetic wave transmission in the three-layer medium.

Take electromagnetic wave incident on a three-layer medium (Figure 2.13) as an example.

According to the intrinsic impedance (η_4 and η_3) of the medium in second and third layers, the reflection coefficients at the interface are obtained:

$$\Gamma_3 = \frac{\eta_4 - \eta_3}{\eta_4 + \eta_3} \tag{2.86}$$

where η_4 is the intrinsic impedance of the third layer $\eta_4 = \sqrt{\frac{\mu_4}{\varepsilon_4}}$, and μ_4 and ε_4 are the permeability and dielectric constant of the third layer, respectively. η_3 is the intrinsic impedance of the second layer, and μ_3 and ε_3 are the permeability and dielectric constant of the second layer, respectively.

The equivalent impedance (η_{ef2}) of the third and second layers is:

$$\eta_{\text{ef2}} = \frac{\eta_3 \left(e^{j\beta_3 d_3} + \Gamma_3 e^{-j\beta_3 d_3}\right)}{e^{j\beta_3 d_3} - \Gamma_3 e^{-j\beta_3 d_3}} \tag{2.87}$$

where β_3 is the phase constant of electromagnetic waves in the second layer medium $j\beta_3 = j\omega\sqrt{\mu_3\varepsilon_3}$, where μ_3 and ε_3 are the permeability and dielectric constant of the second layer medium, respectively. d_3 is the thickness of the second layer.

The reflection coefficients at interface 2 are obtained from the intrinsic impedance (η_1) of the first layer and the equivalent impedance (η_{ef2}) of the medium at the third and second layers.

$$\Gamma_2 = \frac{\eta_{\text{ef2}} - \eta_2}{\eta_{\text{ef2}} + \eta_2} \tag{2.88}$$

where η_2 is the intrinsic impedance of the first layer, and μ_2 and ε_2 are the permeability and dielectric constant of the first layer, respectively.

The equivalent impedance (η_{ef1}) of the medium at the first, second, and third layers is:

$$\eta_{ef1} = \frac{\eta_2 \left(e^{j\beta_2 d_2} + \Gamma_2 e^{-j\beta_2 d_2}\right)}{e^{j\beta_2 d_2} - \Gamma_2 e^{-j\beta_2 d_2}} \tag{2.89}$$

where β_2 is the phase constant of electromagnetic waves in the first layer medium, where μ_2 and ε_2 are the permeability and dielectric constant of the first layer medium, respectively. d_2 is the thickness of the first layer of the medium.

Finally, the reflection coefficient (Γ_1) at the interface 1 is:

$$\Gamma_1 = \frac{\eta_{ef1} - \eta_1}{\eta_{ef1} + \eta_1} \tag{2.90}$$

where η_1 is the intrinsic impedance of air, and μ_1 and ε_1 are the permeability and dielectric constant of air, respectively.

Equation (2.90) is the empirical equation corresponding to the three-layer wave-transparent model. As long as μ, ε, and thickness d are obtained for each of the three layers of medium, Γ_1 can be found, and the corresponding wave-transparent rate can be calculated.

2.5 Summary

The ε and $\tan\delta$ are important parameters to evaluate the wave-transparent performances of the wave-transparent materials. With the decrease of ε and $\tan\delta$, the lower the loss of electromagnetic waves and the higher the wave-transparent rate. According to the analysis of the Clausius–Mossotti, Maxwell–Garnett, and Lichtenecker equations, it can be deduced that the ε and $\tan\delta$ can be further optimized by adjusting the internal structure polarization and corresponding volume fraction of each component for wave-transparent materials. In addition, the testing methods for dielectric properties (resonant method and nonresonant method) of the wave-transparent materials are suitable for different scenarios. For laminated wave-transparent composites with different structures, the wave-transparent models of single-layer and double-layer are proposed in this chapter, and the calculation equations of the corresponding wave-transparent rate are derived in detail.

References

1 Dong, J., Yang, C., Cheng, Y. et al. (2017). Facile method for fabricating low dielectric constant polyimide fibers with hyperbranched polysiloxane. *Journal of Materials Chemistry C* 5 (11): 2818–2825.

2 Duan, H., Zhuang, C., Mei, F. et al. (2022). Benzyl(4-fluorophenyl)phenylphosphine oxide-modified epoxy resin with improved flame retardancy and dielectric properties. *Advanced Composites and Hybrid Materials* 5 (2): 776–787.

3 Shang, X., Zhai, D., Liu, M. et al. (2021). Dielectric properties and electromagnetic wave transmission performance of aluminium silicate fibreboard at 915 MHz and 2450 MHz. *Ceramics International* 47 (6): 7539–7557.

4 Gong, C., Huang, X., Li, J. et al. (2022). Perfluorocyclobutyl aryl ether-based poly(silylene arylacetylene)s with a low dielectric constant for advanced wave-transparent composites. *European Polymer Journal* 181: 111655.

5 Li, J., Lin, J., Wang, Y. et al. (2022). Cross-linked and rigid polyimide composite foams with prominent fire resistant, thermal insulating, and wave-transparent performances. *ACS Applied Polymer Materials* 4 (10): 6994–7003.

6 Xia, L., Yang, Y., Zhang, X. et al. (2018). Crystal structure and wave-transparent performances of lithium aluminum silicate glass-ceramics. *Ceramics International* 44 (12): 14896–14900.

7 Liu, X., Zhang, D., Ma, R. et al. (2022). The two-tier tissue reinforcing and toughening mechanism caused by dual-granularity Si powders ingredient, and dielectric properties in reaction bonded and sintered reaction bonded Si_3N_4 porous ceramics. *Ceramics International* 48 (16): 22877–22884.

8 Liu, H., Zhu, W., Mao, Q. et al. (2023). Single-crystalline $BaZr_{0.2}Ti_{0.8}O_3$ membranes enabled high energy density in PEI-based composites for high-temperature electrostatic capacitors. *Advanced Materials* 35 (22): 2300962.

9 Meng, X.-S., Zhou, Y., Li, J. et al. (2023). All-organic PTFE coated PVDF composite film exhibiting low conduction loss and high breakdown strength for energy storage applications. *Ploymers* 15 (5): 1305.

10 Tang, L., Tang, Y., Zhang, J. et al. (2022). High-strength super-hydrophobic double-layered PBO nanofiber-polytetrafluoroethylene nanocomposite paper for high-performance wave-transparent applications. *Science Bulletin* 67 (21): 2196–2207.

11 Qi, Q., Zheng, P., Lei, Y., and Liu, X. (2019). Design of bi-modal pore structure polyarylene ether nitrile/SiO2 foams with ultralow-k dielectric and wave transparent properties by supercritical carbon dioxide. *Composites Part B: Engineering* 173: 106915.

12 Jiang, W., Zhang, X., Chen, D. et al. (2021). High performance low-k and wave-transparent cyanate ester resins modified with a novel bismaleimide hollow polymer microsphere. *Composites Part B: Engineering* 222: 109041.

13 Wang, L., Yang, J., Cheng, W. et al. (2021). Progress on polymer composites with low dielectric constant and low dielectric loss for high-frequency signal transmission. *Frontiers in Materials* 8: 774843.

14 Anandraj, J. and Joshi, G.M. (2018). Fabrication, performance and applications of integrated nanodielectric properties of materials—a review. *Composite Interfaces* 25 (5–7): 455–489.

15 Huang, X., Sun, B., Zhu, Y. et al. (2019). High-k polymer nanocomposites with 1D filler for dielectric and energy storage applications. *Progress in Materials Science* 100: 187–225.

16 Li, B., Yang, H., Guo, L. et al. (2023). A dual-band antireflection metasurface for air-ground impedance matching in ground penetrating radar. *Physica Scripta* 98 (3): 035515.

17 Sun, K., Yang, X., Lei, Y. et al. (2023). Core-shell structural design and microwave absorption enhancement of multi-dimensional graphene

oxide@polypyrrole/carbonyl iron fiber nanocomposites. *Journal of Alloys and Compounds* 930: 167446.

18 Xie, P., Li, H., He, B. et al. (2018). Bio-gel derived nickel/carbon nanocomposites with enhanced microwave absorption. *Journal of Materials Chemistry C* 6 (32): 8812–8822.

19 Gu, J., Li, Y., Liang, C. et al. (2018). Synchronously improved dielectric and mechanical properties of wave-transparent laminated composites combined with outstanding thermal stability by incorporating iysozyme/POSS functionalized PBO fibers. *Journal of Materials Chemistry C* 6 (28): 7652–7660.

20 Kuzmin, D.A., Bychkov, I.V., and Shavrov, V.G. (2014). Electromagnetic waves reflectance of graphene—magnetic semiconductor superlattice in magnetic field. *IEEE Transactions on Magnetics* 50 (11): 1–4.

21 Razzaz, F. and Alkanhal, M.A.S. (2021). Evanescent and propagating electromagnetic waves in Bi-isotropic layers. *Waves in Random and Complex Media* 31 (5): 879–890.

22 Hu, Z., Wang, Y., Liu, X. et al. (2022). Rational design of POSS containing low dielectric resin for SLA printing electronic circuit plate composites. *Composites Science and Technology* 223: 109403.

23 Smith, R.S., Tsui, T., and Ho, P.S. (2010). Analysis of ultraviolet curing effect on the dielectric constant and molecular structure of a porous dielectric film. *Journal of Electronic Materials* 39 (10): 2337–2345.

24 Kobayashi, K. (1998). Relationship between electronic molar polarizability and super low dielectric constant in B_2O_3-BeF_2-C glass systems. *Journal of Physics and Chemistry of Solids* 59 (9): 1671–1673.

25 Li, J., Chen, M., and Wang, Y. (2022). Preparation and properties of a fluorinated epoxy resin with low dielectric constant. *Journal of Applied Polymer Science* 139 (19): 52132.

26 Schröder, C. and Steinhauser, O. (2010). Simulating polarizable molecular ionic liquids with Drude oscillators. *The Journal of Chemical Physics* 133 (15): 154511.

27 Hubbard, R.L., Tyler, D.R., and Thompson, B. (2021). An empirically derived model for further increasing microwave curing rates of epoxy-amine polymerizations. *Journal of Applied Polymer Science* 138 (1): 49635.

28 Mousavi, M., Pahlavan, F., Oldham, D. et al. (2016). Multiscale investigation of oxidative aging in biomodified asphalt binder. *The Journal of Physical Chemistry C* 120 (31): 17224–17233.

29 Ryazanov, M.I. and Tishchenko, A.A. (2006). Clausius-Mossotti-Type relation for planar monolayers. *Journal of Experimental and Theoretical Physics* 103 (4): 539–545.

30 Zhu, T., Yu, Q., Zheng, W. et al. (2021). Intrinsic high-k-low-loss dielectric polyimides containing ortho-position aromatic nitrile moieties: reconsideration on Clausius-Mossotti equation. *Polymer Chemistry* 12 (16): 2481–2489.

31 Duan, L., Liu, M., Tang, Y. et al. (2022). Effect of crystal structure of $Ca_2AMg_2Ge_2VO_{12}$ (A = Y, Eu) garnets on microwave dielectric properties. *Materials Today Communications* 33: 104706.

32 Li, E., Wen, Q., Yang, H. et al. (2020). Novel temperature stable $NiSnTa_2O_8$ microwave dielectric ceramics with trirutile structure. *Ceramics International* 46 (5): 6079–6084.

33 Wang, J., Wu, H., Wang, Z. et al. (2022). An ultrafast self-polarization effect in barium titanate filled poly(vinylidene fluoride) composite film enabled by self-charge excitation triboelectric nanogenerator. *Advanced Functional Materials* 32 (35): 2204322.

34 Qiu, J., Gu, Q., Sha, Y. et al. (2022). Preparation and application of dielectric polymers with high permittivity and low energy loss: a mini review. *Journal of Applied Polymer Science* 139 (24): 52367.

35 Duan, C.-g., Mei, W.N., Yin, W.-G. et al. (2004). Simulations of ferroelectric polymer film polarization: the role of dipole interactions. *Physical Review B* 69 (23): 235106.

36 Li, X., Lan, T.H., Tien, C.H., and Gu, M. (2012). Three-dimensional orientation-unlimited polarization encryption by a single optically configured vectorial beam. *Nature Communications* 3 (1): 998.

37 Lv, H., Guo, Y., Wu, G. et al. (2017). Interface polarization strategy to solve electromagnetic wave interference issue. *ACS Applied Materials & Interfaces* 9 (6): 5660–5668.

38 Chen, B., Pan, D.F., Duan, M.L., and An, P.L. (2019). Electric modulation of magnetic anisotropy on the granule/matrix interface of a permalloy/PZT composite. *Physical Review B* 100 (13): 134418.

39 Wu, N., Ding, W., Shi, A., and Zhang, W. (2016). Inhomogeneous dynamic nuclear polarization and suppression of electron polarization decay in a quantum dot. *Physics Letters A* 380 (35): 2706–2712.

40 Hasjim, P.L. and Norris, J.R. (2007). Electron spin polarization of radical pairs in simple molecular wires. *The Journal of Physical Chemistry C* 111 (13): 5203–5210.

41 Passchier, I., Higinbotham, D.W., Jager, C.W. et al. (1998). A Compton backscattering polarimeter for measuring longitudinal electron polarization. *Nuclear Instruments and Methods in Physics Research Section A: Accelerators, Spectrometers, Detectors and Associated Equipment* 414 (2): 446–458.

42 Spirina Jenkins, O. and Hunt, K.L.C. (2003). Nonlocal dielectric functions on the nanoscale: Screened forces from unscreened potentials. *The Journal of Chemical Physics* 119 (16): 8250–8256.

43 Yong-Fang, L., Li-Qing, R., Rui-Qiong, M. et al. (2010). Implementing quantum control for the temporal evolution of the wave function of the quantum state by using a phase-controlled light field. *Acta Physica Sinica* 59 (3): 1671.

44 Wang, Q., Sang, D., Jiao, H. et al. (2017). Ionic transport and dielectric properties in $NaNbO_3$ under high pressure. *Applied Physics Letters* 111 (15): 152903.

45 Wu, C., Li, Z., Treich, G.M. et al. (2019). Dipole-relaxation dynamics in a modified polythiourea with high dielectric constant for energy storage applications. *Applied Physics Letters* 115 (16): 163901.

46 Rui-Xiang, W., Guo-Feng, Z., Zhi-Xing, Q., and Rui-Yun, C. (2019). Dipole orientation polarization property of single-molecule manipulated by external electric field. *Acta Physica Sinica* 68 (12): 128201.

47 Bronnikov, S., Kostromin, S., Asandulesa, M. et al. (2020). Interfacial interactions and interfacial polarization in polyazomethine/MWCNTs nanocomposites. *Composites Science and Technology* 190: 108049.

48 He, D., Wang, Y., Song, S. et al. (2017). Polymer-based nanocomposites employing Bi_2S_3@SiO_2 nanorods for high dielectric performance: understanding the role of interfacial polarization in semiconductor-insulator core-shell nanostructure. *Composites Science and Technology* 151: 25–33.

49 Hu, J.T., Gu, A., Liang, G. et al. (2012). Synthesis of mesoporous silica and its modification of bismaleimide/cyanate ester resin with improved thermal and dielectric properties. *Polymers for Advanced Technologies* 23 (3): 454–462.

50 Ma, B., Wang, Y., Chen, Y., and Gao, Y. (2017). Dielectric property and interfacial polarization of polymer-derived amorphous silicon carbonitride. *Ceramics International* 43 (15): 12209–12212.

51 Jeong, Y.G., Baik, D.H., Jang, J.W. et al. (2014). Preparation, structure and properties of poly(p-phenylene benzobisoxazole) composite fibers reinforced with graphene. *Macromolecular Research* 22 (3): 279–286.

52 Sun, H., Zhang, H., Liu, S. et al. (2018). Interfacial polarization and dielectric properties of aligned carbon nanotubes/polymer composites: the role of molecular polarity. *Composites Science and Technology* 154: 145–153.

53 Omri, M.A., Sanjay, M.R., Triki, A. et al. (2019). Dielectric properties and interfacial adhesion of jute, kenaf and E-glass fabrics reinforcing epoxy composites. *Polymer Composites* 40 (6): 2142–2153.

54 Ignaczak, W., Ladegaard Skov, A., and El Fray, M. (2020). Interfacial polarization in thermoplastic basalt fiber-reinforced composites. *Polymers* 7: 1486.

55 Hu, J.T., Gu, A., Liang, G. et al. (2012). Preparation and properties of maleimide-functionalized hyperbranched polysiloxane and its hybrids based on cyanate ester resin. *Journal of Applied Polymer Science* 126 (1): 205–215.

56 Fang, L., Tao, Y., Zhou, J. et al. (2021). A biobased low dielectric resin derived from vanillin and guaiacol. *Polymer Chemistry* 12 (5): 766–770.

57 Tao, Z., Yang, S., Ge, Z. et al. (2007). Synthesis and properties of novel fluorinated epoxy resins based on 1,1-bis(4-glycidylesterphenyl)-1-(3′-trifluoromethylphenyl)-2,2,2-trifluoroethane. *European Polymer Journal* 43 (2): 550–560.

58 Fang, L., Zhou, J., He, C. et al. (2020). Understanding how intrinsic micro-pores affect the dielectric properties of polymers: an approach to synthesize ultra-low dielectric polymers with bulky tetrahedral units as cores. *Polymer Chemistry* 11 (15): 2674–2680.

59 Gu, J., Dong, W., Xu, S. et al. (2017). Development of wave-transparent, light-weight composites combined with superior dielectric performance and desirable thermal stabilities. *Composites Science and Technology* 144: 185–192.

60 Tang, Y., Gu, J., Yu, Y., and Kong, J. (2015). Preparation of POSS/Quartz fibers/cyanate ester resins laminated composites. *Polymer Composites* 36 (11): 2017–2021.

61 Liu, Z., Fan, X., Cheng, L. et al. (2022). Hybrid polymer membrane functionalized PBO fibers/cyanate esters wave-transparent laminated composites. *Advanced Fiber Materials* 4 (3): 520–531.

62 Lakhtakia, A. (1998). Incremental Maxwell Garnett formalism for homogenizing particulate composite media. *Journal of Electromagnetic Waves and Applications* 17 (4): 276–279.

63 Mallet, P., Guérin, C.A., and Sentenac, A. (2005). Maxwell-Garnett mixing rule in the presence of multiple scattering: derivation and accuracy. *Physical Review B* 72 (1): 014205.

64 Riikonen, S., Romero, I., and García de Abajo, F.J. (2005). Plasmon tunability in metallodielectric metamaterials. *Physical Review B* 71 (23): 235104.

65 Dong, X., Hu, Y., Zhao, J., and Wu, Y. (2010). Method to predict effective dielectric constant of porous silica low dielectric constant materials. *Physica B-Condensed Matter* 405 (17): 3551–3554.

66 Tan, S., Xiong, C., Xu, X., and Tsang, L. (2016). Uniaxial effective permittivity of anisotropic bicontinuous random media using NMM3D. *IEEE Geoscience and Remote Sensing Letters* 13 (8): 1168–1172.

67 Xue, Q. (2001). Study on dielectric properties of oil/water random composites. *Journal of Electrostatics* 50 (3): 169–175.

68 Tarik, Z., Jean-Paul, L., and Michel, V. (1998). Theoretical evidence for 'Lichtenecker's mixture formulae' based on the effective medium theory. *Journal of Physics D: Applied Physics* 31 (13): 1589.

69 Simpkin, R. (2010). Derivation of Lichtenecker's logarithmic mixture formula from Maxwell's equations. *IEEE Transactions on Microwave Theory and Techniques* 58 (3): 545–550.

70 Goncharenko, A.V., Lozovski, V.Z., and Venger, E.F. (2000). Lichtenecker's equation: applicability and limitations. *Optics Communications* 174 (1): 19–32.

71 Xi, Z., Ghita, O.R., and Evans, K.E. (2011). Effect of recyclate PTFE/GF laminate incorporation on the dielectric and mechanical properties of PTFE/GF composites using a novel manufacturing process. *Journal of Applied Polymer Science* 122 (4): 2467–2477.

72 Song, Y.L., Ta, J.X., and Wang, X.Z. (2011). Band gaps of two-dimensional antiferromagnetic photonic crystal. *Journal of Applied Physics* 110 (1): 013525.

73 Sun, H.T., Zhang, L.Y., Yao, X., and Wlodarski, W. (1994). AT-cut quartz resonators for determination of viscoelastic and dielectric properties of water/glycerol. *Sensors and Actuators A: Physical* 43 (1): 208–212.

74 Song, W., Yan, Y., Zhang, M. et al. (2021). Cobalt ions doped PNN-PHT ceramics with excellent piezoelectric properties. *Ceramics International* 47 (22): 32414–32423.

75 Kim, B., Kazmirenko, V., Prokopenko, Y. et al. (2005). Non-resonant, electrodeless method for measuring the microwave complex permittivity of ferroelectric thin films. *Measurement Science and Technology* 16 (9): 1792.

76 Lee, W., Sarkar, T.K., Moon, H. et al. (2013). Effect of material parameters on the resonant frequencies of a dielectric object. *IEEE Antennas and Wireless Propagation Letters* 12: 1311–1314.

77 Linfeng, C., Ong, C.K., and Tan, B.T.G. (1999). Amendment of cavity perturbation method for permittivity measurement of extremely low-loss dielectrics. *IEEE Transactions on Instrumentation and Measurement* 48 (6): 1031–1037.

78 Chen, Y.C. and Li, C.H. (2016). Improving quality factor of Mg_2SnO_4 ceramics by removing moisture content from starting raw materials. *Ceramics International* 42 (8): 9749–9751.

79 Liu, X., Gan, L., and Yang, B. (2021). Millimeter-wave free-space dielectric characterization. *Measurement* 179: 109472.

80 Alvarez, J.O., Peñaranda-Foix, F.L., Catalá-Civera, J.M., and Gutiérrez-Cano, J.D. (2020). Permittivity spectrum of low-loss liquid and powder geomaterials using multipoint reentrant cavities. *IEEE Transactions on Geoscience and Remote Sensing* 58 (5): 3097–3112.

81 Hasar, U.C., Barroso, J.J., Bute, M. et al. (2014). Attractive method for thickness-independent permittivity measurements of solid dielectric materials. *Sensors and Actuators A: Physical* 206: 107–120.

82 Sheen, J. and Weng, C.M. (2016). Modifications of the cavity perturbation technique for permittivity measurements of laminated samples. *IEEE Transactions on Dielectrics and Electrical Insulation* 23 (1): 532–536.

83 Sheen, J. (2007). Amendment of cavity perturbation technique for loss tangent measurement at microwave frequencies. *Journal of Applied Physics* 102 (1): 014102.

84 Robinson, M.P., Fitton, L.C., Little, A. et al. (2019). Dielectric replica measurement: a new technique for obtaining the complex permittivity of irregularly shaped objects. *Measurement Science and Technology* 30 (4): 045902.

85 Sheen, J. (2009). Measurements of microwave dielectric properties by an amended cavity perturbation technique. *Measurement* 42 (1): 57–61.

86 Sten, J.C.E. and Koivisto, P.K. (2009). Determining both the permittivity and the permeability of small samples using cavity perturbation method. *Measurement Science and Technology* 20 (5): 057001.

87 Lujun, H., Lei, X., Mohsen, R. et al. (2021). Pushing the limit of high-Q mode of a single dielectric nanocavity. *Advanced Photonics* 3 (1): 016004.

88 Ji-Fuh, L. and Blair, W.D. (1998). High-Q TE01 mode DR filters for PCS wireless base stations. *IEEE Transactions on Microwave Theory and Techniques* 46 (12): 2493–2500.

89 Cheng, H.F., Joseph, P.T., Lee, Y.C., and Lin, I.N. (2007). Effect of high-Q $Ba_4Ti_{13}O_{30}$ materials on the dielectric properties of (Bax,Sr_{1-x})Tio_3 films for microwave communication. *Journal of Electromagnetic Waves and Applications* 21 (11): 1445–1451.

90 Derkach, V.N., Filippov, Y.F., Plevako, A.S. et al. (2004). Determination of microwave parameters of isotropic mediums by using an open quasi-optical spherical resonator. *International Journal of Infrared and Millimeter Waves* 25 (1): 139–148.

91 Wu, Y., Cui, B., Luo, S. et al. (2013). A unique Ka-band measurement system based on quasi-optical dielectric resonator technology for studying small superconducting samples. *IEEE Transactions on Applied Superconductivity* 23 (3): 9000204.

92 Ge, S., Follonier, S., Knoesen, A., and Miller, R.D. (2000). Characterization of thin-film low-dielectric constant materials in the microwave range using on-wafer parallel-plate transmission lines. *IEEE Microwave and Guided Wave Letters* 10 (5): 183–185.

93 Farcich, N.J., Salonen, J., and Asbeck, P.M. (2008). Single-length method used to determine the dielectric constant of polydimethylsiloxane. *IEEE Transactions on Microwave Theory and Techniques* 56 (12): 2963–2971.

94 Requena-Perez, M.E., Albero-Ortiz, A., Monzo-Cabrera, J., and Diaz-Morcillo, A. (2006). Combined use of genetic algorithms and gradient descent optmization methods for accurate inverse permittivity measurement. *IEEE Transactions on Microwave Theory and Techniques* 54 (2): 615–624.

95 Ruan, X. and Chan, C.H. (2019). Terahertz free-space dielectric property measurements using time- and frequency-domain setups. *International Journal of RF and Microwave Computer-Aided Engineering* 29 (9): e21839.

96 Hasar, U.C., Kaya, Y., Ozturk, H. et al. (2022). Improved method for permittivity determination of dielectric samples by free-space measurements. *IEEE Transactions on Instrumentation and Measurement* 71: 1–8.

97 Chang, T., Zhang, X., Yang, C. et al. (2017). Measurement of complex terahertz dielectric properties of polymers using an improved free-space technique. *Measurement Science and Technology* 28 (4): 045002.

98 Bourdier, A. and Gond, S. (2001). Dynamics of a charged particle in a linearly polarized traveling electromagnetic wave. *Physical Review E* 63 (3): 036609.

99 Umul, Y.Z. (2021). Physical optics approach to wave diffraction by a perfect electromagnetic conductor half-plane. *Optical and Quantum Electronics* 53 (5): 221.

100 Grokholskyi, Y.M., Sus, B.A., and Sus, B.B. (2017). Radio waves as a stream of particles in vibrational state. *Radio Electronics, Computer Science, Control* 4: 7–10.

101 Sheen, J. (2005). Time harmonic electromagnetic fields in an biaxial anisotropic medium. *Journal of Electromagnetic Waves and Applications* 19 (6): 753–767.

102 Krasnov, V.M. (2001). Penetration of an electromagnetic wave above a reflection point layer. *Radio Science* 36 (6): 1505–1508.

103 He, S., Hu, Y., and Strom, S. (1994). Electromagnetic reflection and transmission for a stratified bianisotropic slab. *IEEE Transactions on Antennas and Propagation* 42 (6): 856–858.

104 Hillion, P. (2000). Diffraction of electromagnetic waves at plane apertures: a different approach. *Journal of Electromagnetic Waves and Applications* 14 (12): 1685–1697.

105 Garfinkle, D. (2022). Gravitational wave memory and the wave equation. *Classical and Quantum Gravity* 39 (13): 135010.

106 Liu, R., Ji, C., Mock, J.J. et al. (2009). Broadband ground-plane cloak. *Science* 323 (5912): 366–369.

3

Polymer Matrix

3.1 Introduction

The viscosity and service cycle of polymer matrix will directly affect the impregnation of reinforced fibers, the layup, and storage of prepregs, then further affect the wave-transparent performances, mechanical properties, and heat resistances of the polymer-matrix wave-transparent composites [1].

At present, the commonly used polymer matrix for wave-transparent composites mainly includes epoxy resins, polyimide (PI) resins, phenolic (PF) resins, bismaleimide (BMI) resins, silicone resins, polytetrafluoroethylene (PTFE) resins, unsaturated polyester (UP) resins, and cyanate ester (CE) resins [2–4]. Table 3.1 shows the primary physical and chemical properties of common polymer matrix. The chemical structure, physical and chemical properties, and modification methods of above common polymer matrix will be introduced in detail thereafter.

3.2 Common Polymer Matrix

3.2.1 Epoxy Resins

Epoxy resins (ER) refer to the class of polymer matrix containing two or more epoxy groups in the molecular structures. The chemical structures of common ER are shown in Table 3.2 [5, 6]. There are many kinds of ER, which can be roughly divided into five categories according to molecular structures: (1) Glycidyl ethers, such as bisphenol A epoxy resins, novolac epoxy resins, and aliphatic polyol glycidyl ether epoxy resins; (2) Glycidyl lipids, such as diglycidyl phthalate, diglycidyl tetrahydrophthalate, and triglycidyl trimellitate; (3) Glycidyl amines, such as tetraglycidyl methyl diphenylamine containing four epoxy groups and triglycidyl *p*-aminophenol containing three epoxy groups; (4) Alicyclic epoxy resins, such as bis(2,3-epoxycyclopentyl) ether and dicyclopentadiene dioxide; and (5) Aliphatic epoxy resins, such as epoxidized polybutadiene resins [7–9]. The first three ER are formed by the polycondensation of epichlorohydrin and compounds containing active hydrogen atoms, such as phenols, alcohols, organic carboxylic acids, and

Polymer Matrix Wave-Transparent Composites: Materials, Properties, and Applications, First Edition.
Junwei Gu, Yusheng Tang, Jie Kong, and Jing Dang.

Table 3.1 Primary physical and chemical properties of common polymer matrix.

Resins Type	Density (g/cm^3)	Flexural strength (MPa)	Flexural modulus (GPa)	ε (10^6 Hz)	tan δ (10^6 Hz)
Epoxy	1.30	97	3.8	3.0	0.020
PI	1.36–1.43	170	3.8	3.2	0.007
PF	1.30	92	3.5	3.2	0.020
BMI	1.30	150	3.7	3.0	0.014
Organic silicon	—	85	—	3.0–5.0	0.003–0.050
PTFE	2.20	90	—	2.1–2.3	0.0003–0.0004
UP	1.29	85	3.2	3.0	0.018
CE	1.29	80	2.8	2.8–3.2	0.002–0.008

Table 3.2 Chemical structures of common epoxy resins.

Types	Chemical structures
Glycidyl ethers	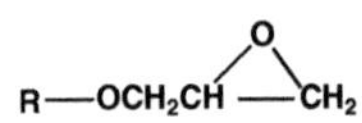
Glycidyl esters	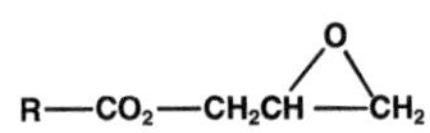
Glycidyl amines	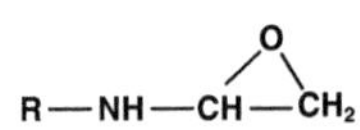
Linear aliphatics	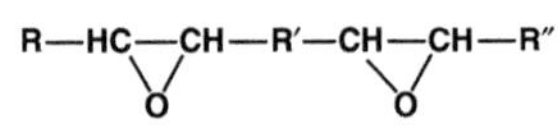
Alicyclics	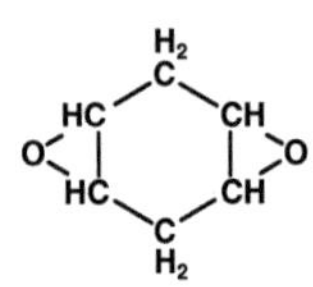

amines. The latter two ER are obtained by epoxidizing olefins with double or triple bonds with peracetic acids or hydrogen peroxide at low temperatures [10, 11].

The three-membered ring of epoxy groups is in the same plane, and the electronegativity of O atom is larger than that of C atom, which can form two reactive centers [12]. Under the introduction of curing agent and appropriate temperature, ER can form highly cross-linked three-dimensional network structures based on the ring-opening reaction of epoxy groups [13]. Epoxy resins, with excellent mechanical properties, outstanding wave-transparent performances, good heat resistance, and chemical stabilities, are widely used to prepare anticorrosion coatings, adhesives, electronic packaging materials, and aerospace materials [14, 15].

Epoxy resin monomers are usually high-viscosity liquids or brittle solids at room temperature, which are cured with the curing agents to form three-dimensional cross-linked networks. Depending on the type of curing reaction, the corresponding curing agents can be divided into reactive and catalytic curing agents [16, 17]. Active hydrogen atoms of reactive curing agents can open the three-membered ring of epoxy groups to form the curing networks through a step-by-step polymerization reaction (Figure 3.1) [18]. Common reactive curing agents include amine curing agents (aliphatic amines, alicyclic amines, aromatic amines, low molecular weight polyamides, etc.), acid anhydride curing agents (aromatic acid anhydrides, alicyclic acid anhydrides, etc.), and polyhydric phenols curing agents (PF resins containing phenolic hydroxyl groups) [19, 20]. Catalytic curing agents can realize the curing reaction of ER by anionic and cationic polymerization mechanisms without participating in constructing ER curing networks. Common anion-catalyzed curing agents mainly include tertiary amines and imidazole containing five-membered nitrogen heterocycles [21, 22]. And cationic-catalyzed curing agents are boron trifluoride and amines or ethers to form complexes, such as the boron trifluoride-ethylamine complex [23–25].

However, high cross-link density of ER results in their high brittleness and poor impact resistance [26–28]. There are several main methods to toughen ER. (1) Inorganic nanoparticles mainly include silica, montmorillonite (MMT), carbon nanotubes (CNTs), and graphene [29, 30]. Bok et al. [31] used fumed silica to modify the bisphenol A epoxy resins to obtain NC resins. When the amount of fumed silica was 5 wt%, the fracture toughness (K_{IC}) of NC resins was increased by 76% compared to that of pure ER. Wang et al. [32] used graphene oxide (GO) to modify

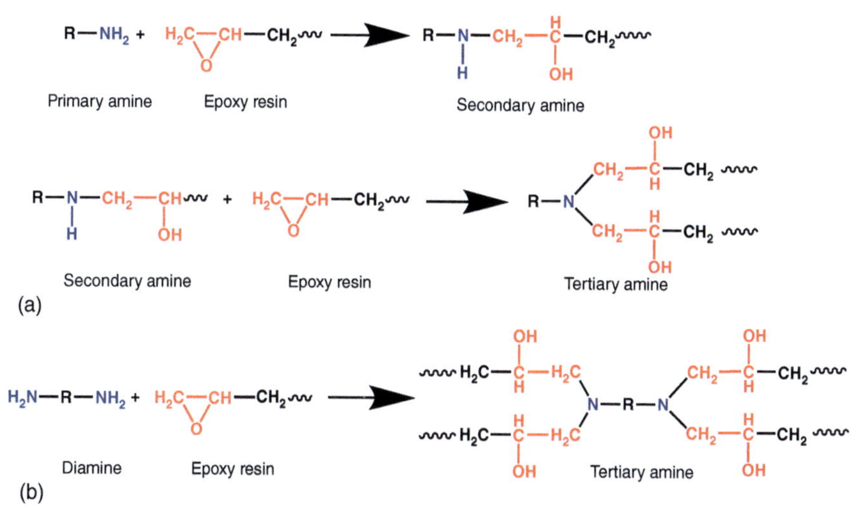

Figure 3.1 Schematic diagram of the curing reaction mechanism for epoxy resins with amine curing agents.

bisphenol A diglycidyl ether (DGEBA) resins (GO/DGEBA). When the amount of GO was 0.1 wt%, the K_{IC} and breaking elongation of GO/DGEBA resins were increased by 79.8% and 33.8%, respectively, compared with those of pure DGEBA resins. (2) Rubber elastomers include carboxyl-terminated liquid nitrile rubber (HTBN), amino-terminated liquid nitrile rubber (ATBN), and hydroxyl-terminated polybutadiene rubber (HTPB). Zhao et al. [33] utilized carboxy butyronitrile nano-rubber particles to modify bisphenol F epoxy resins. When the amount of carboxylated nitrile nano-rubber particles was 15 wt%, the K_{IC} of modified ER was increased by 132.2% at room temperature compared to that of pure ER. Zeng et al. [34] functionalized ethylene glycidyl methacrylate-vinyl acetate solid rubber (EGVA) with amino groups (EGVA-NH_2) and then prepared EGVA-NH_2/epoxy resins by simple blending. When the amount of EGVA-NH_2 was 5 wt%, the impact strength and K_{IC} of EGVA-NH_2/epoxy resins were increased by 130% and 158%, respectively, compared with those of pure ER. (3) Thermoplastic resins mainly include polysulfone (PSF), polyether ketone (PEK), PI, and polyphenylene ether (PPO). Sun et al. [35] toughen ER by PSF. When the amount of PSF was 15 wt%, the impact strength of modified ER increased significantly by about 65.3% compared to that of pure ER. Lee et al. [36] modified DGEBA resins by polyetherimide (PEI). When the amount of PEI was 2 wt%, the K_{IC} and critical strain energy release rate (G_{IC}) of PEI/DGEBA resins were increased by 83% and 194%, respectively, compared with those of pure DGEBA resins. (4) Hyperbranched polymers are mainly including polyester, polysiloxane, polyamide, PI, and polyethyleneimine. Chen et al. [37] synthesized a new hyperbranched polysiloxane (HAHBPs) via one-pot method using diethylene glycol and (3-aminopropyl)triethoxysilane (KH-550) to modify ER (HAHBPs/E-51). When the amount of HAHBPs was 20 wt%, the impact strength of HAHBPs/E-51 resins was increased by 148% compared to that of pure ER. Zhang et al. [38] synthesized a novel bio-based hyperbranched polymer (ITA-HBP) to modify ER (ITA-HBP/EP). The impact strength, fracture toughness, and fracture energy of ITA-HBP/EP resins were significantly improved due to the in-situ strengthening and toughening effect of ITA-HBP. When the amount of ITA-HBP was 10 wt%, the impact strength, $K_{IC,}$ and G_{IC} of ITA-HBP/EP resins were increased by 133.2%, 78.7%, and 124.7%, respectively, compared to those of pure ER.

Additionally, designing and synthesizing ER or curing agents with flexible chain segments can also endow epoxy resins with good toughness [39–41]. Zhang et al. [42] modified ER by first reacting toluene diisocyanate trimer (TDI-T/EP). When the molar ratio of BPA: TDI-T was 1:0.5, the breaking elongation of TDI-T/EP reached 48.02%, significantly higher than that of pure ER (5.12%). Moreover, the different toughening methods would produce different modification effects and impact the dielectric and thermal properties.

3.2.2 Phenolic (PF) Resins

PF resins are a class of resins obtained from phenolic and aldehyde monomers by addition-condensation reaction under the action of catalyst [43–45]. Common phenolic monomers include phenol, resorcinol, and bisphenol A [46].

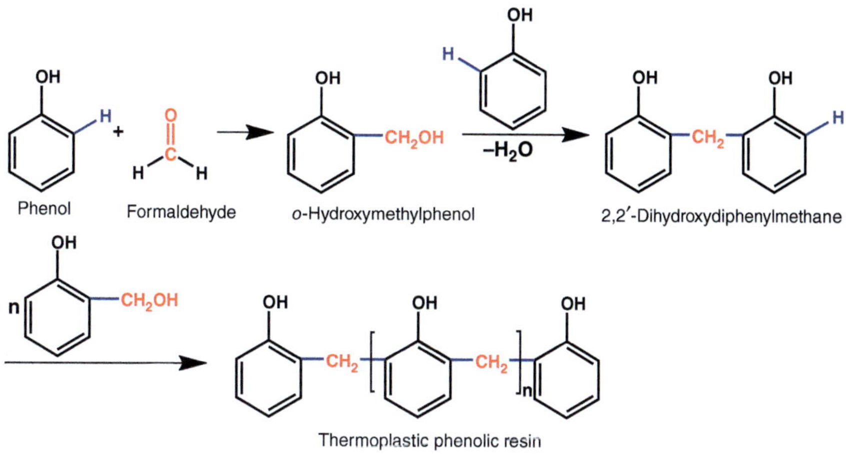

Figure 3.2 Schematic diagram of the synthesis route for thermoplastic phenolic resins.

Common aldehyde monomers include formaldehyde, acetaldehyde, and furfural. PF resins made from phenol and formaldehyde are the most widely used [47]. In the presence of the acidic catalyst and the corresponding molar ratio of phenol to formaldehyde is greater than 1, the thermoplastic PF resins would be obtained (the corresponding synthesis route can be seen in Figure 3.2) [48, 49]. However, in the presence of the alkali catalyst and the molar ratio of phenol to formaldehyde is less than 1, the thermosetting PF resins are then obtained (the synthetic route is shown in Figure 3.3) [50, 51].

In the synthesis process of thermoplastic PF resins, the formaldehyde forms carbocation, reacting with phenol to form *p*-hydroxymethylphenol and *o*-hydroxymethylphenol via addition reaction, which are then reacted with phenol at the ortho-para position by dehydration condensation, to generate 2,6-bis(hydroxymethyl)phenol, 2,4-bis(hydroxymethyl)phenol, and 2,4,6-tris(hydroxymethyl)phenol [52, 53]. Afterward, carbon cation from formaldehyde further reacts with the above three isomers and phenol via condensation polymerization.

However, excessive phenol would increase molecular chains, making each molecular chain containing about 5–6 benzene ring structures without hydroxymethyl, which would not cure even if further heating, finally generating thermoplastic phenolic resins [54]. However, the molecular chain structure of ordinary thermoplastic PF resins possesses one methylene and two phenols, resulting in high density of benzene ring structures and large molecular steric hindrance, leading to poor impact resistances [55, 56]. Simultaneously, ordinary thermoplastic PF resins are easily oxidized due to phenolic hydroxyl and methylene, resulting in poor heat resistance. Therefore, common thermoplastic PF resins are usually less used alone.

Thermosetting PF resins contain a large number of benzene rings. They are highly cross-linked after curing, which have excellent heat resistance, ablation resistance, and flame-retardant properties, and are widely used in the fields of construction, transportation, microelectronics, and aerospace [56, 57]. Table 3.3 shows the specifications and properties of different thermosetting phenolic resins.

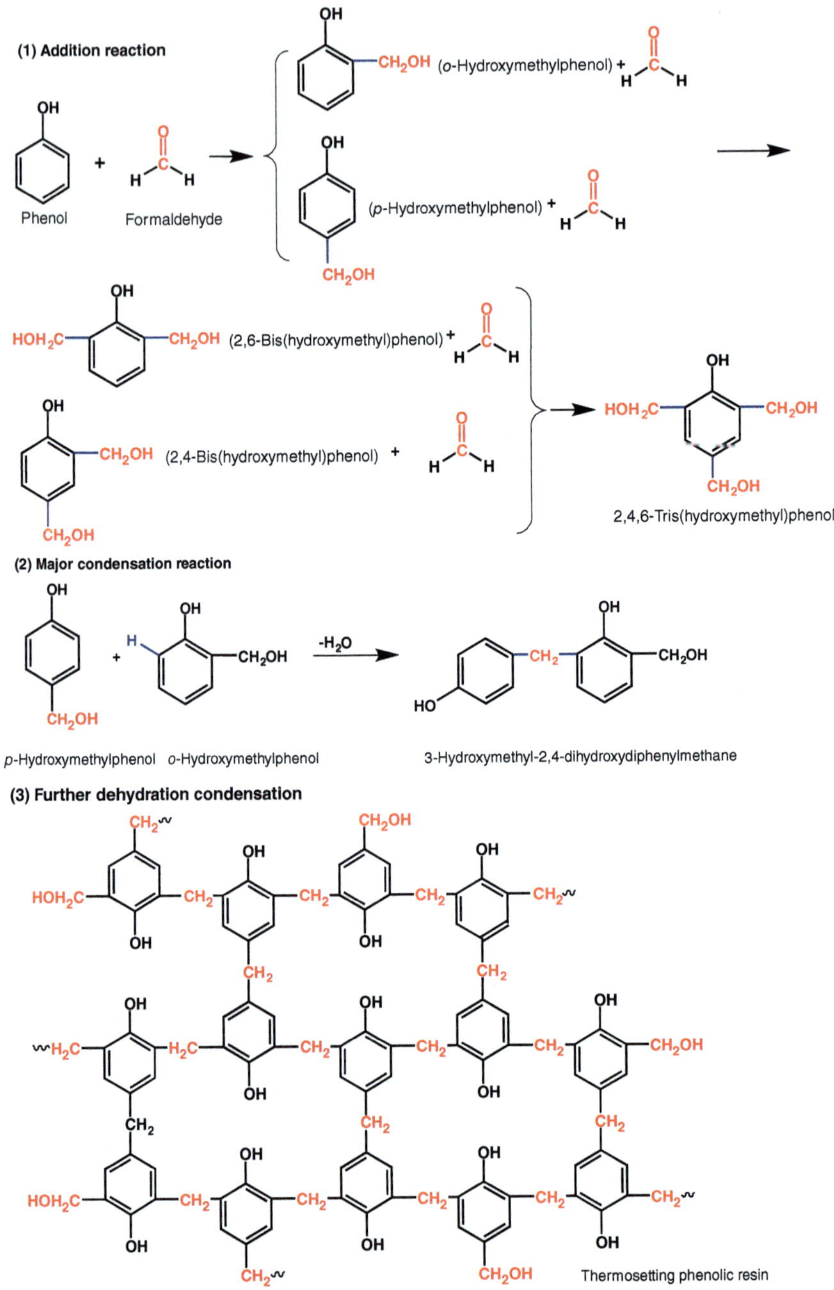

Figure 3.3 Schematic diagram of the synthesis route for thermosetting phenolic resins.

Table 3.3 Specification and properties of common thermosetting phenolic resins.

Types	Resin grade	Appearance	Softening point (°C)	Solid content	Main feature
Tert-butylphenol formaldehyde resin	2402	Irregularly shaped solid	80–120	—	Good water resistance, corrosion resistance, and electrical insulation properties
Cyclic acetophenone formaldehyde resin	2408	Light yellow to brownish-yellow transparent amorphous solid	75–120	—	Good compatibility and adhesion, and can improve gloss and vibrancy of coatings.
Phenol formaldehyde resin	201	Red-brown liquid	—	45–55	Good water resistance and high adhesion
Phenol formaldehyde resin	202	Red-brown liquid	—	72–78	High resistance to abrasion and adhesion
Cresol-phenol formaldehyde resin	2126	Dark brown transparent liquid	—	38–42	High heat curing temperature and good toughness
Butanol etherified cresol formaldehyde resin	284	Red-brown transparent liquid	—	48–52	High adhesion and good corrosion resistance

However, the rigid structure of pure PF resins after curing is too dense and lacks flexible groups, which makes the cured pure PF resins too brittle to stress cracks [58, 59]. Depending on the modification mechanism, there are two main toughening methods of PF resins: chemically reactive toughening and physically blending toughening [60–62]. Chemically reactive toughening means adding a reactive toughening agent to react with phenolic hydroxyl and hydroxymethyl in the PF resins [63, 64]. The flexible groups or long chains in the toughening agents are introduced into the network structures of the PF resins, which would reduce the crosslinking density to improve the toughness of PF resins [65, 66]. Commonly used reactive toughening agents include cashew phenol, polyurethane prepolymer, polyvinyl acetal, carboxy nitrile rubber, and reactive silicone resins. Cardona et al. [67] adopted cashew phenol to prepare the modified phenolic resins (CPF). When the amount of cashew phenol was 40 wt%, the K_{IC} of CPF resins was 1.66 MPa/m^2, higher than that of pure PF resins (0.89 MPa/m^2). Yu et al. [68] prepared nano carboxylic nitrile rubbers to toughen phenolic resins (XNBRL-PF). The functional group (–COOH) of XNBRL could chemically react with $-CH_2OH$ of PF resins to improve the dispersion of rubber particles and the interfacial interaction between rubber particles and PF matrix. When the amount of XNBRL was 6 wt%, the impact strength of XNBRL-PF was increased by 88%. Physical blending type toughening refers to adding nonreactive toughening agents to PF resins by physical blending [69, 70]. Commonly used nonreactive toughening agents include thermoplastic resins, rubber elastomers, and inorganic nanoparticles. He et al. [71] utilized the self-assembled hybrid particles of GO and zirconia (ZrO_2) nanoparticles to modify PF (ZrO_2@GO/PF) resins. When the mass fraction of ZrO_2@GO was 0.1 wt%, the impact strength of ZrO_2@GO/PF resins was 14.74 kJ/m^2, higher than that of pure PF resins (6.85 kJ/m^2). Jin et al. [72] synthesized a new polyether imidazole ionic liquid (PIIL) to modify phenolic resins. When the dosage of PIIL was 30 wt%, the impact strength of the modified PF resins was 9.51 kJ/m^2, higher than that of the pure PF resins (2.10 kJ/m^2).

3.2.3 Bismaleimide (BMI) Resins

BMI resins are one kind of resins derived from PI resins. Monomers with maleimide (MI) groups were prepared from binary amine and maleic anhydride through a two-step reaction of amidation and dehydration cyclization [73, 74]. The corresponding synthesis route is shown in Figure 3.4. BMI resins with different chemical structures can be obtained by reacting different kinds of binary amines with maleic anhydride, mainly divided into aliphatic bismaleimides and aromatic bismaleimides [75–77]. The most widely used BMI is diphenylmethane-type bismaleimides (BDM).

The C=C bonds of imide ring in BMI resins with two strong electron-withdrawing carbonyl adjacent can react with the active hydrogen compounds (diamine, polyphenol, amide, hydrazide, and cyanuric acid) via addition reaction, unsaturated compounds via copolymerization reaction [78, 79]. At the same time, the self-polymerization reaction is carried out with catalyst or heating conditions between C=C bonds. BMI resins contain the benzene ring and imide heterocyclic ring with high crosslinking density, resulting in excellent heat and moisture and radiation resistances, low ε and $\tan\delta$, and good mechanical properties, which are widely used in microelectronics and aerospace fields [80–82]. However, the high crosslinking density and strong structural rigidity of curing BMI resins lead to brittleness and poor impact strength.

3.2.3.1 Modified BMI Resins by Diamine Chain Extension

Diamine chain extension can increase the spacing of crosslinking points in the curing networks of BMI resins to improve the toughness of cured BMI resins [83–85]. Liu et al. [86] modified BDM resins by amino-terminated poly(*o*-phenylene dimethyl acetonitrile) sulfone (PPENS-DA) resins and 3,3′-diallyl-4,4′-bisphenol

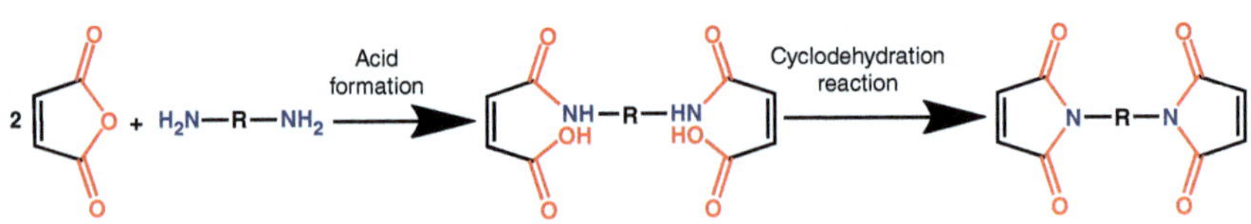

Figure 3.4 Schematic diagram of the synthetic route for BMI resins.

(DABP) curing agents to prepare PPENS-DA/DABP/BDM resins. When the mass fraction of PPENS-DA was 10 wt%, the impact strength of PPENS-DA/DABP/BDM resins was 4.98 kJ/m^2, increased by 36% compared to that of DABP/BDM resins. Liu et al. [87] synthesized chain-expanding aromatic amine (PM_2) with phenolphthalein skeleton to modify BDM resins. When the mass fraction of PM_2 was 0.05 wt%, the impact strength of modified BDM resins was increased by 121.3% compared to that of pure BDM resins.

3.2.3.2 Modified BMI Resins by Allyl Compounds

Allyl compounds toughening is a common method for modifying BMI resins via the Diels–Alder reaction [88–90]. Commonly used allyl-based tougheners include the diallyl bisphenol A resins, allyl epoxy resins, and linear allyl PF resins [91, 92]. Zhou et al. [93] synthesized star-shaped allyl-functionalized phosphonitrile small molecules (HECP) by nucleophilic substitution reaction, which were then performed to obtain HECP/BMI modified resins. When the mass fraction of HECP was 5 wt%, the impact strength of HECP/BMI resins was increased by 195% compared to that of pure BMI resins. Qin et al. [94] synthesized allyl-functionalized hyperbranched polyimide (AT-PAEKI) to modify bisphenol A-based BMI (BPA-BMI) resins. When the mass fraction of AT-PAEKI was 16 wt%, the K_{IC} of the modified BPA-BMI resins was increased by 14.6% compared to that of the pure BPA-BMI resins.

3.2.3.3 Modified BMI Resins by Rubber Elastomers

Rubber elastomers with reactive end groups, such as terminally carboxylic nitrile butadiene rubber (CTBN), terminally vinyl nitrile rubber (VTBN), and core–shell rubber, are introduced into BMI resins networks to produce microscopic phase separation for toughening purposes [95, 96]. When micro-cracks generated by BMI resins under external load pass through the rubber area, the large deformation process generated in the rubber area can absorb a large amount of energy to prevent the development of cracks [97–99]. Liu et al. [100] dispersed carboxylic nitrile rubber (VP-501) nano powder into ER (CYD-128) to modify BMI resins (VP-501/CYD-128/BMI). When the amount of VP-5011 was 5 wt%, the impact strength of VP-501/CYD-128/BMI resins was increased by 64.4% compared to that of CYD-128/BMI resins. Wang et al. [101] utilized the vinyl rubber to modify the BMI resins. When the amount of vinyl rubber was 5 wt%, the impact strength of modified BMI resins was increased by 17.3% compared to that of pure BMI resins.

3.2.3.4 Modified BMI Resins by Thermoplastic Resins

The addition of thermoplastic resins can form a molecular semi-interpenetrating network structure in the BMI resins, which combine the excellent performance characteristics of thermoplastic resins (better toughness and low water absorption) and BMI resins (good chemical resistance and dimensional stability performance) [102, 103]. Currently, high-performance thermoplastic resins commonly used to toughen BMI resins include polyether sulfone (PES), PEI, and polyaryletherketone (PAEK). Chen et al. [104] prepared *nano*-SiO_2/PES/BMI multiphase composites.

When the amounts of PES and *nano*-SiO_2 were 4 and 1.5 wt%, the dielectric constant and dielectric loss of *nano*-SiO_2/PES/BMI composites were 4.55 and 0.0029, respectively. The corresponding volume resistivity and breakdown strength were 1.74×10^{14} Ω m and 29.11 kV/mm, increased by 68.9% and 35.9%, respectively. Babkin et al. [105] used commercial PEI (Ultem 1000P) to modify BMI resins (Ultem 1000P/BMI). When the mass fraction of Ultem 1000P was 15 wt%, the K_{IC} of Ultem 1000P/BMI resins was increased by 116.6% compared to that of pure BMI resins. Han et al. [106] modified BMI resins with the PAEK containing naphthalene biphenyl units (PPEK). When the mass fraction of PPEK was 10 wt%, the impact strength of PPEK/BMI resins was increased by 39% compared with that of pure BMI resins.

3.2.3.5 Modified BMI Resins by Thermoset Resins

Modified BMI resins by thermosetting resins are commonly obtained via the copolymerization reaction of functional groups between two kinds of resin to form the polymer interpenetrating network structure, improving the mechanical properties [106, 107]. Huang et al. [73] used bisphenol A benzoxazine (BOZ) as a toughening agent, diallyl bisphenol A (DABPA), and 3-ethynyl aniline (*m*-APA) as chain extension toughening agents to prepare BOZ/BAD resins. When the mass fraction of BOZ was 30 wt%, the impact strength of BOZ/BAD resin was increased by 53.8% compared to that of BAD resins. Zhang et al. [108] synthesized hyperbranched epoxy resins (AHEP) to modify BMI resins (DEP). When the amount of AHEP was 10 wt%, the impact strength of DEP resins was increased by ten times compared with that of pure BMI resins. Wang et al. [109] modified BMI resins with novel hydroxyl-terminated benzoxazine (Boz-BPA) and CE resins. When the mass fraction of Boz-BPA and CE resins was 10 and 40 wt%, the impact strength and flexural strength of Boz-BPA/CE/BMI modified resins were 14.3 kJ/cm^2 and 128.6 MPa, increased by 53.7% and 22.1%, respectively. And the corresponding 5% thermogravimetric loss temperature was 365 °C, exhibiting the good heat resistance.

3.2.3.6 Modified BMI Resins by Flexible Long-Chain Segments

BMI resins can also be generally toughened by introducing flexible long-chain segments. Zou et al. [110] introduced flexible *n*-dodecyl amine (DDA) into BMI resin monomers to prepare BMI-DDA resins. When the mass fraction of DDA was 1 wt%, the breaking elongation and K_{IC} of BMI-DDA resins increased by 32.3% and 24.3% compared to those of pure BMI resins. Jiang et al. [111] synthesized a new BMI monomer (MADQ) with the aliphatic chain structure to prepare MADQ/BMI resins. When the mass fraction of MADQ was 3 wt%, the impact strength of MADQ/BMI resins was increased by 90.1% compared with that of pure BMI resins.

3.2.3.7 Other Modification Methods

In addition, various modification methods can synergistically toughen BMI to expand the application as a polymer matrix for wave-transparent composites. Han et al. [112] synthesized a novel amino-functionalized metal-organic framework (sN-MIL) modified BMI resins to prepare sN-MIL/BMI resins. When the mass fraction of sN-MIL was 0.3 wt%, the impact strength of sN-MIL/BMI resin was increased by 70% compared to that of pure BMI resins. Shi et al. [113] synthesized a novel hyperbranched unsaturated polyester (HBP) to modify BMI resins. When the amount of HBP was 20 wt%, the impact strength of modified BMI resins was increased by 211.1% compared with that of pure BMI resins.

3.2.4 Silicone Resins

As an organic–inorganic hybrid polymer matrix, the main chain structure of silicone resins is Si—O—Si bonding [114, 115]. Silicone resins usually have four basic structural units, and the details of common silicone resins are shown in Table 3.4. Tri-functional or tetra-functional structural units are essential components for synthesizing silicone resins, and monomer units with functionalities of 1 or 2 can be used to regulate the cross-linking and flexibility of silicone resins, etc. [116, 117]. Different silicone resins can be prepared by adjusting the content of each unit.

Depending on the curing method, silicone resins can be divided into condensation, addition molding, and free radical silicone resins. Among them, condensation and additive silicone resins are the two most widely used types. Condensation-type silicone resins are prepared by removing small molecules such as the water or hydrogen between silicone hydroxyl groups or Si–H bonds under heating or catalyst conditions, which endow their good thermal stability and excellent bonding properties [118, 119]. However, small molecules produced during curing would affect the processing performances (easy to foam and not easy to store). Besides, the molecular weight of cured condensation-type silicone resins is difficult to control. Additive silicone resins are cured by the addition reaction between the unsaturated bonds contained in the silicone resins and the silicon hydrogen bonds to form cross-linked curing networks. No small molecules are released while curing additive silicone resins, which is not easy to foam. However, the catalyst of additive silicone resins is easy to oxidation failure [120, 121]. Free radical silicone resins are initiated by organic peroxides for free radical reaction, which has the advantages of low curing temperature and long storage period. Still, the decomposition of peroxide initiator would adversely affect the properties of silicone resins [122, 123].

Silicone resins have highly cross-linked three-dimensional structures, which combine the advantages of both organic and inorganic materials. Compared with other polymer matrix, silicone resins have lower ε and $\tan\delta$ values in different temperatures and frequencies [124, 125]. Meanwhile, the bonding energy of Si–O

Table 3.4 Basic structural unit of silicone resins.

Expressions	Functionality	R/Si ratio	Marking
$R_3SiO_{1/2}$	1	3	M
R_2SiO	2	2	D
R_2SiO	3	1	T
SiO_2	4	0	Q

Table 3.5 Fiber-reinforced silicone resins wave-transparent composites.

Types	Density	Flexural strength (MPa)	Thermal conductivity (W/(m K))	ε (10 GHz)	$\tan\delta$ (10 GHz)
3D glass fiber/SiO_2 (Russia)	1.5–1.6	10–25	0.36	3.0	0.01
Quartz fiber/SiO_2 (China)	1.78	13.2	—	2.7–2.9	0.008
2.5D glass fiber/SiO_2 (China)	1.6–1.65	25–34.6	0.22	2.8–3.4	0.007–0.01

bond (416 kJ/mol) is larger than that of C–C (332 kJ/mol) and C–O (326 kJ/mol), which is less likely to break at higher temperatures. Besides, the alkane groups attached to Si atoms can undergo oxidation reactions at higher temperatures, resulting in Si–O–Si bonds that are highly cross-linked and more stable [126, 127]. Russia has successfully applied silicone resins as wave-transparent composite matrix on missiles, space shuttles, and space stations (Table 3.5). At the same time, the performance index of quartz fiber-reinforced silicone resins wave-transparent composites developed by the First Research Institute of China Aerospace Industry Corporation (703 Institute) and the 2.5D glass fiber-reinforced silicone resins wave-transparent composites prepared by the of the Fourth Academy of Chinese Aerospace Science and Technology (43 Institute) have reached the international advanced levels.

However, Si–O bonds in silicone resins are highly polarized and prone to bond breaking under the action of electrophilic or nucleophilic reagents, making them less resistant to reagents and environmental performances [128]. Common modification methods of silicone resins include filling and structural modification [129, 130]. Inorganic fillers, such as metal oxides and polyhedral oligomeric sesquisiloxane (POSS), can significantly improve the high-temperature resistance of silicone resins [131, 132]. Huang et al. [133] prepared an ultrafine TiO_2 nanoparticle (S-TiO_2) modified by titanate coupling agents, which were then performed to prepare the S-TiO_2/silicone composites by solvent mixing method. The titanate protective layer on the surface of S-TiO_2 could effectively prevent the aggregation of S-TiO_2 nanoparticles and significantly improve the compatibility between S-TiO_2 nanoparticles and silicone resins, thus significantly improving the thermal stability of S-TiO_2/silicone resins. When the mass fraction of S-TiO_2 was 5 wt%, the T_{50} (the temperature corresponding to 50% weight loss) of S-TiO_2/silicone resins was enhanced to 632 °C, higher than that of pure silicone resins (532 °C). Huang et al. [134] introduced ZrO_2 into silicone resins through in situ sol–gel reaction. When the mass fraction of ZrO_2 was 5 wt%, the residual carbon rate of ZrO_2 particle hybridized silicone nanocomposites was increased by 9 wt%. Dong et al. [135] obtained phenyl polyhedral oligomeric sesquisiloxane (phenyl POSS) by hydrolysis of phenyltrimethoxysilane, which was then performed to prepare the phenyl POSS/silicone composites by simple blending. The phenyl POSS/silicone composites exhibit excellent thermal stability, good adhesion, and toughness. However, the filling modification method generally suffers from poor compatibility between inorganic powders and silicon resins, resulting in agglomeration.

The structural modification of silicone resins can be divided into the main chain and side chain structure modification [136]. The modification of main chain structures usually involves the introduction of heteroatoms, rigid aromatic rings, or aromatic heterocycles to inhibit the internal rotation and the relative slip of main chain in favor of improving the mechanical properties [137, 138]. Jiang et al. [139] synthesized a new boron-doped silicone resin (BSR) by hydrolytic condensation to enhance the thermal degradation temperature of silicone resins. The incorporation of boron in the main chains substantially increased the thermal degradation temperature of silicone resins. Yang et al. [140] introduced a benzylidene support structure on the backbone of silicone resins to inhibit the diffusion of oxygen, which led to a significant improvement in thermal stability. Meanwhile, under the inert atmosphere, the benzylidene units would undergo a chain-branching reaction to inhibit the degradation of silicone resins.

Side chain modification for silicone resins is usually carried out in three ways. The first way is realized by silicone hydrogen addition reaction with vinyl and allyl via silicone hydrogen bonding; the second way is obtained by reaction of silicone hydroxyl group of silicone oligomer with other intermediates containing reactive groups; the third way is nucleophilic substitution reaction of halogen at the end of haloalkyl of silicone intermediate side chains [141, 142]. Wang et al. [143] constructed silicone rubber networks containing epoxy groups by silicon–hydrogen addition reaction between the synthesized vinyl-containing epoxy prepolymers and vinyl-capped silicone oils. The modified silicone rubbers had homogeneous "island" phase separation structures, and the vinyl-containing epoxy prepolymers in the silicone rubber matrix presented good dispersion, resulting in greatly improving the adhesion, tensile properties, and heat resistances of modified silicone rubbers.

3.2.5 Polytetrafluoroethylene (PTFE) Resin

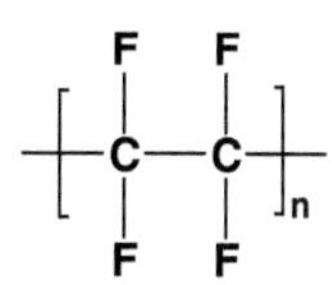

Figure 3.5 Chemical structure formula of PTFE resins.

PTFE resins are completely symmetrical, nonpolar linear polymers with fluorine atoms attached to carbon atoms (Figure 3.5), and their molecular chains form a spatial helical conformation with ultra-low ε and $\tan\delta$ values [144]. The ε of PTFE resins is stable at 2.1, and $\tan\delta$ stays around 1×10^{-4} under a wide temperature and frequency range [145]. In addition, PTFE resins have good chemical and environmental resistances, thermal stabilities, and low moisture absorption (<0.01%). Above 400 °C, no carbon residues are produced on the surface of PTFE resins, and the ablation products are all volatile compounds, which have less effect on the dielectric properties [146–148]. However, PTFE resins have disadvantages, such as high coefficient of linear expansion, poor mechanical properties, and difficulties in secondary processing and molding.

The mechanical properties and processing performance of PTFE resins can be improved by reinforcement, filling, and blending [149–151]. Jiang et al. [152] studied the properties effect of SiO_2/PTFE resins filled with different silicon dioxide (SiO_2) (2 and 15 μm) particle sizes. The small size of SiO_2 (2 μm) could fill the pores between the large size of SiO_2 (15 μm) and facilitate the formation of dense microstructures. When the mass fraction of SiO_2 was 31 wt% for both 2 and 15 μm, the modified PTFE resins maintained lower ε (2.99) and tan δ (0.002) values, and the corresponding thermal expansion coefficient was 18.68 ppm/K, lower than that of pure PTFE resins (26.02 ppm/K). Cheng et al. [153] utilized poly(methyl methacrylate) (PMMA) and sodium dodecyl sulfate (SDS) to surface treat the halloysite nanotubes (F-HNT), which were then to prepare the F-HNT/PTFE composites by physical blending method. The tensile strength of modified F-HNT/PTFE was significantly increased by 2.4–4.0 MPa compared to that of pure PTFE resins. Yu et al. [154] used a covalent grafting method to graft the octavinyl-POSS into poly(methyl methacrylate/butyl methacrylate) (PMB), which were then blended with PTFE resins to prepare POSS-PMB/PTFE composites. The mechanical properties and wear resistances of POSS-PMB/PTFE composites were significantly improved when the amount of octavinyl-POSS was 10 wt%. In addition, the introduction of Si–O–Si bonds in the POSS structures improved the POSS-PMB/PTFE composites in UV, electrical radiation, and atomic oxygen resistances.

PTFE resins have been performed as polymer matrix for wave-transparent composites in wireless communication and aerospace applications [155, 156]. Some American companies, such as Arlon, Taconic, and Rogers, have developed a series of glass fibers reinforced PTFE-based wave-transparent composites, such as DiClad522, CuClad233LX, IsoClad933, AD255, RF35A2, and Avcoat8029.

3.2.6 Unsaturated Polyester (UP) Resins

UP resins (Figure 3.6) are polyesters with unsaturated bonds in the molecular chains and are usually made by condensation of unsaturated diacids with diols [157–159]. Appropriate amounts of vinyl monomers, initiators, and accelerators are usually added into UP resins to form the thermosetting resins with three-dimensional cross-linked structures after curing [160, 161]. The advantages of UP resins are simple and easy to obtain raw materials for production, good product properties to meet the variety needs of molding processes, better sound insulation and light transmission properties, and excellent chemical resistances [162–164].

Where B and A denote the aryl or alkyl in saturated dibasic acid and dihydric alcohol, respectively.

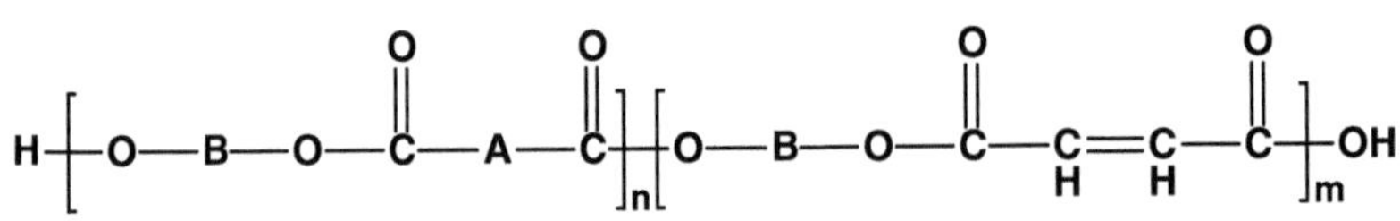

Figure 3.6 Molecular structure formula of the typical UP resins.

Vinyl-like monomers are usually added into linear UP resins for using as crosslinkers and diluents. While reducing the viscosity of the resins system, the linear UP resins are cross-linked by copolymerization between ethylene-based monomers and double bonds to produce the thermosetting resins with the three-dimensional curing structures [165, 166]. After adding vinyl monomer, the curing process of UP resins mainly occurs in the following four reactions: (1) Vinyl monomers play the role of cross-linking agent. The cross-linking reaction occurs between the double bond of different UP resins linear molecular chains to further form the three-dimensional network structures; (2) Vinyl monomers and the double bond in the same UP resins molecular chain produce intramolecular bonding to form the intramolecular ring structures; (3) Ethylene-based monomers interact with the double bonds in the same UP resins molecular chain, resulting in intramolecular bonding and the formation of intramolecular ring structures; (4) Ethylene-based monomers undergo homo-polymerization by themselves to produce the soluble oligomers [167–169].

There are four reactions during the curing of UP resins, affecting the curing reaction behaviors of UP resins [170, 171]. Yang et al. [172] investigated the effects of vinyl monomer (styrene) concentration and molecular structure of unsaturated polyesters on the kinetic and rheological properties of the cross-linking reaction. The conversion rate during actual gelation in UP resin curing was higher than that obtained by extrapolation of the Flory gelation theory. It could be attributed to the large number of intramolecular ring-forming reactions during the curing process and, consequently, the formation of particulate gel. Yang et al. [173] observed the morphological changes in the cross-linking copolymerization reaction of UP resins. The macromolecular radicals in UP resins produced cross-linking reactions between intramolecular reactive groups to form particulate gel with few microns.

The curing and cross-linking process of UP resins can be divided into five stages [174].

(1) Initiation stage: The free radicals generated by the decomposition of initiator react with ethylene-based monomers and unsaturated double bonds in UP resins to produce the corresponding radicals. Most free radicals are in the growth state during the curing with a long survival period. There is little reaction heat at the initiation stage, and the initiation phase ends after the polymerization inhibitor is consumed completely.
(2) Microgel formation stage: The free radicals are linked to adjacent UP resin molecules to obtain particle gels with a diameter of about 10 μm. In the microgel formation stage, the diameter and composition of gels do not change significantly, while the concentration of free radicals gradually increases until it stabilizes at a certain value, and only a small amount of reaction heat is released.
(3) Transition stage: Vinyl free radicals in the microgel react with other reactive groups to form the micelles, and the number of micelles gradually increases with increasing time. In addition, the concentration of free radicals increases slowly until the whole reaction system reaches the gel point. In the vicinity of gel point,

the shear modulus of cured system increases, and the heat of reaction accumulates. At the transition stage, there are two conditions of phase separation and no phase separation due to the different ratios of vinyl monomer and UP resins. If phase separation occurs, the transition stage time is longer; if phase separation does not occur, the transition stage time is shorter.

(4) Macrogel stage: The microgels accumulate gradually with increasing time until they form network structures and the whole system reaches the gel point. At the macrogel stage, most gels produce aggregation to form micelle structures. Since the reaction occurs inside the particles, the molecular weight and the conversion rate of the reactive groups of UP resins increase, and the released reaction heat increases significantly.

(5) Post-gel stage: At the beginning of this stage, the concentration of free radicals in the reaction system tends to increase linearly at a certain rate. By the end of the reaction, the concentration of free radicals slowly tends to the limit value. If the UP resin is cured at room temperature, the final conversion rate is about 30%, and after post-curing treatment, the conversion rate exceeds 70%. At this stage, the heat release of reaction is significant.

Molecular structures of UP resins can be designed according to the properties requirements of final composites [175]. For example, the structure of the binary acid and dihydric alcohol determines the mechanical properties of UP resins. The longer the chain length of the dihydric alcohol or saturated acids is, the better flexibility of the UP resins is [176]. The greater proportion of unsaturated acids, the higher density of crosslinking points of UP resins, which increases their heat resistances and hardness [177, 178]. At the same time, UP resins can be cured at room temperature, and no volatiles escape during the curing process.

However, the poor mechanical properties, low–high-temperature resistances, short storage cycles, and high curing shrinkage of UP resins have limited their wider application [179, 180]. Toughening UP resins by elastomers are the main industrial modification method. Huang et al. [181] prepared the core–shell rubbers (CSRs) using a two-step polymerization method and then investigated the effect of CSRs on the fracture properties of UP resins under 110 °C curing conditions. The soft core of CSRs was rubber-like poly (*n*-butyl acrylate), and the hard shell was made from methacrylate, ethylene glycol dimethacrylate, and various concentrations of glycidyl methacrylate. The fracture toughness and modulus of polyacrylate toughened UP resins increased with the increasing of glycidyl methacrylate content, but the impact strength and tensile strength decreased. Li et al. [182] used the *co*-condensation reaction of the hydroxyl groups at both ends of high-*cis*-terminal polybutadiene (HTPB) with UP resins. Then, they investigated the mechanical and thermal properties of the modified UP resins at room and low temperatures. The breaking elongation, tensile strength, and impact strength of the modified UP resins at room temperature were respectively increased by 124%, 94%, and 132%, and the corresponding breaking elongation, tensile strength, and impact strength of modified UP resins were increased by 220%, 186%, and 167% at low temperature when the amount of HTPB (molecular weight 2000 g/mol) was 5 wt% compared with that of the pure UP resins.

At present, UP resins can be cured at room temperature, and the corresponding molding process is simple and low-cost, which is suitable for the production of large-scale or large radomes [183–185]. UP resins used as polymer matrix for wave-transparent composites internationally are usually phthalic acid type (referred to as *o*-phenylene types), isophthalic acid type (referred to as *m*-phenylene types), bisphenol A type and vinyl ester type, etc. For example, Nan-gatuck Chemical Company (USA) developed UP resins with trade names Vibrinl35 and Vibrinl36, which were used as polymer matrix for Boeing's Bomarc missile radome.

3.2.7 Cyanate Ester (CE) Resins

CE monomer contains two or more cyanate groups (-OCN), which can form triazine ring network structures via cyclization reaction under the action of heat and catalyst (nucleophilic reagent) (Figure 3.7) [186]. The triazine ring structures are highly symmetrical, which is a benefit to weakening the dipole polarization. Therefore, the CE resins have low ε (2.8–3.2) and $\tan\delta$ (0.002–0.008) values, which can be stable over a wide range of temperatures (0–220 °C) and frequencies (10^6–10^{11} Hz). Meanwhile, the cured CE resins present excellent mechanical properties, outstanding heat resistances (the initial thermal decomposition temperature >280 °C), good processing properties, and low moisture absorption rate (<1.5%) and thermal expansion coefficient, which have gradually become a new generation of high-performance polymer matrix for radomes.

CE resins exist in three states: solid, semi-solid, and liquid at room temperature. They can be soluble in tetrahydrofuran, acetone, chloroform, butanone, etc. According to the structure of CE monomers, CE resins are mainly divided into bisphenol A CE resins, bisphenol F CE resins, dicyclopentadiene CE resins, phenolic CE resins, and bisphenol E CE resins (Table 3.6) [187–189].

The carbon atoms in the -OCN group of CE resins monomer have strong electrophilicity and are easily attacked by nucleophiles. Therefore, the -OCN groups can react with the water, phenolic compounds and other hydrogen donor compounds, transition metal complexes, imidazole, and other substances to form highly symmetric triazine ring cross-linking network structures with low polarity and high

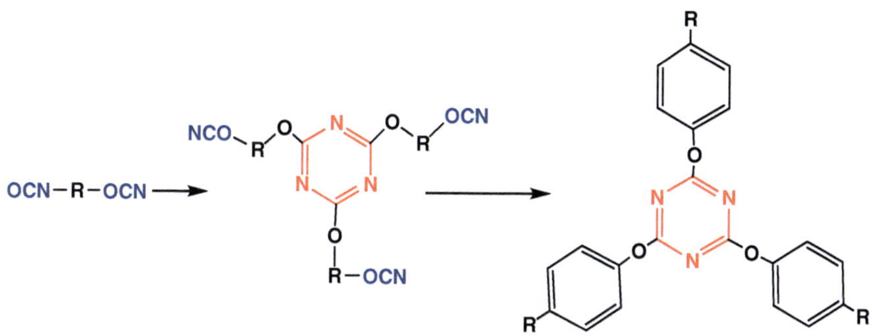

Figure 3.7 Schematic diagram of triazine ring network structures via cyclization reaction.

Table 3.6 The monomer structures and basic properties of CE resins.

CE monomer structures	Melting point (°C)	Glass transition temperature (°C)	Water absorption (%)	ε (1 MHz)
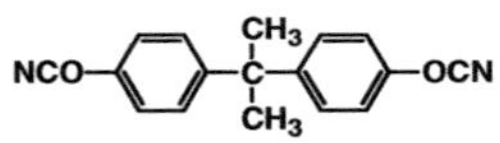	79	289	2.5	2.91
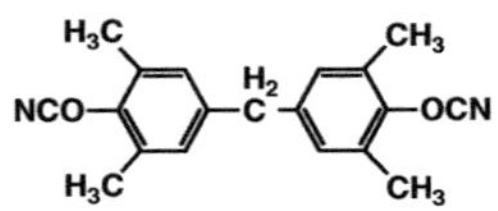	106	252	1.4	2.75
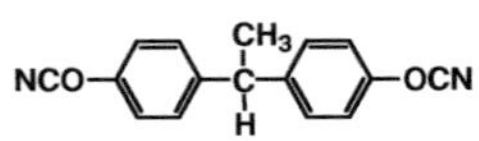	29	258	2.4	2.98
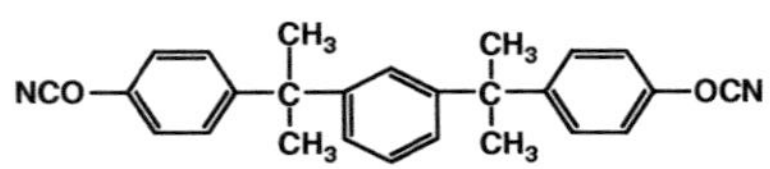	68	192	0.7	2.64
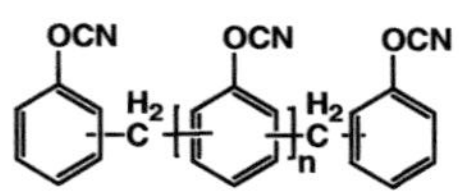	Resinous	>350	3.8	3.08

rigidity after curing [190]. Additionally, the ether bonds between the triazine rings and aromatic rings are benefit to enhancing the intramolecular rotation. The Poisson ratio, tensile strength, and flexural strength of cured CE resins are 0.4, 70–90, and 80–100 MPa, respectively. Compared with ER, CE resins have relatively higher modulus and strength, and the corresponding fracture toughness is between ER and BMI resins.

Meanwhile, the formed cross-linked networks of cured CE resins contain a large number of aromatic rings in favor of endowing excellent heat resistance. The T_g of CE resins is about 240–290 °C, and the initial decomposition temperature (T_d) is about 420–450 °C. The service life of the cured CE resins is up to 25 000 hours in the range of 162–180 °C. The heat resistance of CE resins is better than that of epoxy and BMI resins. The curing system of CE resins does not contain ester and amide bonds, which are easy to hydrolyze, and the ether bond on the molecular chains is almost not affected by water. The annular structure with a large steric hindrance can increase the resistance of water diffusion, endowing good humidity and heat resistance. In addition, CE resins present good adhesion to metal, mainly attributed to the fact that the CE resins can form covalent bonds with functional groups on the metal surfaces.

However, the high cross-linking density of the triazine ring network structures leads to the obvious brittleness of CE resins. Meanwhile, highly crosslinked polymer networks restrict the segment migration of molecular chains.

Physical blending or chemical copolymerization of CE resins by thermosetting polymers is an important way to increase toughness. The addition of thermosetting polymers into CE resins can effectively reduce the crosslink density to endow different branching degrees and microscopic network structures [191–193]. The adopted thermosetting resins mainly include epoxy, BMI, and PF resins. The copolymerization reaction between CE resins and ER is described as follows. Self-polymerizate of CE monomers forms triazine ring cross-linked structures, and then cyanate groups react directly with epoxy groups to form an oxazolidinone structure. Simultaneously, the ER undergo the etherification reaction (Figure 3.8). The modified CE resins form a large number of regular triazine ring structures while introducing ether bonds into the cross-linked networks through *co*-curing reaction between the ER and CE resins [194]. As a result, the toughness of modified CE resins is improved, and the overall mechanical properties are enhanced. Lu et al. [195] modified CE resins by *co*-polymerization of ER and phenolic resin (PR). The addition of ER and PR could improve the toughness while maintaining the thermal stability of CE resins. When the mass ratio of ER:PR:CE was 15:15:70, the flexural and impact strengths of modified CE resins were respectively increased from 113.6 MPa and 5.2 kJ/m^2 (pure CE resins) to 134.5 MPa and 16.7 kJ/m^2. The addition of ER and PR had little effect on the thermal stability and dielectric properties of the modified CE resins. BMI resins can be directly melt-blended with CE resins to obtain *co*-blended system. The –OCN group in CE resins can react with the unsaturated double bond on the maleimide ring of BMI resins to form benzoxazine (BT) resins with a triazine ring,

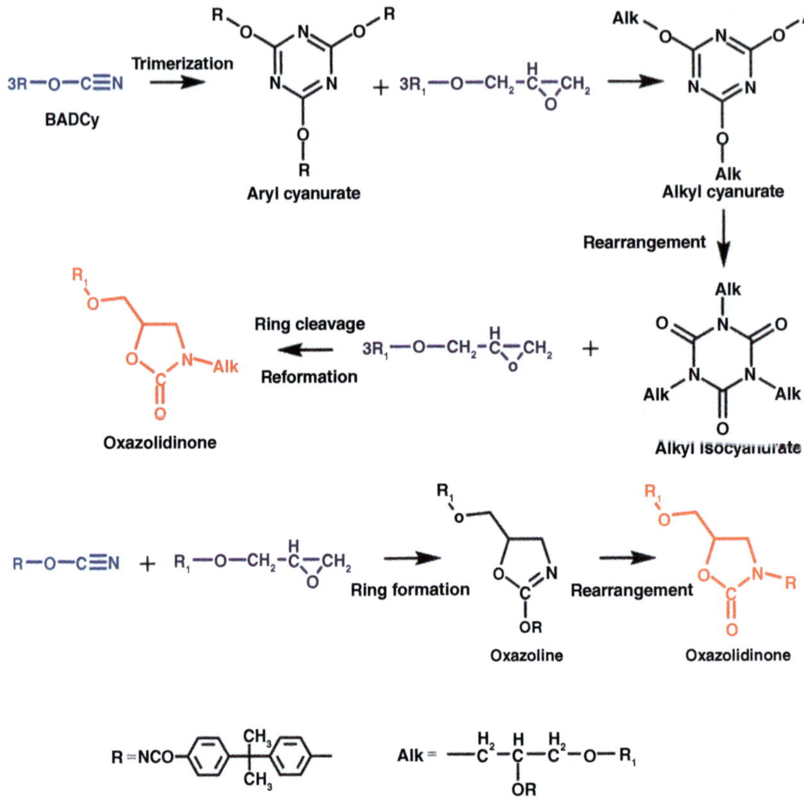

Figure 3.8 Copolymerization between epoxy resins and CE resins.

pyrimidine, and pyridine structure [196–198]. Wan et al. [199] prepared the modified bisphenol A type cyanate (BADCy) resins by copolymerization of BMI resins. The BMI/BADCy resins had good mechanical properties and insulation properties. When the mass fraction of BMI was 20 wt%, the volume resistivity of BMI/BADCy resins was 7.8×10^{15} Ω cm. The corresponding impact strength, flexural strength, tensile strength, and breaking elongation were, respectively, 16.2 kJ/m^2, 157 MPa, 76.6 MPa, and 4.9%, which were 1.59, 1.52, 1.04, and 1.1 times of pure BADCy resins. However, the ε values of modified BADCy increase gradually with the increasing BMI content.

Thermoplastic polymer toughening is a very attractive method of toughening CE resins. The obtained blends possess enhanced toughness while retaining intrinsic properties of the CE resins [200]. The corresponding toughening mechanism of thermoplastic polymers-modified CE resins is usually explained by phase separation theory. The initial compatibility between CE resins and thermoplastic polymers shows the homogeneous phase. As the concentration of thermoplastic polymers increases, microphase separation gradually occurs, forming a two-phase structure with CE resins as continuous phase and thermoplastic polymers as dispersed phase. When the concentration of thermoplastic polymers increases, the continuous phase

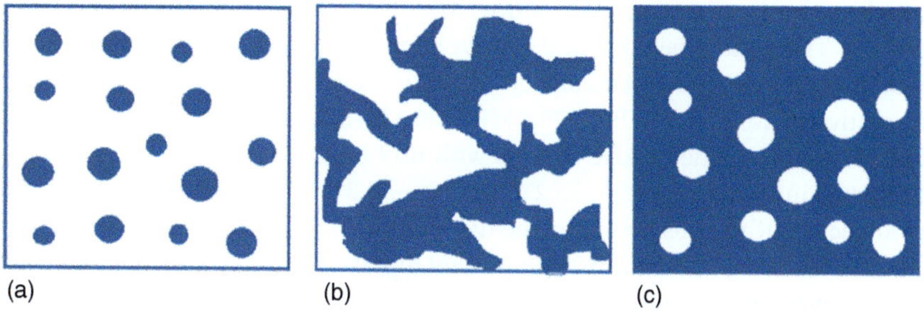

Figure 3.9 Schematic diagram of phase separation structures when the concentration of thermoplastic particles increases: sea-island structures (a), double continuous structures (b), and reverse rotation (c).

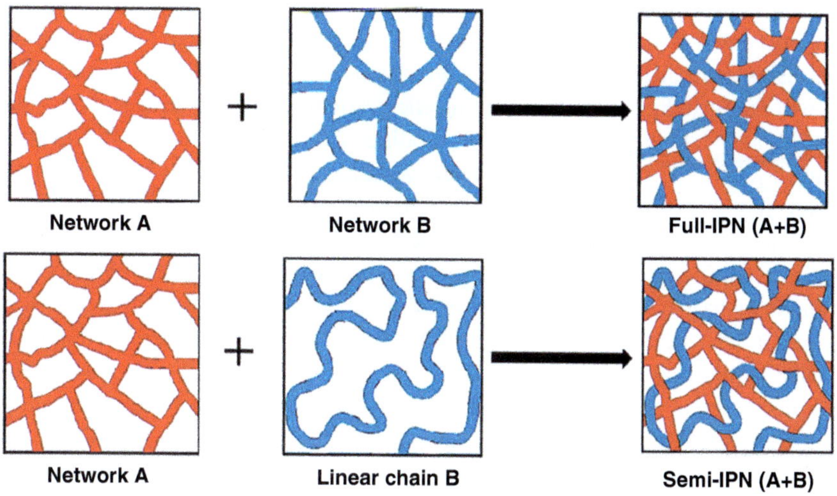

Figure 3.10 Schematic diagram of IPN and semi-IPN structures.

structure will be formed. When the concentration of thermoplastic polymers is higher than that of CE resins, the opposite transformation occurs (Figure 3.9) [201]. This two-phase structure can effectively prevent the propagation of micro-cracks and improve the toughness of the final materials.

The interpenetrating network (IPN) or semi-interpenetrating network (semi-IPN) structures can be formed by combining two incompatible polymers in the form of the networks (Figure 3.10) to further improve the comprehensive properties of modified CE resins. For example, the obtained materials possess the toughness of thermoplastic polymers and the strength and high-temperature resistances of thermosetting polymers [202–204]. Thermoplastic polymers used to toughen CE resins mainly include PES, PAEK, polyester, PEI, and PSU.

PES has good mechanical properties, low shrinkage and hygroscopicity, and good creep resistances. Wang et al. [205] modified BADCy resins with phenolphthalein polyether sulfone (PES-C) and epoxy resins. The addition of PES-C caused the copolymerization reaction and induced the phase separation during the curing process. The tensile, bending and impact strength, and fracture toughness of the modified BADCy resins (Amount of PES-C and epoxy resins of 20 and 13.1 wt%, respectively) were 81.8 MPa, 178.8 MPa, 26.1 kJ/m^2, and 2.54 MPa $m^{1/2}$, higher than those of modified BADCy resins with no PES-C (Amount of epoxy resins of 17.5 wt%), increased by 24.3%, 25.2%, 183.7%, and 225.6%, respectively. Meanwhile, modified BADCy resins presented the higher T_g values (236.9 °C).

PAEK resins are one of the most important high-performance engineering materials, which possess excellent heat, radiation, creep, impact and chemical resistances, dimensional stability, and low ε value. Li et al. [206] prepared phenolphthalein polyether ketone/epoxy resins/CE (PEK-C) polymers with semi-interpenetrating networks by hot melt method. When the mass fraction of PEK-C was 15 wt%, the K_{IC} and G_{IC} of fracture toughness were increased by 20% and 50%, respectively.

In addition, Iijima et al. [207] synthetized the aromatic polyester (PEP) to improve the brittleness of CE resins. Compared with pure CE resins, the K_{IC} of modified CE resins increased by 120% after adding 20 wt% PEP. The modified CE resins with 20 wt% PSu had the optimal toughening effect, and the corresponding K_{IC} value was 1.45 MPa $m^{1/2}$, higher than that of PEI-modified CE resins with the same mass fraction (K_{IC} of 1.10 MPa $m^{1/2}$). Besides, the T_g value of PSU-modified CE resins was lower than that of PEI-modified CE resin at the same PSu mass fraction.

It can be deduced that the thermoplastic polymers have a better toughening effect and can retain higher T_g. However, the ratio between thermoplastic polymers and cyanate matrix should be fully considered to obtain excellent comprehensive properties. In addition, introducing thermoplastic polymers with excessive molecular weight would lead to a sharp increase in melt viscosity, which is not conducive to the processing performance of CE resins [208]. Rubber elastomers possess excellent toughness, which can be used as a good toughening modifier [209]. The active end groups of rubbers can react with the CE resins to introduce flexible chain segments into the cross-linking networks. During the curing process of CE resins, the rubber elastomers can precipitate from the CE resins to further form two-phase structures. Under the condition of external force, rubber particles play the roles on the stress concentration points, which can consume energy to significantly improve the fracture toughness of the modified CE resins [210–212]. Feng et al. [213] used CTBN to improve the toughness of CE resins. The toughness of the modified CE resins was related to the content of CTBN. When the content of CTBN was 10 phr, the impact strength of modified CE resins increased by 200%, but the corresponding energy storage modulus of modified CE resins decreased.

3.3 Design and Preparation of Polymer Matrix with Low Dielectric Constant

The existing ε and $\tan\delta$ values of polymer matrix are still far from the practical application requirements of high-performance wide-band polymer matrix wave-transparent composites [214, 215].

3.3.1 Epoxy Resins

As one of the most important electronic materials, epoxy resins have been widely used in electronic coatings, electronic packaging, and electronic substrates because of their excellent adhesive properties, low density, high mechanical strength, and good heat and chemical resistance [216]. However, the ε and tan δ values of conventional epoxy resins are relatively high (ε for 3.7–3.9 and tan δ for 0.018–0.020), which cannot meet the requirements of advanced microelectronic materials [217–219].

3.3.1.1 Structural Modification of Epoxy Resins with Low ε

The ε value of epoxy resins can be reduced by introducing low-polarity structures (such as low-polarity C–F, C–Si, and O–Si bonds) or structurally symmetrical nonpolar bulk groups into the epoxy resins molecules [220, 221]. Na et al. [222] synthesized the new fluorinated epoxy monomer of 1,5-bis(4-fluorobenzoyl)-2, 6-diglycidyl ether naphthalene (DGENF) by three-step reaction of Friedel–Crafts acylation, demethylation, and nucleophilic reaction (Figure 3.11). Compared with the bisphenol A epoxy resins (BPA-epoxy), bisphenol F epoxy resins (BPF-epoxy), and 3,3′-5,5′-tetramethylbiphenyl bisphenol diglycidyl ether epoxy resins (TMBPER), the ε value of DGENF resins was decreased to 2.97 (1 MHz), due to nonpolar naphthalene groups and low polarizability C–F bonds. Kong et al. [223] synthesized the two new epoxy monomers containing cycloaliphatic hydrocarbons, 1,4-bis(4-(N,N-diglycidylamino)phenoxy) cyclohexane (Cyh epoxy) and 1,3-bis(4-(N,N-diglycidylamino)phenoxy)adamantane (Ada epoxy), then to obtain Cyh/MHHPA and Ada/MHHPA epoxy resins, respectively. The ε values of the Ada/MHHPA were 3.39 (1 MHz), lower than that of the Cyh/MHHPA (3.63), due to the introduction of a large free-volume adamantane structure.

3.3.1.2 Curing Agent with Low ε for Epoxy Resins

Epoxy resin curing agent is an important part of cross-linking networks of cured epoxy resins [224]. Na et al. [225] introduced the 4-trifluoromethylphenyl benzimidazole (4-TFMBI) as curing agent into 3-trifluoromethylphenyl hydroquinone epoxy resins (3-TFM) to prepare 3-TFM/4-TFMBI epoxy resins (Figure 3.12). Compared with 3-TFM/MeHHPA/2-MI, 3-TFM/MeHHPA/4-TFMBI, and 3-TFM/2-MI, the 3-TFM/4-TFMBI had the lowest ε (3.38, 10 MHz). This was due to low polarizability $-CF_3$ groups in 4-TFMBI. Meng et al. [226] successfully synthesized 3,5-bis(benzoyloxy)benzoic acid capped polyaryl ether ketone containing tetramethylbiphenyl (TMPK) by polycondensation and esterification reaction to cure dicyclopentadienyl epoxy (DCPD) resins. Compared with 3,5-bis(benzoyloxy) benzoic acid capped polyaryl ether ketone (BPAPK) and benzene-1,3,5-tribenzoate (TBB) curing agents, the cured DCPD/TMPK epoxy resins had the lowest ε (3.15, 1 MHz), lower than those of DCPD/BPAPK (3.20) and DCPD/TBB (3.32).

3.3.1.3 Epoxy Composites with the Addition of Low ε Compounds, Fillers, or Fibers

The addition of low ε compounds, fillers, or fibers would endow the corresponding epoxy composites with low ε values [227].

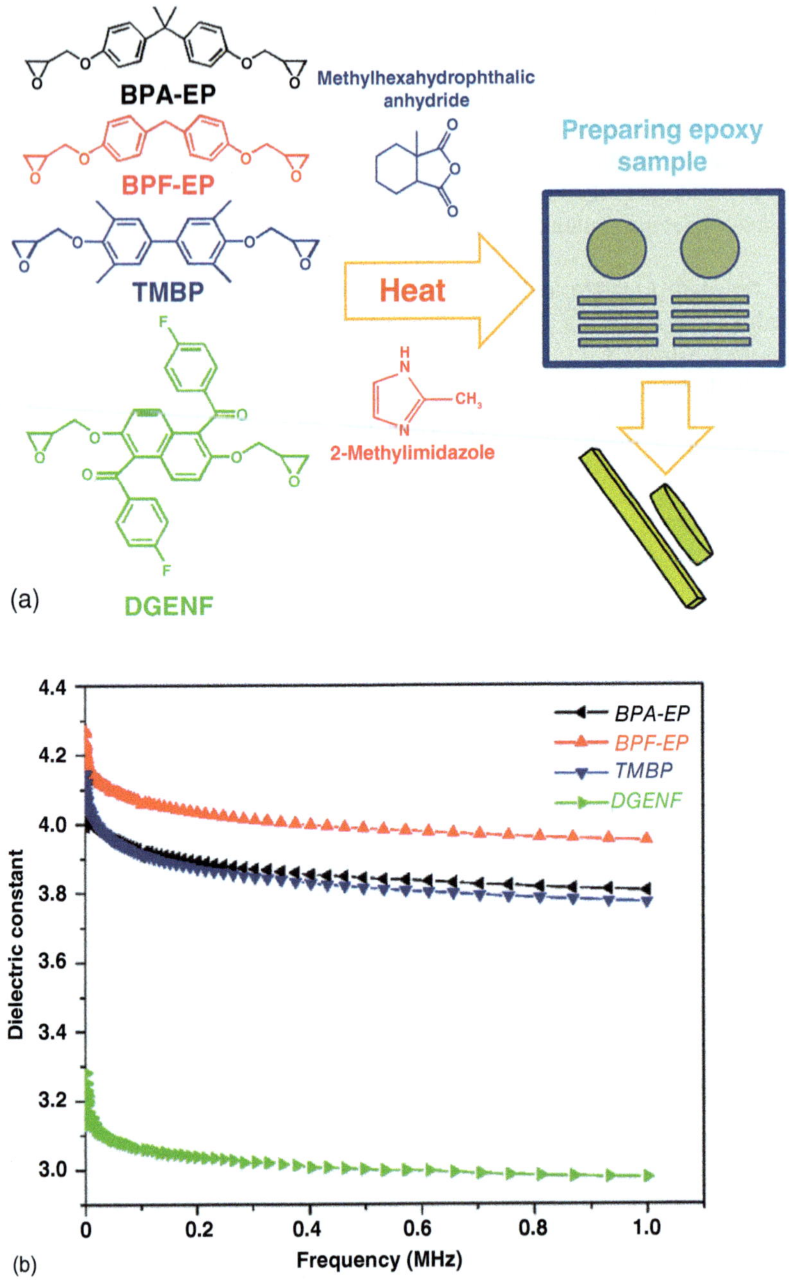

Figure 3.11 Schematic diagram of preparation for BPA-EP, BPF-EP, TMBP, and DGENF resins (a) and their dielectric properties (b). Source: [222]/Royal Society of Chemistry.

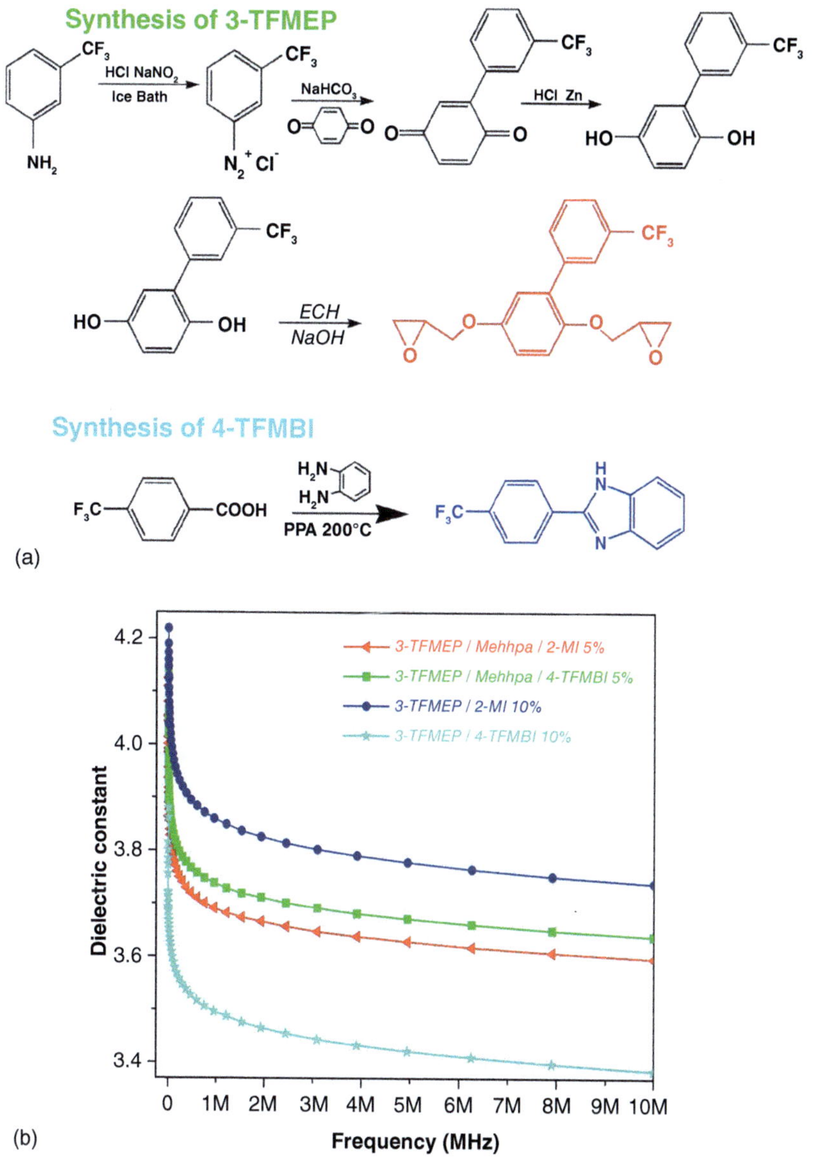

(a)

(b)

Figure 3.12 Synthesis route of TFMEP monomer and 4-TFMBI curing agent (a) and their dielectric properties (b). Source: [225]/with permission of Elsevier.

Zhang et al. [228] prepared the PAEK containing naphthalene and phenolic hydroxyl groups (HPAEK) by polycondensation and demethylation reaction. The phenolic hydroxyl group in HPAEK could react with the epoxy group in the biphenyl group-containing epoxy resins (TMBPER) to prepare porous membranes (ES-HPAEK-TMBPER) by electrostatic spinning technique (Figure 3.13). The ε values of ES-HPAEK-TMBPER porous membrane were 1.9 (1 MHz), lower than that of the TMBPER membrane (3.77), which could be attributed to the large free volume

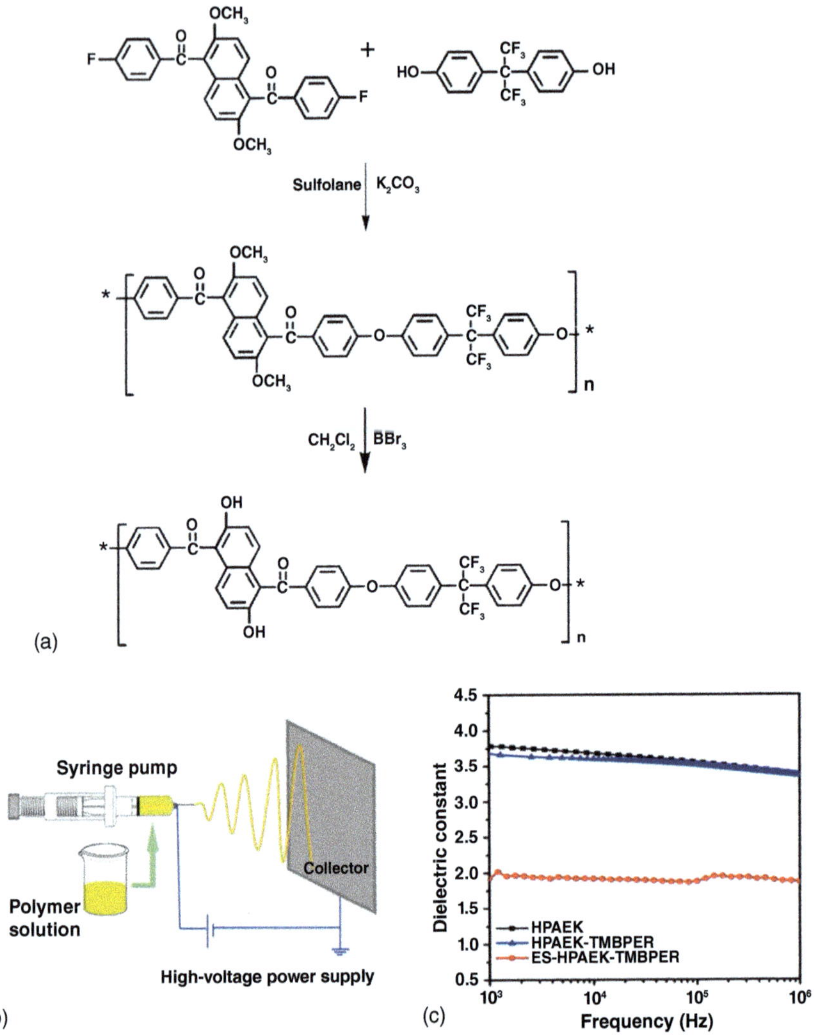

Figure 3.13 (a) Synthetic route of HPAEK, (b) Schematic preparation of ES-HPAEK-TMBPER membrane based on the electrostatic spinning technique, and (c) their dielectric properties. Source: [228]/with permission of Elsevier.

of naphthalene groups and low polarizability $-CF_3$ groups in HPAEK. Yung et al. [229] prepared modified epoxy resins by introducing hollow glass microspheres (HGM). When the volume fraction of HGM was 51.3%, the ε value of the modified epoxy resins was 2.84 (1 MHz), significantly lower than that of the pure epoxy resins (3.98). This was due to the introduction of nanopores into epoxy resins via HGM, which reduced the polarization density, thus decreasing the ε values.

Zhang et al. [230] used polyhedral sesquisiloxane zwitterion (BZPOSS) containing benzoxazine to modify the DCPD epoxy resins to prepare poly(DCPD/BZPOSS) films (Figure 3.14). When the mass fraction of BZPOSS was 10 wt%, the ε and $\tan\delta$ of

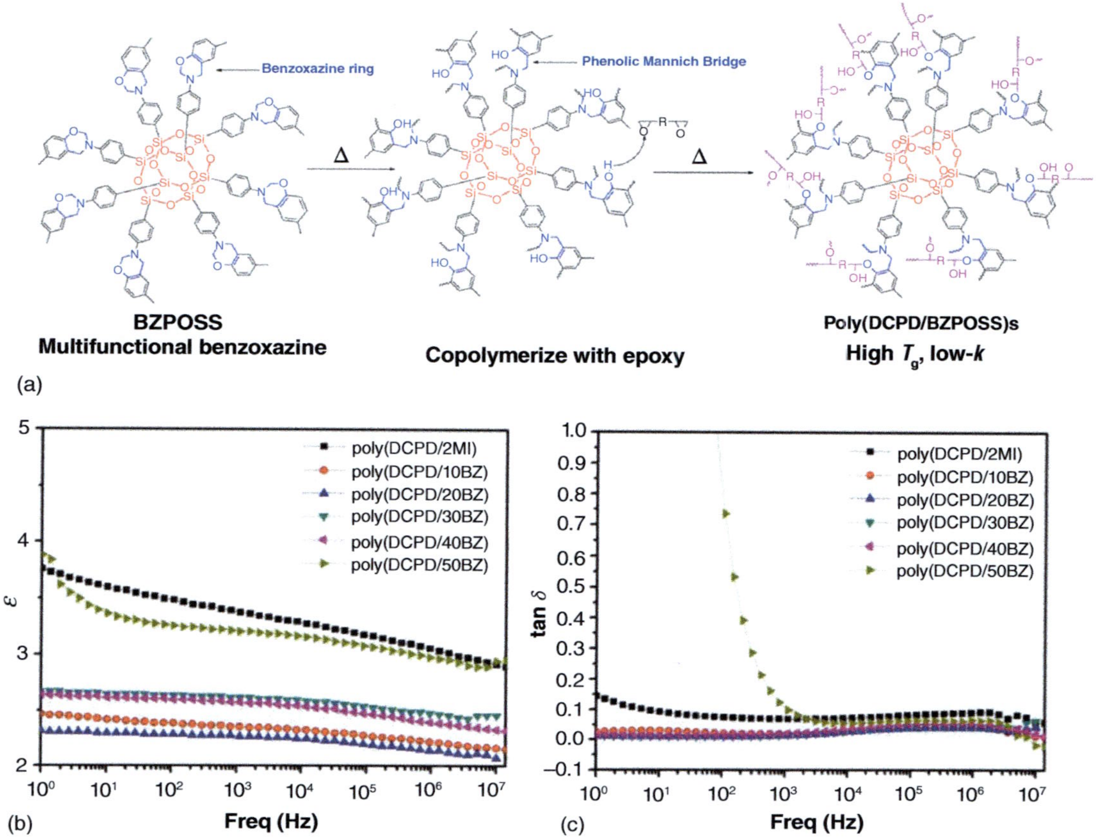

Figure 3.14 Schematic diagram of the preparation for BZPOSS-modified DCPD resins containing benzoxazine (a) and their dielectric performances (b–c). Source: [230]/with permission of Elsevier.

poly(DCPD/BZPOSS) films were, respectively, 2.21 and 0.0402 (1 MHz), which were significantly lower than those of poly(DCPD/MI) films (3.04 and 0.0903), because the introduction of microporous BZPOSS increased the free volume to decrease ε and $\tan\delta$.

3.3.2 Phenolic Resins

PF resins have excellent heat and environmental resistances and good mechanical properties [231, 232]. However, the ε (8.0) and $\tan\delta$ (0.05) of pure PF resins are too high, which limits their wider application as polymer matrix for wave-transparent composites [233]. The modified PF resins have reduced ε to 4.5–5.0 and $\tan\delta$ to 0.015–0.030 [234]. Relevant Chinese research institutions have developed PF resins such as boron phenolic resins, phenol triazine resins, and S-157 phenolic resins. Among them, S-157 resins have been widely used in supersonic shore ship missiles.

Zhou et al. [235] obtained new phenolic resins after functionalization of commercial phenolic resins by using thermally cross-linked-OCF=CF_2 groups as side chains. The phenolic resin had a very low ε, within 0.1–20 MHz, of about 2.52. Lei et al. [236] utilized the epoxy POSS (EPOSS) to modify phenolic resin (EN) by mixing with hexamethylenetetramine (HMTA) to obtain ENH resins. When the amount of EPOSS was 48.88 wt%, the ε of ENH resins was 4.11 (1 MHz), lower than that of pure EN resins (6.42). Mu et al. [237] used silane coupling agent KH-550 to modify the low molecular weight PF resins. When the amount of KH-550 was 5 wt%, the ε and $\tan\delta$ of modified PF resins were, respectively, 2.43 and 0.0168, compared with the unmodified low molecular weight PF resins. This was because the polar phenolic hydroxyl groups on the phenol in the modified PF resin effectively combined with the hydroxyl groups produced by the hydrolysis of KH-550 after KH-550 modification, resulting in a lower number of polar groups to decrease the ε and $\tan\delta$ values.

3.3.3 Bismaleimide Resins

The ε and $\tan\delta$ values of BMI resins are, respectively, 3.1–3.5 and 0.005–0.02, lower than those of epoxy and phenolic resins, which have been rapidly developed as a polymer matrix for radomes [238, 239]. However, the ε of pure BMI resins must be further reduced to meet the demand for high-performance radomes [240]. Wang et al. [241] synthesized a series of novel fluorinated bismaleimide (FBMI) resins by a three-step method. Compared with the commercially available BMI resins, the FBMI resins exhibited lower ε values of 2.49 (500 MHz). It was attributed that introducing low polarizability C–F bonds and large free volume $-CF_3$ groups could reduce the molecular polarizability and the number of polarized molecules of FBMI resins. Devaraju et al. [242] synthesized new bismaleimide (AEBMI) copolymerized by the octa(aminophenyl) sesquisiloxane (OAPS) to prepare OAPS–AEBMI nanocomposites. When the mass fraction of OAPS was 5 wt%, the ε of the OAPS-AEBMI resins was 3.19 (1 MHz), lower than that of the AEBMI resins (4.10), which could be attributed to the introduction of nano porous

OAPS, resulting in increasing the free volume of OAPS–AEBMI, finally to decrease the ε values.

Bismaleimide-triazine (BT) resins are high-performance polymer matrix made by mixing BMI and CE resins. Due to the low ε and good thermal properties, BT resins have been widely used in microelectronics and aerospace applications [243, 244]. Therefore, the development of a new generation of BT resins with low ε is important for the preparation of high-frequency printed circuit boards (PCBs) and aerospace composites. Li et al. [245] synthesized the fluorinated zirconium-based metal–organic skeleton (F4-UiO-66) by reflux method to prepare F4-UiO-66/BT resins (Figure 3.15). When the mass fraction of F4-UiO-66 was 0.5 wt%, the ε and $\tan\delta$ values of F4-UiO-66/BT resins were, respectively, 2.98 and 0.005 (1 MHz), lower than those of BT resins with ε (3.33) and $\tan\delta$ (0.016). This was because the molecular polarization of F4-UiO-66/BT resins was reduced by the introduction of low polarizability triazine ring and C–F bonds, and the introduction of nanopore structure increased the free volume to reduce the polarization density inside F4-UiO-66/BT resins, resulting in lower ε and $\tan\delta$.

Currently, polymer matrix wave-transparent composites based on modified BMI resins as matrix in Europe and America include Hexcel P550 and Hexcel F855. The grades 4501A, 4501B, and 4503 BMI resins developed by Northwestern Polytechnical University (China) have been used as the matrix of polymer-matrix wave-transparent composites because of their excellent mechanical properties, wave-transparent performances and heat resistances.

3.3.4 Silicone Resins

Silicone resins have low ε and $\tan\delta$, but their mechanical strength and heat resistances are poor [246, 247]. Researchers have focused on the modification of silicone resins aiming at synergistic enhancement of wave-transparent performances, mechanical properties, and heat resistance. Zhu et al. [248] blended the methyl hydrosiloxane (MHSis) with the poly (methyl phenyl vinyl siloxane) (PSi) resins to prepare the MHSis/Psi resins. The heat resistances of the MHSis/PSi resins were significantly improved, and the corresponding thermal decomposition temperature was from 355.7 (pure PSi resins) to 382.7 °C, while ensuring the lower ε (2.95) and $\tan\delta$ (0.0045). In addition, Zhan et al. [249] analyzed the effect of linear (LPHS) and branched (TPHS) polysiloxane curing agents on the properties of vinyl-terminated polysiloxane (MPSR) resins. The rubber platform modulus ($T_g + 30$ °C) of TPHS/MPSR resins was 75.97 MPa, higher than that of LPHS/MPSR resins (19.29 MPa). Meanwhile, the ε and $\tan\delta$ values of TPHS/MPSR resin were, respectively, 3.15 and 0.0052, lower than those of LPHS/MPSR resins (ε for 3.28 and $\tan\delta$ for 0.0068). Ren et al. [250] prepared the VHR resins with excellent tensile strength (1.34 MPa), low ε (3.33), and $\tan\delta$ (0.002) by hydrosilation of vinyl terminally functionalized poly(dimethylsiloxane) (V), poly(vinyl functionalized silicone resin) (R), and crosslinker (H) at high temperature. Zhang et al. [251] utilized the poly(methylphenyl vinyl siloxane) (PMPVSi) and hydrogen-functionalized hyperbranched polysiloxane (EHFHPSi) to prepare the ASiR resins. When the

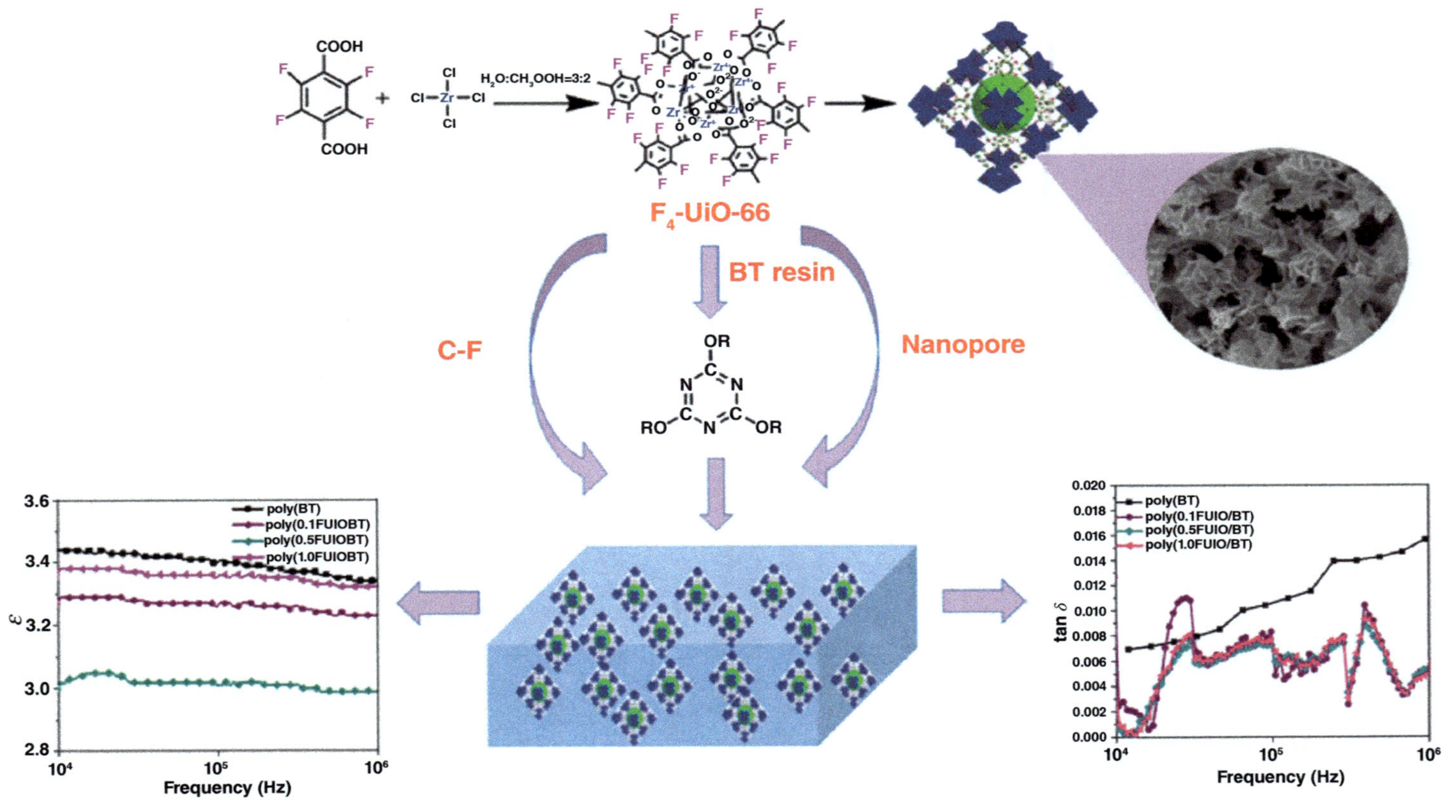

Figure 3.15 Schematic preparation and dielectric properties of the F4-UIO-66 modified BT resins. Source: [245]. Xiaodan Li et al. 2021/Reproduced with permission from Elsevier.

mass ratio of PMPVSi:EHFHPSi was 2:1, the ASiR resins had the lowest ε (2.96) and $\tan\delta$ (0.0003) as well as excellent thermal stability (the initial decomposition temperature of 468 °C).

3.3.5 Polytetrafluoroethylene Resins

Modifying PTFE resins has focused on improving their mechanical properties and maintaining low ε and $\tan\delta$ values. Xu et al. [252] prepared the PTFE resins with good dimensional stability and low ε value by soft molding. The coefficient of linear expansion for PTFE resins was reduced from 1.69×10^{-4} (pure PTFE resins) to 1.31×10^{-4} m/K, and the corresponding ε values were 2.05, and the $\tan\delta$ value was less than 10^{-3}. Pan et al. [253] studied the surface modification of boron nitride by KH-550 (hBN-KH-550), affecting the processing and dielectric properties of hBN/PTFE resins. The hBN-KH-550 had excellent interfacial compatibility with PTFE resins, and the hBN-KH-550/PTFE composites had good processing properties. When the mass fraction of hBN-KH-550 was 20 wt%, the ε values of hBN-KH-550/PTFE composites was 2.51, lower than that of hBN/PTFE composites (2.57), and the $\tan\delta$ value was in the range of 0.004–0.015.

3.3.6 Unsaturated Polyester Resins

The modification of dielectric properties for unsaturated polyester resins has focused on the optimal designing of the molecular structure of synthetic UP resin monomers and selecting vinyl-based curing agents [254, 255]. Hanemann et al. [256] prepared the containing styrene radicals UP resins with low ε (2.6), $\tan\delta$ (0.009), and excellent heat resistance (the thermal decomposition temperature is 260 °C), which was expected to be used in the fields of aviation, aerospace, and communication services. In addition, Canelas et al. [257] synthesized the polybutadiene containing UP resins with lower ε (3.4), $\tan\delta$ (0.009), and higher fracture toughness. The in-plane fracture toughness of the polybutadiene/UP resins increased from 0.14 (pure UP resins) to 0.21 $mN/m^{3/2}$. Zhou et al. [258] synthesized a new end-group, benzyl alcohol hyperbranched polysiloxane (Vi-HPSi), and copolymerized with UP resins to prepare Vi-HPSi/UP resins. The ε value of Vi-HPSi/UP resin was unaffected, the $\tan\delta$ decreased from 0.0315 (pure UP resins) to 0.0196, and the impact strength increased from 8.2 (pure UP resins) to 23.5 kJ/m^2. Chen et al. [259] doped N-phenyl maleimide into UP resins to prepare modified UP resins with high heat resistance. The modified UP resins had lower ε (3.63), and the 5% weight loss temperature (T_5) was 336 °C, higher than that of pure UP resins (327 °C). Sethuraman et al. [260] developed an organic–inorganic hybrid nanocomposite based on vinyl silane functionalized SBA-15 reinforced UP resins. The ε value of SBA-15/UP resin was 2.4, lower than that of the pure UP resins (4.3), and the SBA-15/UP resins still had good heat resistances.

3.3.7 Cyanate Resins

CE resins have ultra-low ε and $\tan\delta$ values, good heat resistances, and excellent processing properties [261]. Some CE resin systems, such as 5575-2 from BASF,

Arcoy-10 from Ciba-Geigy, and XU-71787 from Dow, have been applied to polymer matrix wave-transparent composites. However, with the increasing radar transmitting power and application frequency band, the wave-transparent performances of existing CE resins cannot meet the practical application requirements of full waveband [186, 262].

3.3.7.1 CE Composites Blending with the Addition of Low ε Fillers

The addition of fillers with low ε values and/or micro and nano-cavity structures into the CE resins can decrease the corresponding ε values of CE composites. The relative fillers include the POSS [128], SiO_2 [263], and synthetic hollow microspheres [264].

POSS with the silicon-oxygen core based on $(SiO_{1.5})_n$ is an organic–inorganic hybrid material with nanocavity structures. The incorporation of POSS into the curing networks of CE resins can improve their mechanical and thermal properties while reducing the ε and $\tan\delta$ values. Jiao et al. [265] synthesized a novel hybrid functional nanoparticle POSS-MPS based on the aminopropyl functionalized mesoporous silica (AP-MPS) and glycidyl POSS (G-POSS) to modify CE resins. When the mass fraction of POSS-MPS was 4 wt%, the ε and $\tan\delta$ values of POSS-MPS/CE composites were, respectively, 2.78 and 0.008, lower than those of pure CE resins (ε for 3.27 and $\tan\delta$ for 0.012). The corresponding flexural strength, flexural modulus, impact strength, and T_g values were increased by 14.3%, 4.9%, 57.5%, and 8.7%. Zhang et al. [266] utilized amino POSS (NH_2-POSS) to modify CE (POSS-NH_2/CE) resins. When the mass fraction of NH_2-POSS was 1 wt%, the ε and $\tan\delta$ of POSS-NH_2/CE resins were, respectively, 2.82 and 0.0032, lower than those of pure CE resins (ε for 3.05 and $\tan\delta$ for 0.0080). The T_g value of POSS-NH_2/CE resins (1 wt% NH_2-POSS) was increased from 268.0 (pure CE resins) to 311.0 °C. Ariraman et al. [267] synthesized hydroxy POSS from renewable eugenol and then obtained POSS-OCN by cyanogen bromide esterification, which was copolymerized with BADCy resins to prepare POSS-OCN/BADCy resins. The ε value of POSS-OCN/BADCy resins was reduced to 1.81 at 1 MHz when the mass fraction of POSS-OCN was 30 wt%.

SiO_2 fillers possess the advantages of non-toxicity, cheapness, and heat resistance, and are considered very excellent low-epsilon fillers [268]. However, there are some difficulties in applying SiO_2 to a polymer matrix due to the poor interfacial compatibility between SiO_2 and polymer matrix [269]. Wang et al. [270] grafted polymer molecular chains on the surface of silica nanoparticles (f-SiO_2) by covalent grafting method, which were then used to prepare f-SiO_2/BADCy nanocomposites. Meanwhile, compared with pure BADCy resins, the bending strength, impact strength, and fracture toughness of the f-SiO_2/BADCy nanocomposites were, respectively improved by 31%, 35%, and 35% when the mass fraction of f-SiO_2 was 4 wt%. In addition, the thermal stability of the f-SiO_2/BADCy nanocomposites was also better than that of pure BADCy resins.

In addition, synthetic hollow microspheres are also performed to reduce the ε and $\tan\delta$ values of CE resins. Tang et al. [271] incorporated the cyclodextrin microspheres (M-CDP) into CE resins to prepare M-CDP/CE resins. The ε value of M-CDP/CE resins decreased from 3.40 (pure CE resins) to 2.92 when the mass fraction of M-CDP was 5 wt%. The initial thermal decomposition temperature of M-CDP/CE resins increased from 368 (pure CE resins) to 376 °C.

Porous materials, such as molecular sieves and sesquisiloxane (SSQ), can also prepare CE resins with low ε values. Ariraman et al. [272] prepared the cyclohexyl-branched fatty chain bridged cyanate (BCC) by reduction reaction and modified CE resin (BCC/FMCM-41) by the *co*-blending with glycerol-based silane modified mesoporous material MCM-41. When the mass fraction of FMCM-41 was 10 wt%, the ε value of BCC/FMCM-41 resin was 1.98 at 1 MHz. Zhang et al. [273] mixed the BADCy resins and methyl sesquisiloxane (Me-SSQ) to prepare the Me-SSQ/BADCy resins. The ε value of Me-SSQ/BADCy resin was 2.78 at 1 MHz when the mass fraction of Me-SSQ was 20 wt%.

3.3.7.2 Structural Modification of CE Resins with Low ε

Introducing less polar groups and chemical bonds into CE molecular chains can reduce ε values by reducing polarizability. Pei et al. [274] prepared a series of compounds containing C–Si, Si–O, and C–F to modify CE resins to decrease the ε and $\tan\delta$ values of CE resins while effectively enhancing their resistance to moisture and heat. Liu et al. [275] prepared the new 4,4′-dicyanate based-2,2′-trifluoromethylbiphenyl (TFMBDCy) resins. The cured TFMBDCy resins had lower ε and $\tan\delta$ values, excellent mechanical properties, and heat resistances than cured CE resins. The tensile and impact strengths were 64 MPa and 20 kJ/m^2, respectively, and the corresponding ε, $\tan\delta$, and T_g values were 2.6, 0.0025, and 372 °C, respectively.

Introducing nonpolar bulk groups into CE resin structure can increase the free volume and reduce the polarization density of the cured CE resins, which is beneficial to decreasing the ε value of CE resins. Adamantane is a highly symmetrical structure polycyclic hydrocarbon with a nonpolar and large volume structure, which can reduce the internal polarization density and the packing density of polymer chains to further decrease the ε of CE resins [276]. Tang et al. [277] synthesized a compound containing adamantane (AEAF) to prepare AEAF-*co*-BADCy resins (Figure 3.16). The dipole density of AEAF-*co*-BADCy resins could be reduced by introducing adamantane. When the mass fraction of AEAF was 6 wt%, the real part (ε'_r) and imaginary part (ε''_r) of ε values and $\tan\delta$ values were, respectively, 2.49, 0.012, and 0.0048 (10 GHz), lower than those of pure BADCy resins (2.93, 0.019, and 0.0065). Shieh et al. [278] synthesized the 2,6-dimethylphenol dicyclopentadiene CE (DCPDCY) resins from the dicyclopentadiene, 2,6-dimethylphenol, and cyanogen bromide. The cured DCPDCY resins had a low ε value (2.67) at 10 GHz, which was due to the introduction of nonpolar bridging ring structures and the four bulky $-CH_3$ groups in favor of increasing the free volume.

Besides, the liquid crystal unit orientation can hinder the movement of molecules in the electric field, which is also beneficial to reducing the ε and $\tan\delta$ values [279]. Su et al. [276] used the bisphenol A diglycidyl ether (DGEBA) and 4,4′-bis(2,3-epoxypropoxy)biphenyl (BP) to, respectively, prepare DGEBA/CE and BP/CE resins. Compared with the DGEBA/CE resins (ε for 2.943 and $\tan\delta$ for 0.02), the BP/CE resin had lower ε (2.638) and $\tan\delta$ (0.001) at 10 kHz due to the presence

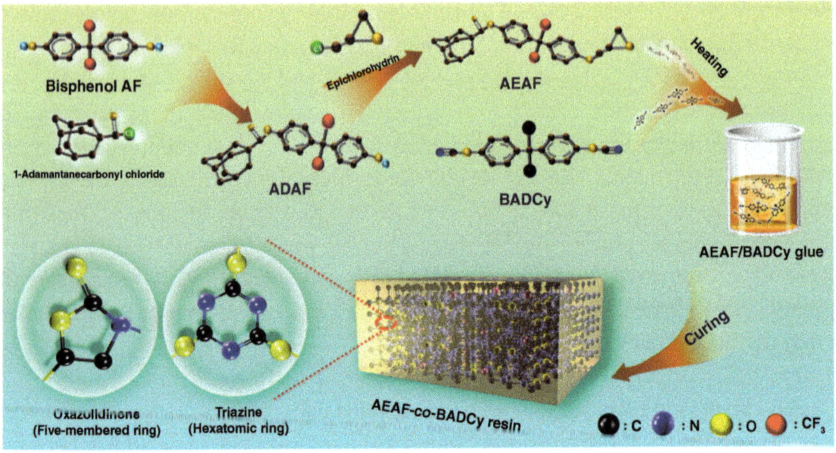

Figure 3.16 Schematic diagram of preparation for AEAF-*co*-BADCy resins. Source: [277]/with permission of Elsevier.

of liquid crystal units in the network structures. Liu et al. [280] utilized the liquid crystal hyperbranched polysiloxane (LCPSi) modified epoxy resin blended with CE (eCE) resin to prepare LCPSi/eCE resins. When the mass fraction of LCPSi was 1.5 wt%, the ε and $\tan\delta$ of LCPSi/eCE resins were about 78% and 45% compared to those of eCE resins (1 kHz). This was because the introduction of low polarizability Si–O bonds in LCPSi could reduce the polarity of LCPSi/eCE resins, finally to further decrease the ε and $\tan\delta$ values.

The introduction of thermoplastic resins with low ε values can also reduce the ε and $\tan\delta$ values of CE resins. PEI has excellent mechanical properties, superior wave-transparent performance, and high service temperature (up to 170–180 °C). In addition, low coefficient of thermal expansion and high resistivity are important properties of PEI. Wen et al. [204] synthesized fluorinated terminated acetimidates (FETI) to modify BADCy resins with the interpenetrating polymer network structures. When the FETI-19 (oligomer with a polymerization degree of 19) was 30 wt%, the tensile strength and impact strength of modified BADCy resins were 95 MPa and 37 kJ/m^2, and the corresponding ε and T_g were 3.15 and 312 °C, respectively. Gu and coworkers [281] synthesized the fluorinated linear random copolymers P(PFS-*co*-GMA) containing epoxy groups by RAFT polymerization using pentafluorostyrene (PFS) and glycidyl methacrylate (GMA) to prepare modified BADCy (*m*-BADCy) resins (Figure 3.17). When the mass fraction of P(PFS-*co*-GMA) was 15 wt%, the ε and $\tan\delta$ values of *m*-BADCy resins were, respectively, 2.59 and 0.0053 (1 MHz), lower than those of pure BADCy resins (ε for 2.97 and $\tan\delta$ for 0.0090). It was attributed that the low polarizability C–F bond in PFS and the ability of P(PFS-*co*-GMA) could increase the intermolecular spatial site resistance with a larger free volume, thus decreasing the ε and $\tan\delta$ values. Wang et al. [282] synthesized the fluorinated linear polyaryletherketone (EFPAEK) to prepare EFPAEK/BADCy resins (Figure 3.18). When the mass fraction of EFPAEK was 20 wt%, the ε and $\tan\delta$ values of EFPAEK/BADCy resins were, respectively, 2.64

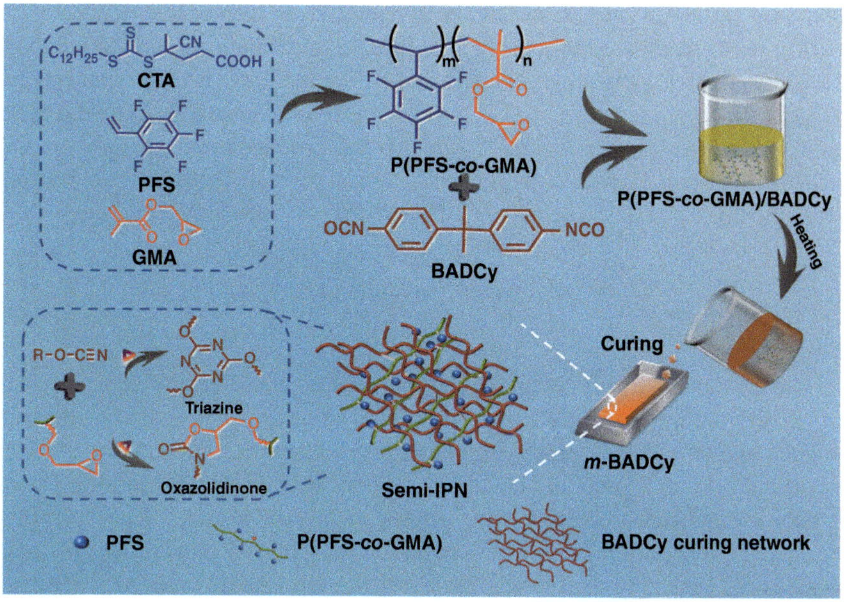

Figure 3.17 Schematic diagram of the preparation for P(PFS-*co*-GMA) modified BADCy resins. Source: [281]/Springer Nature.

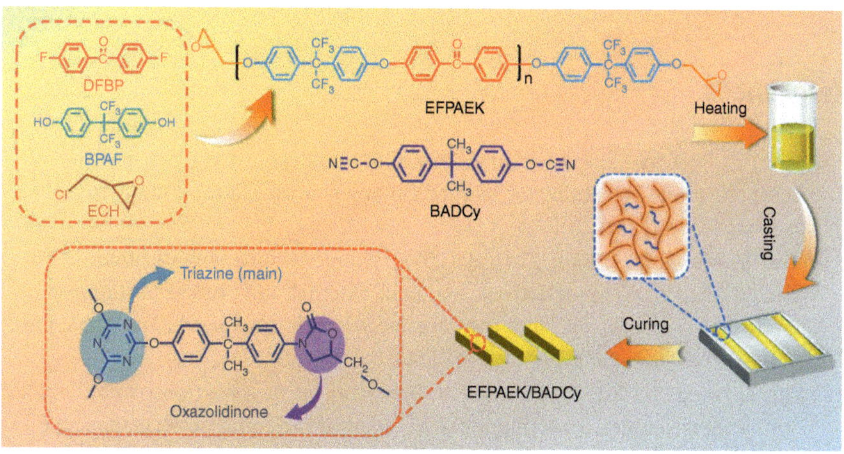

Figure 3.18 Schematic diagram of the preparation for EFPAEK-modified BADCy resins. Source: [282]/Royal Society of Chemistry.

and 0.006 (1 MHz), lower than those of pure BADCy resins (ε for 3.09 and tan δ for 0.008). This was because the strong electronegativity of the $-CF_3$ group in EFPAEK could weaken the polarization effect caused by the external electric field, thus reducing the ε and tan δ values of the EFPAEK/BADCy resins.

Similarly, the introduction of monofunctional –OCN groups into the CE resins can also serve to reduce the ε value. Shieh et al. [283] synthesized a new CE monomer (4CPCY) with monofunctional group-OCN and copolymerized with BADCy resins to obtain the modified BADCy resins. When the mass fraction of 4CPCY was 10 wt%, the ε and $\tan\delta$ values of the modified BADCy resins were, respectively, 2.75 and 0.008 (1 GHz), lower than those of pure BADCy resins (3.03 and 0.0117). This was because the copolymerization reaction of –OCN in 4CPCY with BADCy resin increased the molecular spacing, increasing the free volume inside the BADCy resins, a decrease in the ε and $\tan\delta$ values.

3.4 Summary

This chapter mainly reviews the properties of various commonly used polymer matrix for wave-transparent composites and summarizes the latest research progress and methods on modifying the polymer matrix. In general, the existing polymer matrix still has problems such as poor heat resistance, large ε value, or poor mechanical properties, which make it difficult to meet the demanding comprehensive performance requirements of the polymer matrix for the new radomes. By analyzing the factors that affect the mechanical properties, thermal resistances, and wave-transparent performances of the polymer matrix, researchers can design purposeful solutions to achieve simultaneous improvement of comprehensive properties, which is an important direction for future development of high wave-transparent polymer matrix.

References

1 Guo, Z., Poot, A.A., and Grijpma, D.W. (2021). Advanced polymer-based composites and structures for biomedical applications. *European Polymer Journal* 149: 110388.

2 Dai, Y., Meng, C., Cheng, Z. et al. (2019). Nondestructive modification of aramid fiber based on selective reaction of external cross-linker to improve interfacial shear strength and compressive strength. *Composites Part A: Applied Science and Manufacturing* 119: 217–224.

3 Gu, J., Dong, W., Xu, S. et al. (2017). Development of wave-transparent, light-weight composites combined with superior dielectric performance and desirable thermal stabilities. *Composites Science and Technology* 144: 185–192.

4 Jayalakshmi, C.G., Salunke, A.D., Joshi, M. et al. (2022). Cyanate ester-epoxy blends toward microwave transparent polymer composites through resin film infusion for wideband electromagnetic applications. *Polymer-Plastics Technology and Materials* 61 (2): 145–160.

5 Abbasi, H., Antunes, M., and Velasco, J.I. (2019). Recent advances in carbon-based polymer nanocomposites for electromagnetic interference shielding. *Progress in Materials Science* 103: 319–373.

6 Compton, B.G. and Lewis, J.A. (2014). 3D-printing of lightweight cellular composites. *Advanced Materials* 26 (34): 5930–5935.

7 Hsissou, R., Seghiri, R., Benzekri, Z. et al. (2021). Polymer composite materials: a comprehensive review. *Composite Structures* 262: 113640.

8 Jin, F., Li, X., and Park, S. (2015). Synthesis and application of epoxy resins: a review. *Journal of Industrial and Engineering Chemistry* 29: 1–11.

9 Karger Kocsis, J., Mahmood, H., and Pegoretti, A. (2015). Recent advances in fiber/matrix interphase engineering for polymer composites. *Progress in Materials Science* 73: 1–43.

10 Ma, J., Shang, T., Ren, L. et al. (2020). Through-plane assembly of carbon fibers into 3D skeleton achieving enhanced thermal conductivity of a thermal interface material. *Chemical Engineering Journal* 380: 122550.

11 Mehdikhani, M., Gorbatikh, L., Verpoest, I., and Lomov, S.V. (2019). Voids in fiber-reinforced polymer composites: a review on their formation, characteristics, and effects on mechanical performance. *Journal of Composite Materials* 53 (12): 1579–1669.

12 Ruiz de Luzuriaga, A., Martin, R., Markaide, N. et al. (2016). Epoxy resin with exchangeable disulfide crosslinks to obtain reprocessable, repairable and recyclable fiber-reinforced thermoset composites. *Materials Horizons* 3 (3): 241–247.

13 Wang, C., Zhao, M., Li, J. et al. (2017). Silver nanoparticles/graphene oxide decorated carbon fiber synergistic reinforcement in epoxy-based composites. *Polymer* 131: 263–271.

14 Wang, X. and Wu, P. (2018). Melamine foam-supported 3D interconnected boron nitride nanosheets network encapsulated in epoxy to achieve significant thermal conductivity enhancement at an ultralow filler loading. *Chemical Engineering Journal* 348: 723–731.

15 Yang, X., Fan, S., Li, Y. et al. (2020). Synchronously improved electromagnetic interference shielding and thermal conductivity for epoxy nanocomposites by constructing 3D copper nanowires/thermally annealed graphene aerogel framework. *Composites Part A: Applied Science and Manufacturing* 128: 105670.

16 Yang, X., Guo, Y., Luo, X. et al. (2018). Self-healing, recoverable epoxy elastomers and their composites with desirable thermal conductivities by incorporating BN fillers via in-situ polymerization. *Composites Science and Technology* 164: 59–64.

17 Zhang, R., Shi, X., Tang, L. et al. (2020). Thermally conductive and insulating epoxy composites by synchronously incorporating si-sol functionalized glass fibers and boron nitride fillers. *Chinese Journal of Polymer Science* 38 (7): 730–739.

18 Zheng, Y., Chen, L., Wang, X., and Wu, G. (2020). Modification of renewable cardanol onto carbon fiber for the improved interfacial properties of advanced polymer composites. *Polymers* 12 (1): 45.

19 Jiang, S., Zha, S., Xia, L., and Guan, R. (2015). Synthesis and characterization of diphenylsilanediol modified epoxy resin and curing agent. *Journal of Adhesion Science and Technology* 29 (7): 641–656.

20 Kim, W.G. and Lee, J.Y. (2002). Contributions of the network structure to the cure kinetics of epoxy resin systems according to the change of hardeners. *Polymer* 43 (21): 5713–5722.

21 Kim, W.G. and Lee, J.Y. (2002). Curing characteristics of epoxy resin systems that include a biphenyl moiety. *Journal of Applied Polymer Science* 86 (8): 1942–1952.

22 Tüzün, F.N., Gürü, M., and Akyüz, Y. (2007). Comparison of the curing kinetic behavior for two epoxy resin systems containing EPIKOTE 828–EPIKURE 3090 and DURATEK KLM 606A–DURATEK KLM 606B. *Polymer Composites* 28 (6): 762–770.

23 Wang, X. and Qi, D.T. (2013). Studies on the flexible epoxy resin curing agent. *Asian Journal of Chemistry* 25 (15): 8825–8826.

24 Wang, X., Zheng, S.R., and Wang, R.M. (2013). Synthesis and properties of a flexible curing agent. *Asian Journal of Chemistry* 25 (8): 4256–4260.

25 Wang, Y., Lin, S., Tu, Y. et al. (2016). Synthesis and properties of a novel bis-phenylbiguanide as latent-curing agent for epoxy resin. *Fine Chemicals* 33 (2): 224–230.

26 Yuanqin, X., Hongxia, D., Hua, P. et al. (2009). Synthesis of a waterborne epoxy curing agent and its properties. *Journal of Hunan University Natural Sciences* 36 (11): 53–56.

27 Zhang, C., Sun, G., Zhang, L., and Na, H. (2007). Curing properties of the novel curing agent of epoxy resin. *Journal of Jilin University (Engineering and Technology Edition)* 37 (2): 353–356.

28 Zhang, J., Qiu, Q., Yek, W.Y. et al. (2012). Preparation and application of a new curing agent for epoxy resin. *International Journal of Polymeric Materials and Polymeric Biomaterials* 61 (7): 520–531.

29 Liu P, Sheng X, Li L, Hao Z. Influence of toughened epoxy resin dosage on performance of asphalt mixture. *New Chemical Materials* 2019;47(9):245-8,52.

30 Xiong, J., Huang, Y., Wang, Q., and Zeng, S. (2004). Development of research on toughening epoxy resin. *Plastics* 33 (3): 58–61.

31 Bok, G., Lim, G., Park, K., and Kim, Y. (2021). Mechanical properties and fracture toughness of fumed silica epoxy composites containing glycidyl terminated polysiloxanes. *Ceramics International* 47 (18): 25738–25743.

32 Wang, X., Jin, J., and Song, M. (2013). An investigation of the mechanism of graphene toughening epoxy. *Carbon* 65: 324–333.

33 Zhao, Y., Chen, Z., Liu, Y. et al. (2013). Simultaneously enhanced cryogenic tensile strength and fracture toughness of epoxy resins by carboxylic nitrile-butadiene nano-rubber. *Composites Part A: Applied Science and Manufacturing* 55: 178–187.

34 Zeng, Y., Wang, Q., Xu, P. et al. (2022). Toughening of epoxy resin with balanced performance by in situ constructing interfacial bonding and elastic segments into epoxy network. *Materials Chemistry and Physics* 277: 125529.

35 Sun, Z., Xu, L., Chen, Z. et al. (2019). Enhancing the mechanical and thermal properties of epoxy resin via blending with thermoplastic polysulfone. *Polymers* 11 (3): 461.

36 Lee, S., Kang, M., Kim, S. et al. (2021). Roles of small polyetherimide moieties on thermal stability and fracture toughness of epoxy blends. *Polymers* 13 (19): 3310.

37 Chen, K., Zhang, T., Cui, Y. et al. (2018). OH- and NH2-terminated hyperbranched polysiloxanes: controlled synthesis, characterization, toughening, and reinforcing epoxy resin. *Macromolecular Chemistry and Physics* 219 (18): 1800200.

38 Zhang, J., Mi, X., Chen, S. et al. (2020). A bio-based hyperbranched flame retardant for epoxy resins. *Chemical Engineering Journal* 381: 122719.

39 Huang, W., Shan, G., and Weng, Z. (2007). Toughening of bisphenol A and carboxyl-terminated butadiene-acrylonitrile to epoxy resin. *China Synthetic Rubber Industry* 30 (2): 119–123.

40 Liu, S., Wang, Y., Zhu, L., and Li, X. (2010). Toughening epoxy resin with polyether polyol. *Journal of Functional Polymers* 23 (2): 144–148.

41 Xuan, Z., Yi, J., Duan, Z., and Zhang, Q. (2004). Research on toughening modification of epoxy resin. *Fiber Reinforced Plastics/Composites* (1): 24–27.

42 Zhang, X., Zhao, T., Cao, T. et al. (2022). Study on the compounding of a new type of trimer epoxy resin curing agent. *Journal of Applied Polymer Science* 139 (24): 52368.

43 Chen, J. (2017). Study on mechanical and thermal properties of nano carbon particles reinforced phenolic building foam. *China Plastics Industry* 45 (6): 144–147.

44 Hu, Y., Zou, W., and Peng, J. (2014). Advances on heat and humidity aging resistance of phenolic resins. *China Plastics Industry* 42 (4): 7–11.

45 Kimura, H., Ohtsuka, K., Matsumoto, A. et al. (2015). High performance phenolic resin based on untreated natural herbaceous lignin. *Polymers and Polymer Composites* 23 (8): 525–534.

46 Lian, H.Q. and Wu, X. (2004). Synthesis and photochromism property of three naphthacenequinone phenolic resins. *Chemical Journal of Chinese Universities* 25 (9): 1740–1742.

47 Ma, C., Sung, S., Wang, F. et al. (2001). Thermal, mechanical, and morphological properties of novolac-type phenolic resin blended with fullerenol polyurethane and linear polyurethane. *Journal of Polymer Science Part B: Polymer Physics* 39 (20): 2436–2443.

48 Ma, L., Zhou, X., and Li, X. (2004). Modification progresses of phenolic resin. *Plastics* 33 (5): 39–42. 94.

49 Poduval, M.K.K. and Kim, T.H. (2014). Synthesis of triazole-functionalized phenolic resin and its inherent flame retardant property. *Bulletin of the Korean Chemical Society* 35 (11): 3249–3253.

50 Tang, K., Zhang, A., Ge, T. et al. (2021). Research progress on modification of phenolic resin. *Materials Today Communications* 26.

51 Wang B, Lin M, Li Q, Yang Z, Li M. The influence of the synthesis condition of phenolic resin on the performances of phenolic resin /acrylamide microspheres. *Applied Chemical Industry* 2017;46(4):620-4,8.

52 Wang, Y., Zhai, B., and Wang, S. (2003). Preparation of resin from phenolic residuum. *Environmental Protection of Chemical Industry* 23 (3): 153–156.

53 Yaohua, Y. and Limei, C. (2006). Study progress in phenolic resin/thermoplastics blending. *Plastics* 35 (5): 9–14.

54 Zhao, Y., Xu, R., Xiao, Y. et al. (2022). Mechanical performances of phenolic modified epoxy resins at room and high temperatures. *Coatings* 12 (5): 643.

55 Huang, R. and Wang, J. (2013). Study on process conditions for preparation of ester-curing alkaline phenolic resin by two-step method. *Foundry* 62 (6): 500–503.

56 Liu, T., Zeng, L., Ye, X., and Zhao, J. (2010). Research on preparation and thermal properties of phenolic resin modified by bismalemide. *Journal of Wuhan University of Technology* 32 (7): 12–14. 27.

57 Liu, Y., Liu, S., Li, Z., and Yu, J. (2012). Curing kinetics of allyl phenolic resin. *Journal of Northeastern University Natural Science* 33 (4): 551–554.

58 Liu, D., Liu, Y., and Feng, X. (2015). Study on decomposition kinetics of cardanol modified phenolic resin. *China Plastics Industry* 43 (4): 10–14.

59 Zhang, Z., Zhao, P., Sun, G. et al. (2019). Effect of concentration of phenolic resin on impregnation efficiency of carbon/carbon composites prepared by liquid impregnation-carbonization method. *Journal of Solid Rocket Technology* 42 (6): 771–778.

60 Ge, T., Hu, X., and Wang, D. (2020). Modification and properties of phenolic resin based wear-resistant composite. *Plastics* 49 (2): 69–72. 84.

61 Ji, B., Xue, X., Wang, Y. et al. (2018). Effects of different scale modifiers on the main mechanical properties of Calamus sp. *Journal of Anhui Agricultural University* 45 (6): 1039–1043.

62 Sun, M., Gu, Z., Zhang, W., and Gu, Y. (2004). Study on reaction characteristics of benzoxazine/phenolic mixed resins. *Aerospace Materials & Technology* 34 (6): 33–37.

63 Jian, Z., Shuhua, Q.I., and Li, C. (2007). Study of eco-compatible phenolic resin modified by polyphenylene ether. *China Plastics Industry* 35 (2): 13–16.

64 Li, M., Gu, Y., Wang, Q. et al. (2017). Modification of phenolic foam with epoxy resin. *China Plastics* 31 (11): 66–71.

65 Lihua, Y., Lin, H., and Jianwei, Y. (2007). Research on the heat-stability of the nano-particle SiC modified phenolic resin. *Polymer Materials Science & Engineering* 23 (3): 148–151.

66 Liu, C. and Mo, G. (2015). Preparation and characterization of a novel spiro-phenolic resin. *New Chemical Materials* 43 (11): 129–131.

67 Cardona, F., Kin-Tak, A.L., and Fedrigo, J. (2012). Novel phenolic resins with improved mechanical and toughness properties. *Journal of Applied Polymer Science* 123 (4): 2131–2139.

68 Yu, Z., Li, J., Yang, L. et al. (2012). Synthesis and properties of nano carboxylic acrylonitrile butadiene rubber latex toughened phenolic resin. *Journal of Applied Polymer Science* 123 (2): 1079–1084.

69 Qin, Z., Sun, Y., Wu, X., and Liu, G. (2014). Study on copolymerization of epoxy-terminated polydimethylsiloxane and phenolic resin. *China Plastics Industry.* 42 (12): 37–39. 82.

70 Sun, L., Yang, J., and Yan, J. (2022). The structure and properties of water-based silicone blended phenolic resin and its application in oil filter paper-based materials. *Korean Journal of Chemical Engineering* 39 (8): 2201–2210.

71 He, Y., Duan, R., Zhang, Q. et al. (2021). Reinforce the mechanical toughness, heat resistance, and friction and wear resistance of phenolic resin via constructing self-assembled hybrid particles of graphite oxide and zirconia as nano-fillers. *Advanced Composites and Hybrid Materials* 4 (2): 317–323.

72 Jin, X., Guo, L.Y., Deng, L., and Wu, H. (2017). Study on epoxy resin modified by polyether ionic liquid. *IOP Conference Series: Materials Science and Engineering* 213: 012037.

73 Huang, F., Huang, F., Zhou, Y., and Du, L. (2010). Thermal properties of new bismaleimide resins containing hydrogen silsesquioxane and diallyl bisphenol A. *Journal of Applied Polymer Science* 117 (5): 2932–2936.

74 Huang, F., Zhou, Y., Hang, X. et al. (2009). Octahydridrosilsesquioxane-containing bismaleimide resins. *Aerospace Materials & Technology.* 39 (1): 22–24.

75 Huang, Z.X., Mei, Q.L., and Zhou, Z.F. (2001). Research on the fire resistance of BMI/DDM resin toughened by CTBN. *Journal of Wuhan University of Technology Materials Science Edition* 16 (1): 75–78.

76 Jia, Y., Liu, Z., Li, S., and Wang, X. (2019). Curing reaction of hyperbranched polysiloxane/bismaleimide resin. *Thermosetting Resin* 34 (5): 11–16.

77 Kim, W.G. and Nam, T.Y. (1996). Curing characteristics of o-cresol novolac epoxy resin modified by bismaleimide. *Journal of Polymer Science Part A: Polymer Chemistry* 34 (6): 957–962.

78 Li, L. and Chen, J. (2009). Benzoxazin/bismalelmide resin curing and more toughness glass fabric laminated materials. *The Journal of Engineering Plastics Application* 37 (6): 20–23.

79 Suzuki, M. and Takahashi, A. (1997). Mechanical and viscoelastic property of cured bismaleimide-biscyanamide resin. *Kobunshi Ronbunshu* 54 (1): 25–30.

80 Qian, L., Hongxia, Y., Fen, F., and Pingsen, L. (2007). Developments of toughening modification of bismaleimide resins. *China Plastics* 21 (1): 13–17.

81 Wang, Y., Li, J., Zhou, X. et al. (2020). Chemical degradation of carbon fiber reinforced bismaleimide resin composites to recover carbon fiber and nitrogen doped carbon materials. *Composites Science and Technology* 199: 108342.

82 Wu, G., Cheng, Y., Xie, Q. et al. (2014). Synthesis of a bismaleimide/cyanate ester copolymer containing phenolphthalein functional group with excellent dielectric properties and thermally stable. *Journal of Polymer Research* 21 (12): 615.

83 Chandran, M.S., Krishna, M., and K S, Rai KS. (2010). Preparation and characterization of chain-extended bismaleimide/carbon fibre composites. *International Journal of Polymer Science* 2010: 987357.

84 Sava, M. and Grigoras, C.V. (2005). Bismaleimide monomers and polymers with ester and ether units: synthesis and properties. *Journal of Macromolecular Science, Part A* 42 (8): 1095–1108.

85 Sipaut, C.S., Padavettan, V., Rahman, I.A. et al. (2014). An optimized preparation of bismaleimide–diamine co-polymer matrices. *Polymers for Advanced Technologies* 25 (6): 673–683.

86 Liu, C., Jia, H., Li, N. et al. (2021). Enhanced toughness and thermal properties of bismaleimide resin based on the synergistic effect of reactive amino-terminal poly(phthalazinone ether nitrile sulfone) and bisallyl bearing diphenol group. *Polymers for Advanced Technologies* 32 (3): 1205–1213.

87 Liu, S., Xiong, X., Chen, P. et al. (2018). Bismaleimide-diamine copolymers containing phthalide cardo structure and their modified BMI resins. *High Performance Polymers* 30 (5): 527–538.

88 Hu, R., Wang, R., Qiang, X., and Ma, R. (2013). Preparation and research of new allyl compounds modified bismaleimide resin. *Engineering Plastics Application* 41 (9): 10–14.

89 Yang, J., Liang, G., and Li, L. (2006). Study on properties of low dielectric loss resin matrix. *Journal of Applied Polymer Science* 102 (1): 315–319.

90 Zhang, Z., Li, X., Bao, Y. et al. (2021). Bismaleimide resins modified by an allyl ether of bio-based resveratrol with excellent halogen-free and phosphorus-free intrinsic flame retardancy and ultrahigh glass transition temperature. *Polymer Degradation and Stability* 193: 109717.

91 Liang, G., Zhang, Z., and Yang, J. (2007). Study on properties of low dieletric loss resin matrix. *Polymer Bulletin* 58 (5): 1021–1029.

92 Meng, J. and Hu, X. (2005). Synthesis and exfoliation of bismaleimide-organoclay nanocomposites. *Polymer* 46 (7): 2418.

93 Zhou, X., Qiu, S., He, L. et al. (2021). Synthesis of star-shaped allyl phosphazene small molecules for enhancing fire safety and toughness of high performance BMI resin. *Chemical Engineering Journal* 425: 130655.

94 Qin, H., Mather, P., Baek, J., and Tan, L. (2006). Modification of bisphenol-A based bismaleimide resin (BPA-BMI) with an allyl-terminated hyperbranched polyimide (AT-PAEKI). *Polymer* 47 (8): 2813–2821.

95 Chian, K.S., Du, X.Y., Goy, H.A. et al. (2002). Mechanical properties and morphology of poly(ethylene glycol)-side-chain-modified bismaleimide polymer. *Journal of Applied Polymer Science* 86 (3): 715–724.

96 Ohtsuka, K., Kimura, H., Ikeshita, S. et al. (2016). Novel bismaleimide/diallylbisphenol A resin modified with multifunctional thiol containing isocyanuric ring and long-chain aliphatic unit. *High Performance Polymers* 28 (5): 591–599.

97 Subramaniam, K., Das, A., Steinhauser, D. et al. (2011). Effect of ionic liquid on dielectric, mechanical and dynamic mechanical properties of multi-walled carbon nanotubes/polychloroprene rubber composites. *European Polymer Journal* 47 (12): 2234–2243.

98 Sun, S., Guo, M., Yi, X., and Zhang, Z. (2017). Phase separation morphology and mode II interlaminar fracture toughness of bismaleimide laminates toughened by thermoplastics with triphenylphosphine oxide group. *Science China Technological Sciences* 60 (3): 444–451.

99 Yuan, L., Gu, A., Liang, G., and Zhang, Z. (2008). Microcapsule-modified bismaleimide (BMI) resins. *Composites Science and Technology* 68 (9): 2107–2113.

100 Liu, N., Wang, H., Ma, B. et al. (2022). Enhancing cryogenic mechanical properties of epoxy resins toughened by biscitraconimide resin. *Composites Science and Technology* 220: 109252.

101 Wang, D., Wang, X., Liu, L. et al. (2017). Vinyl-terminated butadiene acrylonitrile improves the toughness, processing window, and thermal stability of bismaleimide resin. *High Performance Polymers* 29 (10): 1199–1208.

102 Li, L. and Lan, L. (2002). Study on the curing reaction of modified BMI. *Polymer Materials Science & Engineering* 18 (3): 85–88.

103 Zhang, B.Y., Li, P., and Chen, X.B. (1998). Influence of composition on the propertics of BMI neat resins and BMI/T300 composites. *Polymers & Polymer Composites* 6 (3): 155–159.

104 Wang, R., Yang, L., Zheng, S. et al. (2002). Fourther reduction of undesirable brittleness of bismaleimide resins. *Journal of the Northwestern Polytechnical University* 20 (1): 145–150.

105 Babkin, A.V., Erdni-Goryaev, E.M., Solopchenko, A.V. et al. (2016). Mechanical and thermal properties of modified bismaleimide matrices toughened by polyetherimides and polyimide. *Polymers for Advanced Technologies* 27 (6): 774–780.

106 Han, Y., Liao, G., Xu, Y. et al. (2009). Cure kinetics, phase behaviors, and fracture properties of bismaleimide resin toughened by poly(phthalazinone ether ketone). *Polymer Engineering and Science* 49 (12): 2301–2308.

107 Upadhyaya, P., Roy, S., Haque, M.H., and Lu, H. (2013). Influence of nano-clay compounding on thermo-oxidative stability and mechanical properties of a thermoset polymer system. *Composites Science and Technology* 84: 8–14.

108 Zhang, Z., Wang, X., Chen, D. et al. (2021). Synthesis and characterization of hyperbranched epoxy with terminal ally group and its application of toughen bismaleimide. *Journal of Wuhan University of Technology Materials Science Edition* 36 (5): 720–729.

109 Wang, Y., Kou, K., Wu, G. et al. (2013). Study on synthesis of novel benzoxazine resin and its modificaion of CE/BMI. *Engineering Plastics Application* 41 (10): 22–25.

110 Zou, Q., Xiao, F., Gu, S.Q. et al. (2019). Toughening of bismaleimide resin based on the self-assembly of flexible aliphatic side chains. *Industrial & Engineering Chemistry Research* 58 (36): 16526–16531.

111 Jiang, X., Chu, F., Zhou, X. et al. (2022). Construction of bismaleimide resin with enhanced flame retardancy and mechanical properties based on a novel DOPO-derived bismaleimide monomer. *Journal of Colloid and Interface Science* 614: 629–641.

112 Han, X., Yuan, L., Gu, A., and Liang, G. (2018). Development and mechanism of ultralow dielectric loss and toughened bismaleimide resins with high heat and moisture resistance based on unique amino-functionalized metal-organic frameworks. *Composites Part B: Engineering* 132: 28–34.

113 Shi, J., Zhang, X., Weng, L. et al. (2021). High toughness and excellent electrical performance bismaleimide resin modified by hyperbranched unsaturated polyester of flexible aliphatic side chains. *High Performance Polymers* 33 (6): 695–703.

114 Lork, A. (2004). Silicone resin emulsion coatings: the strength profile of the silicone resin network in paint microstructures. *Surface Coatings International Part B: Coatings Transactions* 87 (1): 41–46.

115 Wang, D., Gui, D., Liu, J., and Yu, H. (2008). Preparation and performance of amino silicone oil modified epoxy resin. *Plastics* 37 (5): 1–3.

116 Xu, Y., Long, J., Zhang, R. et al. (2020). Greatly improving thermal stability of silicone resins by modification with POSS. *Polymer Degradation and Stability* 174: 109082.

117 Zhang, C. (2007). Study on synthesis and thermo-stability of ladderlike polymethyl/phenylsilsesquioxane. *New Chemical Materials* 35 (6): 37–38.

118 Dai, G., Pei, Y., Chu, J. et al. (2022). Progress in modification methods and mechanisms of silicone resins for high temperature resistance. *Polymer Materials Science & Engineering* 38 (3): 174–182.

119 Guo, Y., Liang, G., Qiu, Z., and Liu, A. (2006). Study on silicone modified epoxy resin cured by polyamide. *Aerospace Materials & Technology* 36 (5): 30–33.

120 Han, J., Ye, S., Xu, Z. et al. (2020). Mechanical properties of vinyl MQ silicone reinforced liquid silicone rubber. *Polymer Materials Science & Engineering* 36 (7): 169–175.

121 Huang, P., Zheng, Z., Mao, K. et al. (2021). Effect of heat blockage on ablative thermal protection of silicone resin-carbon fiber fabrics. *Acta Materiae Compositae Sinica* 38 (9): 3045–3055.

122 Huang, W., Huang, Y., and Yu, Y.Z. (2000). Trimethylsilylated silica as rheology modifier for silicone resins. *Chinese Journal of Polymer Science* 18 (1): 51–55.

123 Kong, G., Yu, Q., Li, Y. et al. (2020). Thermal properties of silicone resin modified by silanol based POSS. *Engineering Plastics Application* 48 (2): 98–101.

124 Su, Q., Liu, S., Sun, C. et al. (2020). Study on dispersion property of TiB in silicone resin system. *New Chemical Materials* 48 (10): 92.

125 Wang, G. and Xie, J. (2004). Study on the synthesis of MQ silicone resin by organization of water glass. *Journal of Building Materials* 7 (3): 299–305.

126 Wu, C., Gao, D., Li, W., and Jia, M. (2009). Study of structural characterization and properties of silicone resin. *Paint & Coatings Industry* 39 (4): 22–25.

127 Wu, C.B., Li, W., Gao, D.H., and Jia, M.Q. (2009). Study of resistance of silicone resin to heat and irradiation. *Polymer-Plastics Technology and Engineering* 48 (10): 1094–1100.

128 Wu, Z., Shen, L.Y., Tong, W. et al. (2017). Preparation and properties research of modified silicone resin. *Journal of Testing and Evaluation* 45 (5): JTE20160127.

129 Liu, Y.R., Huang, Y.D., Liu, L. et al. (2006). Effects of methacryl-POSS on the heat of organic silicone resin. *Journal of Solid Rocket Technology* 29 (3): 217–221.

130 Zhang, Y., Zhang, Y., Li, Y. et al. (2014). Synthesis and properties of silicone /polyether amine cured epoxy resin. *Fine Chemicals* 31 (10): 1188.

131 Zhang, Y., Huang, Y., Liu, X., and Yu, Y. (2003). Studies on the silicone resins cured with polymethylsilazanes at ambient temperature. *Journal of Applied Polymer Science* 89 (6): 1702–1707.

132 Zhu, Q., Wang, Z., Zeng, H. et al. (2021). Effects of graphene on various properties and applications of silicone rubber and silicone resin. *Composites Part A: Applied Science and Manufacturing* 142: 106240.

133 Huang, P., Shi, H., Xiao, H. et al. (2017). High performance surface-modified TiO_2/silicone nanocomposite. *Scientific Reports* 7: 5951.

134 Huang, Y., Feng, Y., Sun, X. et al. (2019). Preparation of ZrO_2/silicone hybrid materials for LED encapsulation via in situ sol-gel reaction. *Polymers for Advanced Technologies* 30 (7): 1818–1824.

135 Dong, Y., He, J., and Yang, R. (2019). Phenolic resin/polyhedral oligomeric silsesquioxane (POSS) composites: mechanical, ablative, thermal, and flame retardant properties. *Polymers for Advanced Technologies* 30 (8): 2075–2085.

136 Wang, X., Lu, H., Wang, H., and Feng, S. (2010). Synthesis and photophysical properties of rare earth-containing luminescent silicone resin from cooperative molecular design and assembly. *Journal of Non-Crystalline Solids* 356 (31): 1581–1586.

137 Li, Y., Zhang, T., Jiang, B. et al. (2019). Interfacially reinforced carbon fiber silicone resin via constructing functional nano-structural silver. *Composites Science and Technology* 181: 107689.

138 Pathak, S.S., Sharma, A., and Khanna, A.S. (2009). Value addition to waterborne polyurethane resin by silicone modification for developing high performance coating on aluminum alloy. *Progress in Organic Coatings* 65 (2): 206–216.

139 Jiang, Y., Hao, Z., Luo, H. et al. (2018). Synergistic effects of boron-doped silicone resin and a layered double hydroxide modified with sodium dodecyl benzenesulfonate for enhancing the flame retardancy of polycarbonate. *RSC Advances* 8 (20): 11078–11086.

140 Yang, Z., Han, S., Zhang, R. et al. (2011). Effects of silphenylene units on the thermal stability of silicone resins. *Polymer Degradation and Stability* 96 (12): 2145–2151.

141 Gonzalez, G., Chiappone, A., Dietliker, K. et al. (2020). Fabrication and functionalization of 3D printed polydimethylsiloxane-based microfluidic devices obtained through digital light processing. *Advanced Materials Technologies* 5 (9): 2000374.

142 Wan, F., Pei, X., Yu, B. et al. (2012). Grafting polymer brushes on biomimetic structural surfaces for anti-algae fouling and foul release. *ACS Applied Materials & Interfaces* 4 (9): 4557–4565.

143 Wang, Y., Cao, R., Wang, M. et al. (2020). Design and synthesis of phenyl silicone rubber with functional epoxy groups through anionic copolymerization and subsequent epoxidation. *Polymer* 186: 122077.

144 Chen, G. and Huang, B. (2013). Preparation and application of high performance polytetrafluoroethylene fiber scrim. *New Chemical Materials* 41 (11): 185–186. 95.

145 Gu, A. and Liang, G. (2007). New approach to fabricate densificated continuous fiber reinforced polytetrafluoroethylene composites with significantly improved interfacial bonding and mechanical properties. *Journal of Applied Polymer Science* 104 (6): 3588–3591.

146 Gu, A., Liang, G., and Yuan, L. (2009). Novel preparation of glass fiber reinforced polytetrafluoroethylene composites for application as structural materials. *Polymers for Advanced Technologies* 20 (1): 39–42.

147 Hu, F., Wei, X., Pi, P. et al. (2005). Preparation of copper clad laminates based on polytetrafluoroethylene with blending modification. *Journal of Wuhan University of Technology Materials Science Edition* 33 (11): 31–36.

148 Mognonov, D.M., Kornopol'tsev, V.N., Ayurova, O.Z., and Dashitsyrenova, M.S. (2015). Improving the wear resistance of polytetrafluoroethylene friction surface. *Russian Journal of Applied Chemistry* 88 (6): 1070–1073.

149 Nagatsu, M., Sugiyama, K., Motrescu, I. et al. (2018). Surface modification of fluorine contained resins using an elongated parallel plate electrode type dielectric barrier discharge device. *Journal of Photopolymer Science and Technology* 31 (3): 379–383.

150 Patil, P.D., Isaias, O., Stamboulides, C. et al. (2008). Paste extrusion of polytetrafluoroethylene (PTFE) powders through tubular and annular dies at high reduction ratios. *Journal of Applied Polymer Science* 108 (2): 1055–1063.

151 Tian, P., Kou, K., Ding, M., and Yan, L. (2005). Preparation of stretching porous polytetrafluoroethylene membrane. *New Chemical Materials* 33 (12): 81–83.

152 Jiang, G., Hu, R., Xi, X. et al. (2013). Facile preparation of superhydrophobic and superoleophilic sponge for fast removal of oils from water surface. *Journal of Materials Research* 28 (4): 651–656.

153 Cheng, Z., Chang, X., Liu, Z., and Qin, D. (2018). Surface-modified halloysite nanotubes as fillers applied in reinforcing the performance of polytetrafluoroethylene. *Clay Minerals* 53 (4): 643–656.

154 Yu, C., Ju, P., Wan, H. et al. (2020). Tribological properties of the polyacrylate/PTFE coating modified by POSS in the space environment. *Journal of Applied Polymer Science* 137 (21): 48730.

155 Usami, R., Shirai, O., Kitazumi, Y., and Kano, K. (2016). Electrochemical interpretation of parabolic relation between the hydrophobicity and the permeability of tetraalkylammonium chlorides. *Journal of Electroanalytical Chemistry* 782: 161–167.

156 Zhang, Z., Zeng, Z., and He, J. (2004). Development and performance of modified PTFE dispersing resin. *New Chemical Materials* 32 (3): 45–47.

157 Barcikowski, W., Królikowski, W., and Lenart, S. (2021). Microstructures of unsaturated polyester resins modified with reactive liquid rubbers. *Polimery* 62 (9): 650–657.

158 Brzozowski, Z.K., Szymańska, E., and Bratychak, M.M. (1997). New epoxy-unsaturated polyester resin copolymers. *Reactive and Functional Polymers* 33 (2): 217–224.

159 Budkin, V.A., Soloveva, N.B., Samuilov, Y.D. et al. (1996). Certain features of synthesis of unsaturated norbornene polyester resins. *Russian Journal of Applied Chemistry* 69 (4): 593–596.

160 Cheng, S. and Wen, D. (2003). Study progress in toughening of unsaturated polyester resins. *Fiber Reinforced Plastics/Composites* (3): 49–51.

161 Diao, Z., Gu, A., Liang, G., and Yuan, L. (2010). Preparation and properties of UV-curable heat-resistant unsaturaled polyester resins. *The Journal of Engineering Plastics Application* 38 (4): 18–21.

162 Dinakaran, K. and Alagar, M. (2002). Preparation and characterization of bismaleimide (N,N′-bismaleimido-4,4′-diphenyl methane)–vinyl ester oligomer-modified unsaturated polyester interpenetrating matrices for advanced composites. *Journal of Applied Polymer Science* 86 (10): 2502–2508.

163 Dong, D. and Zhou, J. (2003). Synthesis of water extended unsaturated polyester resin. *Fiber Reinforced Plastics/Composites* (3): 24–26.

164 Duliban, J. (2007). Amine modifiers with an s-triazine ring for unsaturated polyester resins. *Macromolecular Materials and Engineering* 292 (10–11): 1126–1139.

165 Gao, Q.Y., Li, X.H., and Liu, Z.G. (2002). Studies on the UV-curing of unsaturated polyester/polyepoxy resin blocked copolymer resin. *Acta Polymerica Sinica* 2: 245–248.

166 Gawdzik, B., Matynia, T., and Chmielewska, E. (2001). Modification of unsaturated polyester resin with bismaleimide. *Journal of Applied Polymer Science* 82 (8): 2003–2007.

167 Ge, H., Guan, R., Liu, H., and Wang, J. (2004). Study on activated terminated polyurethane rubber modifying unsaturated polyester resin II:activated terminated polyurethane rubber reducing shrinkage of unsaturated polyester resin. *Fiber Reinforced Plastics/Composites* 3: 25.

168 Hegemann, G. (2003). Low viscous unsaturated polyester resin for monomer free UP-resins. *Macromolecular Symposia* 199 (1): 333–342.

169 Hongbo, X., Jun, W., Xiaoli, Y., and Haopeng, C. (2006). Research on unsaturated polyester resin wetting velocity. *Journal of Wuhan University of Technology* 28 (7): 15–17.

170 Ge, H., Wang, J., Liu, H. et al. (2004). Study on activated terminated polyurethane rubber modifying unsaturated polyester resin I: activated terminated polyurethane rubber toughening unsaturated polyester resin. *Fiber Reinforced Plastics/Composites* (1): 21.

171 Liu, C. and Zhou, Y. (2020). Progress in oil-based unsaturated polyester resins. *Chemistry and Industry of Forest Products* 40 (6): 107–117.

172 Yang, Y. and Suspene, L. (1991). Curing of unsaturated polyester resins: viscosity studies and simulations in pre-gel state. *Polymer Engineering and Science* 31 (5): 321–332.

173 Yang, Y.S. and Lee, L.J. (1988). Microstructure formation in the cure of unsaturated polyester resins. *Polymer* 29 (10): 1793–1800.

174 Jia, Z., Li, X., and Zhao, Q. (2010). Effect of artificial weathering on surface properties of unsaturated polyester (UP) resin. *Materials Chemistry and Physics* 121 (1): 193–197.

175 Chen, W., Gu, Y., Xu, Z. et al. (2014). Preparation of double bond-terminated polyester resin for powder coatings. *Paint & Coatings Industry* 44 (10): 19.

176 Duliban, J. (2011). Hydroxyalkyl derivatives of phenyldiamine as modifiers for unsaturated polyester resins, part 1: derivatives of phenyl-1,4-diamine. *Journal of Applied Polymer Science* 120 (1): 311–321.

177 Li, Y., Feng, L., Ji, R., and Zhang, L. (2004). Study on modification of unsaturated polyester resin with bismaleimide. *China Plastics* 18 (7): 64–66.

178 Sharmila, R.J., Alagar, M., and Chozhan, C.K. (2007). Synthesis and characterization of N,N′-bismaleimido-4,4′-diphenyl methane-vinyl ester oligomer modified unsaturated polyester interpenetrating matrices for advanced composites. *High Performance Polymers* 19 (2): 213–230.

179 Wang, W., Shi, J., and Jing, C. (2013). Preparation of unsaturated polyester resin-based photoluminescent composites. *Fiber Reinforced Plastics/Composites* 5: 22–24.

180 Xu, S., Chen, J., He, Y. et al. (2013). Preparation and curing of high temperature resistant unsaturated polyester resin. *Petrochemical Technology* 42 (7): 802–806.

181 Huang, Y., Wu, J., Liang, J. et al. (2008). Toughening of unsaturated polyester resins with core–shell rubbers. *Journal of Applied Polymer Science* 107 (2): 939–950.

182 Li, S., Deng, Y., Fu, Z. et al. (2021). Hydroxyl-terminated polybutadiene based waterborne polyurethane acrylate emulsions: synthesis, characterization, and damping property. *Journal of Applied Polymer Science* 138 (17): 50300.

183 Yang, H. and Lee, L.J. (2002). Effects of resin chemistry on redox polymerization of unsaturated polyester resins. *Journal of Applied Polymer Science* 84 (1): 211–227.

184 Yuan, S., Zhong, J., Wu, T. et al. (2018). Preparation of waterborne unsaturated polyester-modified acrylate hybrid varnish and properties of its film. *Electroplating & Finishing.* 37 (18): 803–810.

185 Zhang, B. and Ye, D. (2014). Preparations and properties of unsaturated polyester coatings modified with hyperbranched unsaturated resins. *Paint & Coatings Industry* 44 (9): 20–25.

186 Li, J., Ren, C., An, D. et al. (2020). Systematic evaluation of cyanate ester/epoxidized cresol novolac copolymer resin system for high temperature power electronic packaging applications. *Polymer* 195: 122454.

187 Anuradha, G. and Sarojadevi, M. (2008). Synthesis and characterization of novel betti type cyanate esters. *Polymer Bulletin* 61 (2): 197–206.

188 Corley, C.A., Guenthner, A.J., Sahagun, C.M. et al. (2014). Di(cyanate ester) networks based on alternative fluorinated bisphenols with extremely low water uptake. *ACS Macro Letters* 3 (1): 105–109.

189 Laskoski, M., Dominguez, D.D., and Keller, T.M. (2006). Oligomeric cyanate ester resins: application of a modified Ullmann synthesis in the preparation of thermosetting polymers. *Journal of Polymer Science Part A: Polymer Chemistry* 44 (15): 4559–4565.

190 Ding, J., Huang, Y., and Han, T. (2016). Functional graphene nanoflakes/cyanate/epoxy nanocomposites: mechanical, dielectric and thermal properties. *Iranian Polymer Journal* 25 (1): 69–77.

191 Bershtein, V., Fainleib, A., Egorova, L. et al. (2015). The impact of ultra-low amounts of amino-modified MMT on dynamics and properties of densely cross-linked cyanate ester resins. *Nanoscale Research Letters* 10 (1): 165.

192 Chen, Y., Li, Z., Liu, Y. et al. (2020). Curing mechanism and mechanical properties of Al_2O_3/cyanate ester–epoxy composites. *Journal of Electronic Materials* 49 (2): 1473–1481.

193 Lei, Y., Xu, M., Jiang, M. et al. (2017). Curing behaviors of cyanate ester/epoxy copolymers and their dielectric properties. *High Performance Polymers* 29 (10): 1175–1184.

194 Yamei, Z., Doudou, Z., and Li, G. (2014). Shape-memory behavior of bisphenol A-type cyanate ester/carboxyl-terminated liquid nitrile rubber coreacted system. *Colloid and Polymer Science* 292 (10): 2707–2713.

195 Lu, S., Zhou, Z., Fang, L. et al. (2007). Preparation and properties of cyanate ester modified by epoxy resin and phenolic resin. *Journal of Applied Polymer Science* 103 (5): 3150–3156.

196 Fan, J., Hu, X., and Yue, C.Y. (2003). Dielectric properties of self-catalytic interpenetrating polymer network based on modified bismaleimide and cyanate ester resins. *Journal of Polymer Science Part B: Polymer Physics* 41 (11): 1123–1134.

197 Hu, X., Fan, J., and Yue, C.Y. (2001). Rheological study of crosslinking and gelation in bismaleimide/cyanate ester interpenetrating polymer network. *Journal of Applied Polymer Science* 80 (13): 2437–2445.

198 Iredale, R.J., Ward, C., and Hamerton, I. (2017). Modern advances in bismaleimide resin technology: a 21st century perspective on the chemistry of addition polyimides. *Progress in Polymer Science* 69: 1–21.

199 Wan, L., Zhang, X., Wu, G., and Feng, A. (2017). Thermal conductivity and dielectric properties of bismaleimide/cyanate ester copolymer. *High Voltage* 2 (3): 167–171.

200 Inamdar, A., Cherukattu, J., Anand, A., and Kandasubramanian, B. (2018). Thermoplastic-toughened high-temperature cyanate application in advanced composites. *Industrial & Engineering Chemistry Research* 57 (13): 4479–4504.

201 Zheng, L., Yuan, L., Guan, Q. et al. (2018). High-k 3D-barium titanate foam/phenolphthalein poly(ether sulfone)/cyanate ester composites with frequency-stable dielectric properties and extremely low dielectric loss under reduced concentration of ceramics. *Applied Surface Science* 427: 1046–1054.

202 Liang, G. and Zhang, M. (2002). Enhancement of processability of cyanate ester resin via copolymerization with epoxy resin. *Journal of Applied Polymer Science* 85 (11): 2377–2381.

203 Meier, C., Parlevliet, P.P., and Döring, M. (2017). Interpenetrating polymer networks formed by cyanate esters and phenylethynyl-terminated imides. *High Performance Polymers* 29 (5): 556–568.

204 Wen, Y., Yan, J., Liu, J., and Wang, Z. (2017). Interpenetrating polymer networks based on cyanate ester and fluorinated ethynyl-terminated imide oligomers. *Express Polymer Letters* 11 (12): 936–945.

205 Wang, G., Wang, R., Fu, G. et al. (2015). Study on phenolphthalein poly (ether sulfone)-modified cyanate ester resin and epoxy resin blends. *Polymer Engineering and Science* 55 (11): 2591–2602.

206 Li, J., Chen, P., Ma, Z. et al. (2009). Preparation and properties of phenolphthalein polyether ketone/epoxy resin/cyanate semi-interpenetrating polymer. *Polymer Science and Engineering* 25 (7): 141–143.

207 Iijima, T., Katsurayama, S., Fukuda, W., and Tomoi, M. (2000). Modification of cyanate ester resin by poly(ethylene phthalate) and related copolyesters. *Journal of Applied Polymer Science* 76 (2): 208–219.

208 Sun, X., Zeng, M., Lu, C. et al. (2010). Modification of epoxy resin by cyanate ester resin and liquid butadiene-acrylonitrile rubbers. *Polymer-Plastics Technology and Engineering* 49 (8): 772–778.

209 Hayes, B.S., Seferis, J.C., and Parker, G.A. (2000). Rubber modification of low temperature cure cyanate ester matrices and the performance in glass fabric composites. *Polymer Engineering and Science* 40 (6): 1344–1349.

210 Borrajo, J., Riccardi, C.C., Williams, R.J.J. et al. (1995). Rubber-modified cyanate esters: thermodynamic analysis of phase separation. *Polymer* 36 (18): 3541–3547.

211 Cao, Z., Mechin, F., and Pascault, J.-P. (1994). Effects of rubbers and thermoplastics as additives on cyanate polymerization. *Polymer International* 34 (1): 41–48.

212 Zeng, M., Sun, X., Yao, X. et al. (2009). Free-volume properties and toughening behavior of cyanate ester resin/carboxyl-randomized liquid butadiene-acrylonitrile rubber composites. *Journal of Materials Science* 44 (16): 4270–4278.

213 Feng, Y., Fang, Z., and Gu, A. (2004). Toughening of cyanate ester resin by carboxyl terminated nitrile rubber. *Polymers for Advanced Technologies* 15 (10): 628–631.

214 Lei, Y., Wang, Y., Xue, J., and Song, Y. (2013). Influence of pyrolysis conditions on fabrication of polymer-derived BN fiber for wave transparent application. *Composites Part B: Engineering* 51: 254–259.

215 Lingesh, B.V., Ravi Kumar, B.N., Rudresh, B.M., and Reddappa, H.N. (2018). Hybridization effect of fibers on mechanical properties of PA66/PP blend-based thermoplastic composites. *Advanced Composites and Hybrid Materials* 1 (4): 766–776.

216 Lee, E.Y., Chae, I.S., Park, D. et al. (2016). The electrical properties for phenolic isocyanate-modified bisphenol-based epoxy resins comprising benzoate group. *Journal of Nanoscience and Nanotechnology* 16 (3): 3009–3013.

217 Liu, Y., Dai, J., Liu, X. et al. (2017). Bio-based epoxy resins derived from eugenol with low dielectric constant. *Journal of Electronic Packaging* 139 (3): 031006.

218 Satia, M.S.D. and Jaafar, M. (2016). Properties of treated calcium copper titanate filled epoxy thin film composites for electronic applications. *Journal of Applied Polymer Science* 133 (16): 43313.

219 Wang, J.Y., Yang, S.Y., Huang, Y.L. et al. (2012). Synthesis and properties of trifluoromethyl groups containing epoxy resins cured with amine for low Dk materials. *Journal of Applied Polymer Science* 124 (3): 2615–2624.

220 Wang, Y., Chen, W., Yang, C. et al. (2007). Novel epoxy nanocomposite of low Dk introduced fluorine-containing POSS structure. *Journal of Polymer Science Part B: Polymer Physics* 45 (4): 502–510.

221 Wang, Z., Zhang, X., Weng, L., and Liu, L. (2019). Low dielectric constant and high toughness epoxy resin based on hyperbranched polyester grafted by flexible chain modified. *Journal of Materials Science: Materials in Electronics* 30 (6): 5936–5946.

222 Na, T., Jiang, H., Zhao, L., and Zhao, C. (2017). Preparation and characterization of novel naphthyl epoxy resin containing 4-fluorobenzoyl side chains for low-k dielectrics application. *RSC Advances* 7 (85): 53970–53976.

223 Kong, L., Cheng, Y., Jin, Y. et al. (2016). Low k epoxy resin containing cycloaliphatic hydrocarbon with high crosslinking density. *Journal of Applied Polymer Science* 133 (21): 43456.

224 Zhang, P., Yao, T., Xue, K. et al. (2022). Low-dielectric constant and viscosity tetrafunctional bio-based epoxy resin containing cyclic siloxane blocks. *Journal of Applied Polymer Science* 139 (20): 52176.

225 Na, T., Jiang, H., Liu, X., and Zhao, C. (2018). Preparation and properties of novel fluorinated epoxy resins cured with 4-trifluoromethyl phenylbenzimidazole for application in electronic materials. *European Polymer Journal* 100: 96–102.

226 Meng, H., Zhang, Q., Lu, M. et al. (2021). Cure kinetics and properties of high-performance epoxy thermosets cured with active ester-terminated poly (aryl ether ketone). *High Performance Polymers* 33 (9): 1047–1060.

227 Zhang, X., Liu, M., Chen, Y. et al. (2022). Epoxy resin/hollow glass microspheres composite materials with low dielectric constant and excellent mechanical performance. *Journal of Applied Polymer Science* 139 (33): e52787.

228 Zhang, Y., Zhao, C., Liu, J., and Na, H. (2018). Preparation and characterization of ultralow dielectric and fibrous epoxy thermoset cured with poly(arylene ether ketone) containing phenolic hydroxyl groups. *European Polymer Journal* 109: 110–116.

229 Yung, K.C., Zhu, B.L., Yue, T.M., and Xie, C.S. (2009). Preparation and properties of hollow glass microsphere-filled epoxy-matrix composites. *Composites Science and Technology* 69 (2): 260–264.

230 Zhang, S., Yan, Y., Li, X. et al. (2018). A novel ultra low-k nanocomposites of benzoxazinyl modified polyhedral oligomeric silsesquioxane and cyanate ester. *European Polymer Journal* 103: 124–132.

231 Chu, W., Dai, L., Chen, J. et al. (2016). Tunable mesoporous phenolic-silica composites templated by poly(ethylene oxide-b-epsilon-caprolactone) block copolymer. *Journal of Nanoscience and Nanotechnology* 16 (9): 9085–9092.

232 Ding, J.P., Tao, Z.Q., Fan, L., and Yang, S.Y. (2009). Synthesis and properties of fluorinated biphenyl-type epoxy resin. *Journal of Applied Polymer Science* 113 (3): 1429–1437.

233 Hwang, H., Hsu, S., and Wang, C. (2008). Synthesis and physical properties of low-molecular-weight redistributed poly(2,6-dimethyl-1,4-phenylene oxide) for epoxy resin. *Journal of Applied Polymer Science* 110 (3): 1880–1890.

234 Shieh, J., Lin, C., Huang, C., and Wang, C. (2006). Synthesis and characterization of novel dihydrobenzoxazine resins. *Journal of Applied Polymer Science* 101 (1): 342–347.

235 Zhou, W., Yoshino, M., Kita, H., and Okamoto, K. (2001). Carbon molecular sieve membranes derived from phenolic resin with a pendant sulfonic acid group. *Industrial & Engineering Chemistry Research* 40 (22): 4801–4807.

236 Lei, Z., Ji, J., Wu, Q. et al. (2019). Curing behavior and microstructure of epoxy-POSS modified novolac phenolic resin with different substitution degree. *Polymer* 178: 121587.

237 Mu, Q., Wei, C., and Feng, S. (2009). Studies on mechanical properties of sisal fiber/phenol formaldehyde resin in-situ composites. *Polymer Composites* 30 (2): 131–137.

238 Huang, P., Gu, A., Liang, G., and Yuan, L. (2011). High-performance hyperbranched poly(phenylene oxide) modified bismaleimide resin with high thermal stability, low dielectric constant and loss. *Journal of Applied Polymer Science* 120 (1): 451–457.

239 Yang, J., Mao, X., Du, L. et al. (2018). Thermally stabilized bismaleimide–triazine resin composites for 10-GHz level high-frequency application. *High Performance Polymers* 30 (7): 833–839.

240 Zhang, Z., Tian, D., Niu, Z. et al. (2021). Enhanced toughness and lowered dielectric loss of reactive POSS modified bismaleimide resin as well as the silica fiber reinforced composites. *Polymer Composites* 42 (12): 6900–6911.

241 Wang, Z., Ho, J., and Shu, W. (2012). Studies of fluorine-containing bismaleimide resins part I: synthesis and characteristics of model compounds. *Journal of Applied Polymer Science* 123 (5): 2977–2984.

242 Devaraju, S., Vengatesan, M.R., Selvi, M., and Alagar, M. (2014). Thermal and dielectric properties of newly developed linear aliphatic-ether linked bismaleimide-polyhedral oligomeric silsesquioxane (POSS-AEBMI) nanocomposites. *Journal of Thermal Analysis and Calorimetry* 117 (3): 1047–1063.

243 Hu, J., Gu, A., Liang, G. et al. (2011). Preparation and properties of high-performance polysilsesquioxanes/bismaleimide-triazine hybrids. *Journal of Applied Polymer Science* 120 (1): 360–367.

244 Zhang, L., Zhang, C., and Qiu, W. (2003). Preparation of copper clad laminates with high perfoamnce BT resin. *Electronic Components & Materials* 22 (9): 27–29.

245 Li, X., Liu, X., Hu, X. et al. (2021). A novel low-dielectric nanocomposite with hydrophobicity property: fluorinated MOFs modified bismaleimide-triazine resin. *Materials Today Communications* 29: 102802.

246 Yuan, Y., Diao, S., Zhao, C. et al. (2020). Preparation of hollow glass microsphere/organic silicone resin composite material with low dielectric constant by in-situ polymerization. *Silicon* 12 (6): 1417–1423.

247 Zhao, C., Diao, S., Yuan, Y., and Wang, M. (2021). Preparation and properties of hollow glass microsphere/silicone rubber composite material with the TtSilicone resin. *Silicon* 13 (2): 517–522.

248 Zhu, M., Gu, A., Liang, G., and Yuan, L. (2013). High-performance transparent solvent-free silicone resins with stable storage and low viscosity based on new hyperbranched polysiloxanes. *High Performance Polymers* 25 (5): 594–608.

249 Zhan, X., Liu, H., Zhang, J. et al. (2014). Comparative study of silicone resin cured with a linear and a branched cross-linking agent. *Industrial & Engineering Chemistry Research* 53 (11): 4254–4262.

250 Ren, P., Liang, G., and Zhang, Z. (2009). Study on glass fabric reinforced polytetrafluoroethylene composites infused with melted cyanate ester resin. *Journal of Reinforced Plastics and Composites* 28 (18): 2221–2230.

251 Zhang, Y., Liu, Z., Zhang, X., and Guo, S. (2021). Sandwich-layered dielectric film with intrinsically excellent adhesion, low dielectric constant, and ultralow dielectric loss for a high-frequency flexible printed circuit. *Industrial & Engineering Chemistry Research* 60 (31): 11749–11759.

252 Xu, F., Yao, L., Zhao, D. et al. (2011). Effect of a surface resin layer covering the radiating patch on performance of a three dimensional integrated microstrip antenna. *Journal of Composite Materials* 45 (15): 1627–1635.

253 Pan, C., Kou, K., Jia, Q. et al. (2017). Improved thermal conductivity and dielectric properties of hBN/PTFE composites via surface treatment by silane coupling agent. *Composites Part B: Engineering* 111: 83–90.

254 Chen, X., Li, K., Zheng, S., and Fang, Q. (2012). A new type of unsaturated polyester resin with low dielectric constant and high thermostability: preparation and properties. *RSC Advances* 2 (16): 6504–6508.

255 Fraga, A.N., Frullloni, E., de la Osa, O. et al. (2007). Relationship between water A absorption and dielectric behavior of glass fiber reinforced unsaturated polyester resin. *Journal of Composite Materials* 41 (4): 393–402.

256 Hanemann, T., Schumacher, B., and Haußelt, J. (2010). Polymerization conditions influence on the thermomechanical and dielectric properties of unsaturated polyester-styrene-copolymers. *Microelectronic Engineering* 87 (1): 15–19.

257 Canelas, D.A., Abbey, K.J., and Fentress, E. (2002). Properties and toughening of heat-resistant thermosets based on unsaturated ester resins. *Journal of Applied Polymer Science* 86 (4): 821–843.

258 Zhou, C., Liang, G., and Gu, A. (2016). Thermally resistant unsaturated polyester resin with low dielectric loss based on special benzyl alcohol terminated hyperbranched polysiloxane for producing high efficiency motors using vacuum pressure impregnation technique. *RSC Advances* 6 (8): 6672–6678.

259 Chen, J., Liu, K., Chen, C. et al. (2010). Efficient gel-type electrolyte with bismaleimide via in situ low temperature polymerization in dye-sensitized solar cells. *Journal of Polymer Science Part A: Polymer Chemistry* 48 (22): 4950–4957.

260 Sethuraman, K., Lakshmikandhan, T., and Alagar, M. (2016). Surface free energy and dielectric properties of vinyltriethoxysilane functionalized SBA-15-reinforced unsaturated polyester nanocomposites. *Polymer Composites* 37 (12): 3433–3441.

261 Ariraman, M., Sasikumar, R., and Alagar, M. (2015). Hybridization of PDMS based cyanate ester and DGEBA for radiation resistant and microelectronics applications. *RSC Advances* 5 (78): 63641–63649.

262 Chidambaram, V., Rong, E.P.J., Lip, G.C., and Daniel, R.M.W. (2013). Cyanate ester-based encapsulation material for high-temperature applications. *Journal of Electronic Materials* 42 (9): 2803–2812.

263 Vengatesan, M.R., Devaraju, S., Dinakaran, K., and Alagar, M. (2012). SBA-15 filled polybenzoxazine nanocomposites for low-k dielectric applications. *Journal of Materials Chemistry* 22 (15): 7559–7566.

264 Devaraju, S., Vengatesan, M.R., Selvi, M. et al. (2013). Mesoporous silica reinforced cyanate ester nanocomposites for low k dielectric applications. *Microporous and Mesoporous Materials* 179: 157–164.

265 Jiao, J., Zhao, L., Wang, L. et al. (2015). Structures and properties of CE/POSS-MPS composites with low dielectric constant. *Engineering Plastics Application* 43 (10): 1–6.

266 Zhang, Z., Liang, G., Wang, X. et al. (2013). Curing behavior and dielectric properties of amino-functionalized polyhedral oligomeric silsesquioxane/cyanate ester resin hybrids. *High Performance Polymers* 25 (4): 427–435.

267 Ariraman, M., Sasikumar, R., and Alagar, M. (2016). Cyanate ester tethered POSS/BACY nanocomposites for low-k dielectrics. *Polymers for Advanced Technologies* 27 (5): 597–605.

268 Zhou, J., Jiang, Y., Wu, G. et al. (2017). Investigation of dielectric and thermal conductive properties of epoxy resins modified by core-shell structured PS@SiO2. *Composites Part A: Applied Science and Manufacturing* 97: 76–82.

269 Ling, W., Gu, A., Liang, G., and Yuan, L. (2010). New composites with high thermal conductivity and low dielectric constant for microelectronic packaging. *Polymer Composites* 31 (2): 307–313.

270 Wang, J., Liang, G., and Zhu, B. (2007). Nano-SiO2/BADCy composites prepared by sol-gel method. *Journal of Aeronautical Materials* 27 (1): 61–64.

271 Tang, Y., Yuan, L., Liang, G., and Gu, A. (2014). High performance low-k cyanate ester resins with a thermally stable cyclodextrin microsphere. *RSC Advances* 4 (31): 16136–16145.

272 Ariraman, M., Sasi Kumar, R., and Alagar, M. (2014). Studies on FMCM-41 reinforced cyanate ester nanocomposites for low k applications. *RSC Advances* 4 (101): 57759–57767.

273 Zhang, Z., Pei, J., Liang, G., and Yuan, L. (2011). Methyl silsesquioxane/cyanate ester resin organic–inorganic hybrids with low dielectric constant. *Journal of Applied Polymer Science* 121 (2): 1004–1012.

274 Pei, S., Han, Y., Guo, Y. et al. (2020). Thermal and water absorption properties of cyanate ester resins modified by fluoride-containing and silicone-containing components. *Polymers for Advanced Technologies* 31 (6): 1245–1255.

275 Liu, J., Meng, Q., Li, L. et al. (2020). A high-performance polycyanurate network derived from 4,4′-biscyanato-2,2′-trifluoromethylbiphenyl. *Polymer Chemistry* 11 (4): 784–788.

276 Su, X. and Jing, X. (2007). Novel adamantane-containing epoxy resin. *Journal of Applied Polymer Science* 106 (2): 737–742.

277 Tang, L., Zhang, J., Tang, Y. et al. (2020). Fluorine/adamantane modified cyanate resins with wonderful interfacial bonding strength with PBO fibers. *Composites Part B: Engineering* 186: 107827.

278 Shieh, J., Hwang, H., Yang, S., and Wang, C. (2005). Synthesis and properties of a cyanate ester containing dicyclopentadiene(II). *Journal of Polymer Science Part A: Polymer Chemistry* 43 (3): 671–681.

279 Guo, H., Lu, M., Liang, L. et al. (2017). Liquid crystalline epoxies with lateral substituents showing a low dielectric constant and high thermal conductivity. *Journal of Electronic Materials* 46 (2): 982–991.

280 Liu, Z., Yuan, L., Liang, G., and Gu, A. (2015). Tough epoxy/cyanate ester resins with improved thermal stability, lower dielectric constant and loss based on unique hyperbranched polysiloxane liquid crystalline. *Polymers for Advanced Technologies* 26 (12): 1608–1618.

281 Zhou, Y., Zhang, J., Qu, C. et al. (2021). Synchronously improved wave-transparent performance and mechanical properties of cyanate ester resins via introducing fluorine-containing linear random copolymer. *Advanced Composites and Hybrid Materials* 4 (4): 1166–1175.

282 Wang, C., Tang, Y., Zhou, Y. et al. (2021). Cyanate ester resins toughened with epoxy-terminated and fluorine-containing polyaryletherketone. *Polymer Chemistry* 12 (26): 3753–3761.

283 Shieh, J., Yang, S., Wu, M., and Wang, C. (2004). Synthesis and characterization of novel low-dielectric cyanate esters. *Journal of Polymer Science Part A: Polymer Chemistry* 42 (11): 2589–2600.

4

Reinforced Fibers

4.1 Inorganic Fibers

Inorganic fibers are usually made from natural inorganic minerals by high-temperature melting and artificial wire drawing [1, 2]. The inorganic fibers for polymer matrix wave-transparent composites mainly include glass fibers, quartz fibers, and so on [3–5].

4.1.1 Glass Fibers

Glass fibers are usually made of leaf stone, quartz sand, limestone, and other ores as raw materials *via* the method of high-temperature melting, wire drawing, and yarn winding [6, 7]. The main components of glass fibers are silica (SiO_2) and metal oxide. Glass fibers are widely used in the fields of aerospace and microelectronics industry [8–10], due to their low dielectric constant (ε) and dielectric loss tangent value ($\tan\delta$), excellent heat resistances, outstanding electric insulation properties, low cost, good dimensional stabilities, processing performances, and so on.

4.1.1.1 Types of Glass Fibers

Glass fibers can be divided into nonalkali glass fibers, medium-alkali glass fibers, and high-alkali glass fibers according to their composition. Also, they can be divided into high-strength glass fibers, high elastic modulus glass fibers, low dielectric glass fibers, and high silicon-oxygen glass fibers according to the using characteristics [11, 12]. Table 4.1 lists the types and performance of the common glass fibers.

4.1.1.2 Classification by Raw Material Composition

4.1.1.2.1 Nonalkali Glass Fibers

Nonalkali glass fibers (E-glass fibers) are composed of aluminum borosilicate, and the content of alkali metal oxide is 0–2 wt%. E-glass fibers were first developed and prepared by Owens Corning company in the 1930s [13, 14]. E-glass fibers present excellent electrical insulation and water resistance and are widely used as reinforced

Polymer Matrix Wave-Transparent Composites: Materials, Properties, and Applications, First Edition.
Junwei Gu, Yusheng Tang, Jie Kong, and Jing Dang.

Table 4.1 Comparison of the properties of common glass fibers.

Fiber types	Properties				
	Density ($g\,cm^{-3}$)	Tensile strength (GPa)	Modulus (GPa)	ε (10^6 Hz)	$\tan\delta$ (10^9 Hz)
E-glass fibers	2.54	3.75	72	6.13	0.0038
S-glass fibers	2.49	4.00	85	5.21	0.0068
M-glass fibers	2.77	3.70	91	7.00	0.0039
D-glass fibers	2.60	2.40	52	4.00	0.0025
High silica glass fibers	2.30	2.50	52	4.00	0.0048

fibers for polymer matrix wave-transparent composites. However, poor acid resistance of E-glass fibers would limit their application in acidic environments [15].

4.1.1.2.2 Medium-Alkali Glass Fibers

Medium-alkali glass fibers (C-glass fibers) are mainly composed of sodium calcium silicate, and the content of metal oxide content is 8–12 wt%. C-glass fibers present low cost and excellent chemical resistance, especially better than E-glass fibers in acid resistance [16, 17]. However, C-glass fibers present poor mechanical and electrical properties, and tensile strength is about 10–20% lower than E-glass fibers [18].

4.1.1.2.3 High-Alkali Glass Fibers

The metal oxide content (mainly sodium and calcium oxide) of high alkali glass fibers (A-glass fibers) is 14–17 wt% [19, 20], and the mechanical properties are much lower than those of E-glass and C-glass fibers. In addition, A-glass fibers are not resistant to water erosion, and the surface should be hydrophobic in practical application [21]. Therefore, A-glass fibers are extremely uncompetitive in the market competitiveness due to the complicated pretreatment process and high cost.

4.1.1.3 Classification by Usage Characteristics

4.1.1.3.1 Glass Fibers with High Strength

The main chemical compositions of glass fibers (S-glass fibers) with high strength are 65 wt% SiO_2, 25 wt% alumina (Al_2O_3), and 10 wt% magnesium oxide (MgO) [22], which were first developed and put into production by Owens Corning company (USA) in the 1950s, the corresponding trade name was S-2® glass fibers.

S-glass fibers need to add flux in the production process to further broaden the forming temperature range and reduce the upper temperature of SiO_2 and other oxide crystallization. Then, the flux can reduce air bubbles in the S-glass fibers, endowing them with high strength, high-temperature resistance, and high wave-transparent performance [23, 24]. S-2glass fibers from AGY Corporation of the United States, T-glass fibers from Nittobo Corporation of Japan, and the S-glass fibers from Nanjing Glass Fiber Research and Design Institute of China are widely used in the fields of aerospace, shipbuilding, and chemical industry, and made into missile engine shell, aerospace aircraft lining, bulletproof armor, and high-pressure containers [25].

4.1.1.3.2 Glass Fibers with High Elastic Modulus

Glass fibers (M-glass fibers) with high elastic modulus possess more metal oxides than S-glass fibers, such as BeO, Y_2O_3, ZrO_2, TiO_2, La_2O_3, and CeO_2. These metal oxides are performed to improve the elastic modulus of the M-glass fibers [26]. Hipertex® (Owens Corning, U. S.) and Vipro® glass fibers (Gibson, U. S.) are M-glass fibers. The elastic modulus of Hipertex glass fibers is approximately 17%, higher than that of ordinary glass fibers [27]. It is worth noting that the drawing temperature of M-glass fibers needs to be increased, and the heat treatment needs to be performed at higher temperatures during molding to avoid a wide range of crystallization area and surface microcracks in the structure [28].

4.1.1.3.3 Glass Fibers with Low ε Values

With the rapid development of information technology, the frequency of electromagnetic waves received and transmitted by electronic components is increasing [29]. In order to reduce the hysteresis and signal attenuation of electromagnetic waves during the transmission process, it is urgent to develop glass fibers with low ε values (D-glass fibers) to meet the market demand. At present, D-glass fibers are composed of SiO_2 and B_2O_3 and contain a small amount of Al_2O_3, CaO, and MgO as fluxes to reduce the forming temperature and processing cost. The main chemical compositions of D-glass fibers developed by Asahi Glass Co., Ltd. (Japan) are 75 wt% SiO_2 and 20 wt% B_2O_3, and the content of other alkali oxides is about 3 wt%. The ε and $\tan\delta$ are 4.3 and 0.14×10^{-4} at 2.45 GHz, which are 2/3 and 1/4 of that of E-glass fibers, respectively. Nittobo (Japan) has developed NE-glass fibers® at low cost, and the components are shown in Table 4.2. The ε and $\tan\delta$ values of NE-glass fibers are similar to those of D-glass fibers, but the forming temperature (1360 °C) is lower than

Table 4.2 The components of NE-glass fibers.

Components	Mass fraction (wt%)
SiO_2	50–56
Al_2O_3	14–18
B_2O_3	24–28
CaO	0–2.5
MgO	0–2.5
Na_2O	0–0.15
K_2O	0–0.15
Li_2O	0–0.15
TiO_2	1–4

that of D-glass fibers (1400 °C), in favor of reducing the difficulty of the production process [30].

4.1.1.3.4 Glass Fibers with High Silica Content

Glass fibers with high silica content are the kind of high purity SiO_2 (SiO_2 content of 95–99 wt%) continuous glass fibers, which present excellent high-temperature resistance (900 °C for a long time), ablation resistance, heat insulation performance, and low ε and $\tan\delta$. The heat resistance of glass fibers with high silica content mainly depends on the content of SiO_2, and the heat resistance increases with the increasing SiO_2 content [31, 32]. When the SiO_2 content reaches 98 wt%, the heat resistance can reach 1200 °C. H.I. Thompson Co. (USA) pioneered the industrial production of glass fibers with high silica under the trade name Refrasil® fibers, which are widely used in the field of thermal ablative materials for spacecraft, high-temperature insulation, and liquid filtration [33].

4.1.1.4 Preparation of Glass Fibers

The fabrication technology of glass fibers is mainly divided into crucible wire drawing and pool kiln wire drawing [34]. The above two methods are to mix silica sand, quartz, boric acid, clay, and other raw materials in different proportions and further to place them into high-temperature furnaces for melting and forming glass melts. The glass melts flow out from the small hole of the spinneret and then are cooled to obtain glass fiber products of different specifications (Figure 4.1) [35].

The raw materials are treated differently during the process of crucible wire drawing and pool kiln wire drawing. The crucible wire drawing method needs to melt the raw materials into glass balls at high temperatures first, and then carry out the following steps (wire drawing.) after secondary melting [36, 37]. The crucible wire

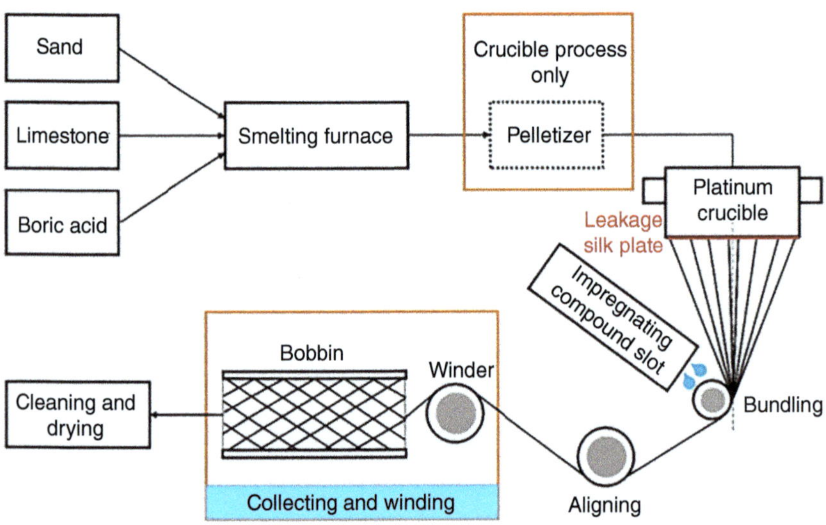

Figure 4.1 Schematic diagram of the preparation process for glass fibers.

drawing method presents the disadvantages of high energy consumption, unstable molding process, and low yield, and has been eliminated by most manufacturers. For the pool kiln wire drawing method, the raw materials are fused in the kiln to make glass molten liquid, which is transported to the porous leakage plate after excluding the bubble, and the glass fibers are formed by high-speed wire drawing [38]. The pool kiln wire drawing method possesses the advantages of a simple process, less energy consumption, stable forming, and high production efficiency. It is suitable for large-scale industrial production. According to statistics, the glass fibers produced by the pool kiln wire drawing method account for over 90% of the world's glass fiber output [39]. In addition, various glass fiber products (no-twist roving and chopped fiber felt) of different lengths, winding, and weaving modes can be made by spinning machine and loom in the post-treatment process (Figure 4.2).

The glass fibers are brittle and wear-resistant and prone to fracture during the process of wire drawing and weaving. In addition, glass fibers are susceptible to the erosion of water vapor in the air after drawing and forming, resulting in a decrease in the tensile strength. In order to further optimize the preparation process of glass fibers, manufacturers usually adopt infiltrating agents to treat the surface of glass fibers after drawing. The infiltrating agents are mixed emulsions composed of surfactant, bonding component, and lubricating component. The first is to facilitate the glass fibers monofilament bunching and subsequent stranding process; the second is to prevent the surface of the original yarn from sticking to

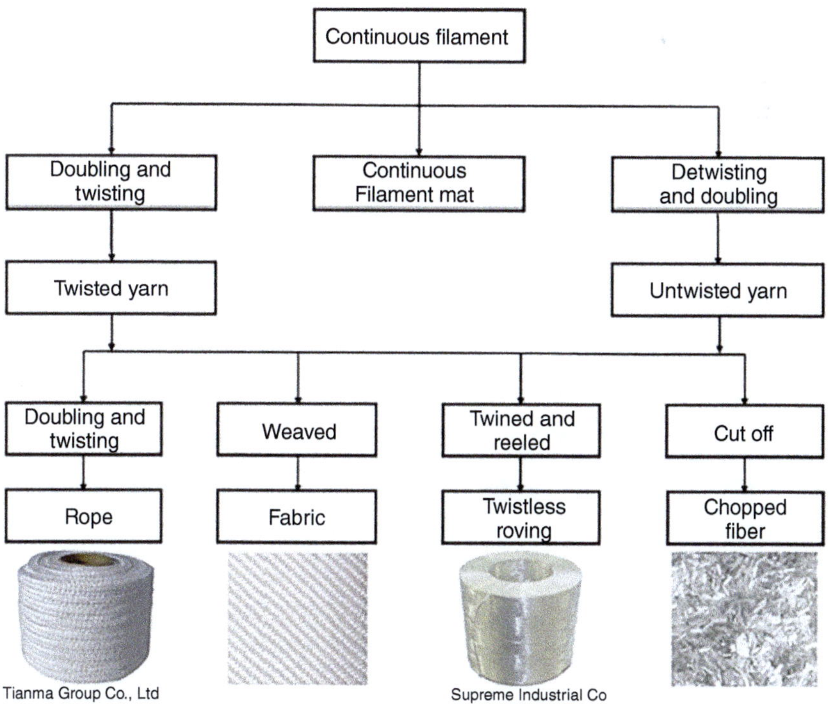

Figure 4.2 Schematic diagram of classification for glass fiber products.

each other during winding; the third is to increase the abrasion resistance of the original yarn, improve the tensile strength, and reduce the abrasion of the glass fibers surface during weaving; finally, it is to protect the glass fibers from the erosion of air and moisture [40–42]. The infiltrating agents are divided into textile-type infiltrating agents and enhanced-type infiltrating agents. Textile-type infiltrating agents contain paraffin and starch-oil type. The enhanced infiltrating agents mainly comprise polyester, polyurethane, and polyacrylate film-forming agents. In addition, coupling agents, lubricants, antistatic agents, and other chemical treatment agents are often used during the preparation process of glass fibers [43, 44].

4.1.1.5 Structure of Glass Fibers

The primary structure of glass fibers is an inorganic ionic polymer with SiO_2 as the skeleton, and the secondary structure is an isotropic amorphous ionic structure [45]. There are two theories about the structure of glass fibers: the irregular network theory [46] and the crystallography theory [47].

4.1.1.5.1 Irregular Network Theory

The structure of glass fibers can be explained by the spatial structure of the three-dimensional (3D) network. Figure 4.3 shows the structure model of sodium-calcium glass fibers. In the network, one O ion can be connected with two Si ions and one can be connected with three O ions to form an oxygen polyhedron structure. Multiple oxygen polyhedrons are connected in irregular directions by common O

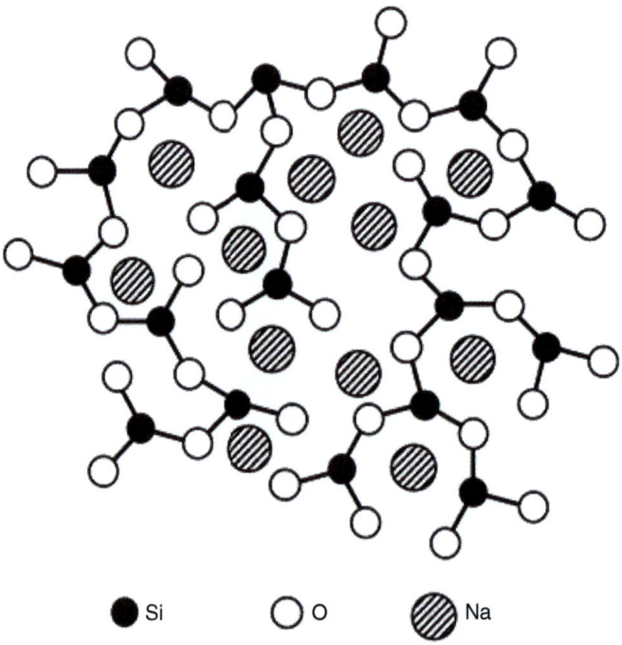

Figure 4.3 Irregular network theory of internal structure model for sodium–calcium glass fibers.

ions at the apex angle and develop into an irregular continuous network structure in three dimensions. The irregular, continuous network of glass fibers can be a 3D skeleton structure, a 2D layered structure, a 1D chain structure, or a coexistence of multiple structures [48].

4.1.1.5.2 Crystallography Theory

According to the Crystallography theory, the internal structure of glass fibers is composed of a discontinuous collection of atoms (i.e. crystal) dispersed in an amorphous medium. For example, in the heating-index test of silicate glass, the index of refraction rises regularly with the increase of temperature. Still, it suddenly decreases at 520–590 °C, and this temperature range is just consistent with the transition temperature of β-quartz $\rightarrow$ α-quartz [49], which indicates the existence of quartz "microcrystalline" structure in the glass. At first, the phenomenon of microcrystal is explained as a microcrystal is a microscopic polyphase composed of crystals with very incomplete lattice, such as $[Si_8O_{20}]^{8-}$, $[Si_3O_9]^{6-}$ and other independent atomic groups [50]. There is a disordered intermediate layer between microcrystals in the glass fibers, which reflects the near-range order and microstructure inhomogeneity of the glass fiber's structure.

4.1.1.6 Properties of Glass Fibers

4.1.1.6.1 Tensile Strength

Glass fibers present excellent tensile strength (2.0–3.8 GPa, special glass fibers up to 4.8 GPa) and low elongation at break (<3%). During the high-temperature drawing of glass fibers, microcracks on the surface of glass fibers can be reduced by increasing the drawing temperature, controlling the pore size of the leakage plate, and treating the glass fibers with infiltrating agents [51]. As the diameter of glass fibers decreases, the size and number of microcracks on the surface of glass fibers decrease, and the corresponding stress concentration decreases, resulting in an increase in tensile strength [52].

4.1.1.6.2 Electrical Insulation and Dielectric Properties

Glass fibers present excellent electrical insulation properties, and the corresponding volume resistivity is about $10^{11} \sim 10^{18}$ Ωm. Therefore, glass fibers are often used in large-scale integrated circuit boards and transformer heat dissipation shells. However, the electrical insulation properties of glass fibers are easily affected by the chemical composition and environmental factors. The increasing alkali content, or humidity and temperature, would result in low resistivity of glass fibers [53]. In addition, glass fibers are widely used in radar radomes, electromagnetic windows, aircraft interiors, and printed circuit boards due to the low ε and tanδ and high wave-transparent performances (>90%) [54].

4.1.1.6.3 Heat-Insulating Properties

The structure of glass fibers is similar to those of amorphous organic polymers, and the glass transition temperature (T_g) is up to 600 °C. Therefore, glass fibers are often used as flame-retardant materials [55]. The glass fibers are easy to create voids during the drawing and molding process so the thermal conductivity (0.034 W/(m K)) is lower than that of SiO_2 and other raw materials (0.7–1.28 W/(m·K)). In addition, glass fibers possess a low thermal expansion coefficient of 2.7×10^{-6}/°C and also can be applied in insulation materials [56].

4.1.1.6.4 Chemical Stability

Aside from hydrofluoric acid, glass fibers present excellent corrosion resistance to other acids, dilute alkalis, salts, and organic solvents. The main reason is that silicate in glass fibers can interact with acid reagents to form a protective film on the surface, which can delay the further erosion and dissolution of acid reagents. However, alkaline metal oxides in glass fibers are soluble in concentrated alkali [57, 58].

4.1.2 Quartz Fibers

The main component of quartz fibers is SiO_2, and the corresponding content can reach more than 99.9%. Quartz fibers present high-temperature resistances, excellent electrical insulation properties, ablative resistances, low ε and $\tan\delta$, and so on, and are widely used in the fields of aerospace, electronic information, and optical fiber communication [59]. At present, quartz fibers manufacturers mainly include Saint Gobain (French, Quartzel®), Joint Processing System (USA, Astroquartz®), NPO Stekloplastic (Russia), Asahi Fiber Glass (Japan) and Feilihua (China), and so on. Compared with glass fibers, quartz fibers possess relatively higher SiO_2 purity, better electrical insulation properties, and lower ε and $\tan\delta$ values but relatively higher cost [60].

The raw material of quartz fibers is quartz glass, made of natural quartz stone and ultra-high purity synthetic quartz stone (Figure 4.4), melted at a temperature above 1750 °C. The energy of the quartz glass melting process is mainly electric energy or hydrogen energy. There is no pollutant emission during the whole preparation

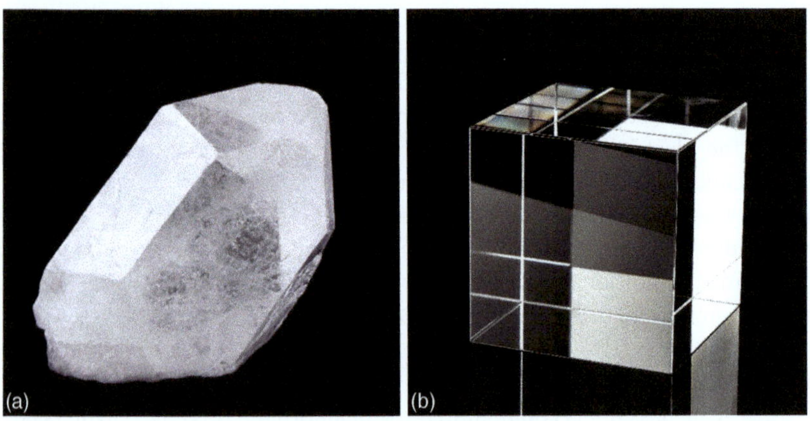

Figure 4.4 Natural quartz stone (a) and ultra-high purity synthetic quartz stone credit: Now Taobao.com (b). Source: (a) Cratervalley/Adobe stock.

process, but relatively high production cost [61, 62]. The main raw materials of ordinary glass are quartz, borax, soda ash, white arsenic, and sodium nitrate, and the melting temperature is 1150–1300 °C. The energy of the ordinary glass melting process is mainly coal, heavy oil, and electricity, resulting in relatively lower production costs.

The properties of quartz fibers mainly depend on the SiO_2 content and processing technology. SiO_2 content of quartz fibers can reach 99.9–99.9999999%, thus giving them low coefficient of thermal expansion, excellent temperature resistances, chemical stabilities, electrical insulation, and radiation resistances, far better than those of ordinary glass fibers [63, 64].

4.1.2.1 Development History and Types of Quartz Fibers

4.1.2.1.1 Development History

From the 1950s to 1960s, in order to meet the development requirements of thermal insulation and communication performance of spacecraft, the Georgia Institute of Technology (the United States) developed quartz ceramic materials with high-temperature resistance and wave-transparent performance, which were made of quartz glass fragments as raw materials, and sintered after grinding, pulping, pouring and molding. Quartz ceramic materials were widely considered and vigorously [65, 66] due to their excellent electrical properties and thermal shock resistances at high temperatures, and the production cost is significantly lower than those of Al_2O_3 ceramic and glass ceramics. In the 1970s, quartz ceramic materials were used in various types of missiles, including the Patriot missile of the United States (Figure 4.5a) and the Aspide missile of Italy (Figure 4.5b).

Figure 4.5 Patriot missile of the United States (a), the Aspide missile of Italy credit: Lockheed Martin (b), and the Trident SLBM of the United States credit: Root Nation (c). Source: (c) Unknown author/Wikimedia Commons/Public domain.

From the 1970s to 1980s, in order to overcome the poor mechanical and erosion resistance of quartz ceramic materials, American General Electric Company and Philco Ford Company developed 3D multidirectional quartz fabric reinforced SiO_2-based composites with different structures (named AS-3DX and Markite3DQ, respectively), which have been successfully applied into the antenna windows and radomes of Trident SLBM (the United States, Figure 4.5c).

Since the 1980s, China has researched the quartz fibers and their reinforced composites. Now, China has broken through a series of technical problems, such as preparation of quartz fibers and high-purity silica sol, structure design of reinforced quartz fabric, cyclic impregnation and moisture-proof treatment, and developed and fabricated a series of wave-transparent materials with excellent mechanical properties, high wave-transparent performances and ablative resistances, which possess puncture structures, three-way orthogonal and shallow bend cross-linking structures [67, 68].

4.1.2.1.2 Types of Quartz Fiber Products

Quartz fiber products can be divided into quartz cotton, quartz yarn, quartz fabric, and quartz sleeves according to size and shape (Figure 4.6). The main types and applications of quartz fiber products are shown in Table 4.3. Quartz fiber products have the advantages of high purity, high-temperature resistance, good heat insulation, corrosion resistance, low ε and $\tan\delta$, and excellent insulation properties, which are widely used in the fields of communication, electronics, and aerospace industries. Quartz fiber products are usually applied in the filling materials for the fiber-drawing furnace seal, the braided materials for the aircraft-enhanced wave-transparent parts, and the high-frequency printed circuit boards [69].

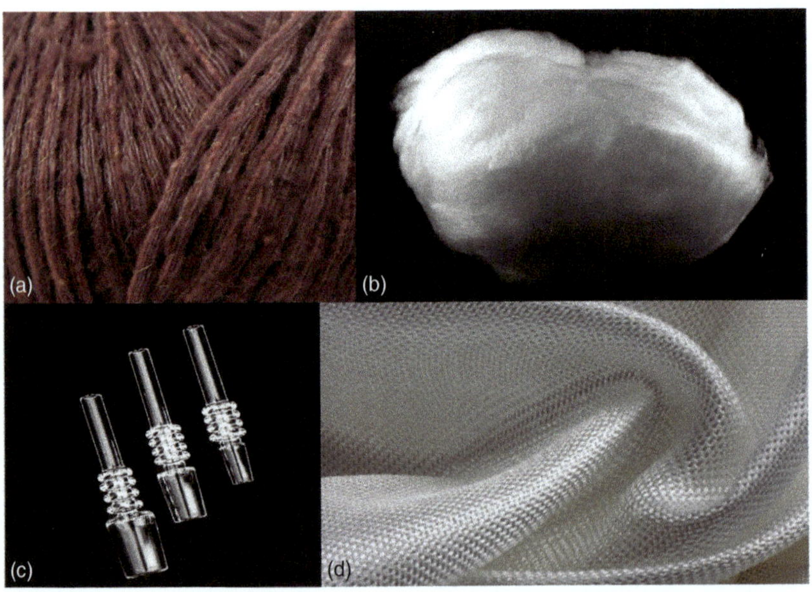

Figure 4.6 Quartz yarn (a), quartz cotton (b), quartz fabric (c) Credit: DHgate.com, and quartz sleeve (d) Credit: MarekPhotoDesign.com/Adobe Stock Photos.

Table 4.3 Types and application of quartz fiber products.

Fibers	Properties		
	Application industries	**Application characteristics**	**Application fields**
Quartz wool	Communication Semiconductor Aerospace	High temperature resistant; Adiabatic	Fiber drawing furnaces Sealing materials Aircraft thermal insulation materials
Quartz yarn	Aerospace	High temperature resistant; Corrosion resistant; Low dielectric	Aircraft wave-transmitting materials
Quartz woven fabric	Aerospace		
Quartz cloth	Aerospace Electronics	Low dielectric; Insulative	Aircraft wave-transmitting materials High-frequency printed circuit boards.
Quartz socket tube	Automotive industry Aerospace Nuke industry	Corrosion resistant Low dielectric Radiation resistant	Auto spare parts Wire conduit tubes Protective jackets

4.1.2.2 Preparation of Quartz Fibers

The preparation process of quartz fibers mainly includes the synthesis of quartz rods and melting and wire drawing of fibers

4.1.2.2.1 Preparation of Quartz Rods

Quartz rods are mainly made of natural crystal or quartz powder by high temperature melting above 1750 °C to get quartz glass, and then smelting and forming. The common preparation process of quartz glass includes electro-melting, flame melting, and high-frequency plasma melting processes [66].

Electro Melting Process There are two main types of electro melting, including two-step and continuous melting processes. Both processes adopt the electric resistance, arc, medium frequency induction, and high-frequency excitation to heat and fuse the quartz powder and further obtain the quartz glass with low hydroxyl content. During the two-step smelting process, the vacuum furnace is used as the heating source to melt the quartz powder to get the quartz glass. The quartz rod with a certain shape is prepared by secondary heating and shaping. The characteristics of the two-step smelting process present a flexible production process, variety of products, and easy to control the quality of products. However, the energy consumption and cost of the process are high.

During continuous smelting process, the quartz powder is sent to the top of the furnace by automatic device, and the quartz glass is continuously pulled out from the bottom of the furnace after high-temperature smelting. The process has the advantages of high automation, short production cycle, and low production cost. The continuous melting process uses tungsten heaters and crucibles, usually protected by helium or hydrogen gas, and so on. It is worth noting that the surface of quartz fibers prepared using hydrogen as a protective gas is relatively rough, and the content of hydroxyl group is high. It can be mainly attributed that SiO_2 can easily combine hydrogen atoms to form Si—OH bonds [70].

Flame Melting Process The flame melting process is the method of producing quartz glass by heating quartz powder and other raw materials with hydrogen and oxygen energy, which contains a one-step melting method and vertical flame melting casting method. During the one-step smelting method, quartz powder is fused by combustion of hydrogen and oxygen on a special lathe to obtain the quartz rods and tubes. However, the obtained products have large dimensional tolerance and rough surfaces. Vertical flame casting method is putting the quartz glass target frame into the furnace to melt the quartz powder and deposit on the target frame through the combustion of hydrogen and oxygen, thereby forming a certain size of quartz rods, resulting in high hydroxyl content.

High-Frequency Plasma Melting Process The high-frequency plasma melting process uses high-frequency plasma as the heat source to melt the quartz powder and deposit it layer by layer on the master tube, which rotates at a constant speed. The outer diameter of the master tube gradually increases, and finally, the quartz glass with a certain size is obtained. Then, the quartz glass is further made into quartz rods. The quartz rod obtained by this method needs to be cold worked to ensure accurate size and processes the advantages of high purity and low hydroxyl content, often used in optical fibers, high wave-transparent performance materials.

4.1.2.2.2 Melting and Wire Drawing of Quartz Fibers

The melting and wire drawing of quartz fibers mainly consists of the following steps (Figure 4.7): first, the pretreated quartz rods (generally 4.4 mm in diameter) are fused at the temperature of 2000–2100 °C, and the fused quartz rods are pulled through the wire leakage plate at a certain speed to form a single fiber. Then, the single fiber (about 10 μm in diameter) is gathered into bundles and subsequently is surface treated by a lubricating tank equipped with an infiltrating agent to prevent surface adhesion and reduce the wear of the quartz fibers [71, 72]. Finally, the treated fibers are wound on the winding tube with a certain speed and dried to obtain the finished quartz fibers.

4.1.2.3 Structure of Quartz Fibers

Quartz fibers follow the structural characteristics of "short range ordered and long range disordered" for glass. The four valence electrons of the Si atom in the SiO_2 configuration form covalent bonds with the four O atoms, and the Si atom is in the

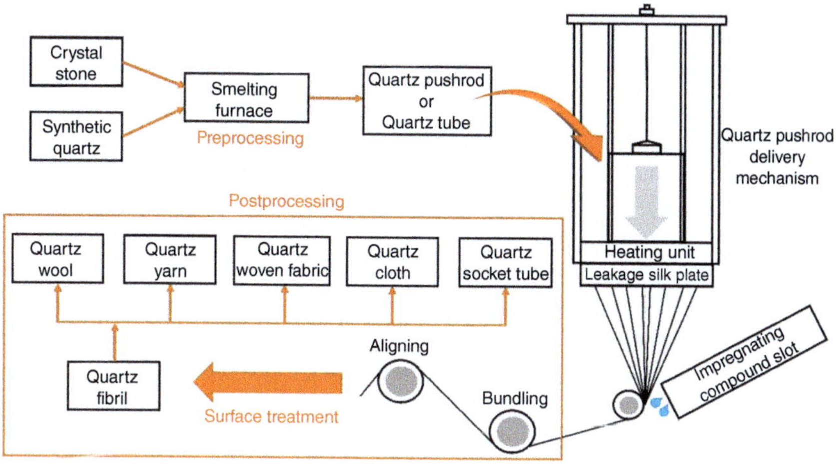

Figure 4.7 Schematic diagram of the preparation process for quartz fibers.

center of the tetrahedron to form the A4 packing. The O atoms form a 3D network of low-density structures near the center of the Si–Si (Figure 4.8) [73]. The irregular network of amorphous SiO_2 phase with high purity endows quartz fibers with excellent heat resistance and dielectric properties. However, internal impurities and surface microcracks of quartz fibers would lead to interfacial phase defects. Therefore, during the process of quartz fiber production, it is necessary to strictly control the purity of SiO_2 to ensure the original structure and properties of quartz fibers at high temperatures and high frequencies [74, 75].

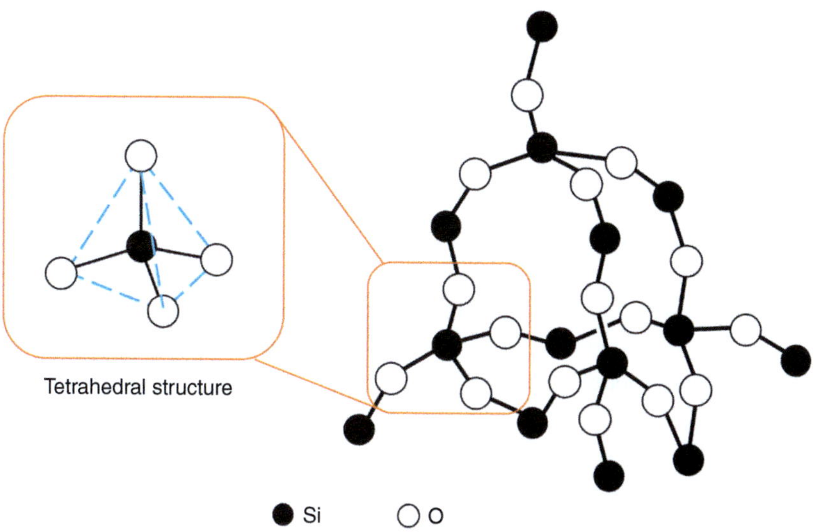

Figure 4.8 Schematic diagram of SiO_2 structure inside quartz fibers.

4.1.2.4 Properties of Quartz Fibers

4.1.2.4.1 Mechanical Property

The tensile strength of quartz fibers is about 6.0 GPa, and the elastic modulus is up to 73.0 GPa. During the process of drawing and forming quartz fibers, the surface treatment with infiltrating agents can further reduce the microcracks and the stress concentration of quartz fibers [76]. In addition, quartz fibers can possess high tensile strength and excellent dimensional stabilities at high temperatures.

4.1.2.4.2 Electrical Insulation and Dielectric Properties

Quartz fibers are composed of high-purity SiO_2 with very few impurities and present a high resistivity of 10^{10}–10^{20} Ωm, showing excellent electrical insulation properties [77]. In addition, quartz fibers have low ε (3.70–3.80) and $\tan\delta$ (0.0002), much lower than ε (4–6.13) and $\tan\delta$ (0.0058–0.0079) of glass fibers. In addition, quartz fibers still have stable ε and $\tan\delta$ at high temperatures (>700 °C) and high frequency (0.3–300 GHz), which are widely used in radar radome, high-frequency printed circuit boards, and so on.

4.1.2.4.3 Heat Resistances

Quartz fibers possess excellent heat resistances; the long-term service temperature can reach 1050 °C, and the short-term service temperature can reach 1200 °C [78]. Quartz fibers are often used in high-temperature furnace sealing materials, aircraft insulation materials, and so on, due to the low coefficient of thermal expansion (5.6×10^{-7} °C). In addition, quartz fibers also have outstanding thermal shock resistances, which can still maintain the original structure and properties after being heated to 1093 °C [79].

4.1.2.4.4 Chemical Stability

Quartz fibers possess excellent chemical stability due to the high stability of SiO_2. Aside from hydrofluoric acid and hot phosphate acid reagents, quartz fibers can resist the damage of weak alkali and organic solvents [80]. In addition, when the quartz fibers crack or fracture, the infiltration of the chemical medium would not cause secondary damage to the internal SiO_2 structure.

4.2 Organic Fibers

Organic fibers are the fibers made of polymers. The reinforced organic fibers for polymer matrix wave-transparent composites mainly include aramid, ultra-high molecular weight polyethylene (UHMWPE) and poly-*p*-phenylene benzobisoxazole (PBO) fibers [81–84].

4.2.1 Aramid Fibers

Aramid fibers are called aromatic polyamide fibers, and the main raw material is synthetic aromatic polyamide [85]. The main chain of aramid fibers is composed of aromatic ring and amide bond. About 86% of the amide groups are directly bonded to the aromatic ring. Because of the unique molecular chain orientation and a large number of amide structures, aramid fibers present high specific strength/specific modulus, excellent thermal stability, low density, and good wear resistance. They are widely used in rocket engine housing, radar radomes, and electromagnetic windows [86, 87].

In the 1940s, with the commercialization of nylon fibers and the advent of polyester fibers, DuPont expected to develop a kind of fiber with ultra-high strength and high-temperature resistances to further occupy the organic fibers market [88]. In the 1950s–1960s, DuPont developed cryogenic solution polymerization technology and reported that inorganic salt solutions could encourage organic solutions to dissolve insoluble polymers, which facilitated the creation of aramid fibers. In 1967, DuPont first commercialized aramid fibers and fabricated the intermediate aramid fibers with high-temperature resistance, whose trade name is Nomex®. In 1972, DuPont developed counterpoint aramid fibers with relatively better mechanical properties than those of Nomex, and the corresponding trade name is Kevlar®. Since then, other countries developed various aramid fiber products in succession. Technora® fibers were developed by Teijin Corporation (Japan) in 1978. Twaron® fibers were fabricated by Akzo Nobel (Netherlands) in 1986. Terlon®, SVM®, and Armoc® fibers have also been launched in Russia. In the 1980s, China launched a series of aramid products, such as Temetar®meta-aramid and Temetarpara-aramid fibers.

4.2.1.1 Types of Aramid Fibers

Aramid fibers can be divided into para-aramid, meta-aramid, and heterocyclic aramid fibers according to the molecular structure. Table 4.4 shows the common aramid fibers and their properties.

4.2.1.1.1 Para-Aramid Fibers

The amide groups are linked to para-phenylene in the molecular chain of para-aramid fibers (Figure 4.9), such as poly-*p*-phenylene terephthamide (PPTA) and poly-*p*-benzamide (PBA) [89–91]. Kevlar from DuPont (USA), Technora from Teijin Corporation (Japan), Twaron from the Netherlands, Armos® from Russia, and Aramid II from China are all para-aramid fibers. The most prominent advantage of para-aramid fibers is high strength and high modulus. The tensile strength of para-aramid fibers is six times, the tensile modulus is two to three times that of steel wire, and the density is only one-fifth that of steel wire.

4.2.1.1.2 Meta-Aramid Fibers

The amide group is connected with interposition phenyl in the molecular chain of meta-aramid fibers (Figure 4.10), and the representative one is poly-*m*-phenylene isophthalamide (MPIA) [92, 93]. There is no obvious conjugation effect between the molecular chain of *m*-phenylene isophthalamide in meta-aramid fibers. The rotational energy within the molecular chain is lower than that of para-aramid fibers, resulting in higher flexibility of the meta-aramid fibers than that of para-aramid

Table 4.4 Common aramid fibers and their properties.

Fibers	Type	Properties			
		Density (g cm^{-3})	Tensile strength (GPa)	Modulus (GPa)	Tensile elongation at break (%)
Aramid I	Poly-*p*-phenylene terephthamide	1.46	2.8	15	–
Aramid II		1.44	2.6	9	3.2
Kevlar49		1.44	3.6	131	3.5
Kevlar 29		1.44	2.8	69	3.6
Kevlar 129		1.44	3.3	94	3.3
Kevlar 149		1.47	2.4	165	1.3
Twaron CT		1.44	2.9	90	3.3
Twaron HM		1.45	2.9	121	2.1
Technora		1.39	3.3	72	4.6
Armos		1.45	4.4	142	3.5
Aramid 1313	Poly-*m*-phenylene-isophthalamide	1.38	0.66	20	20
Nomex		1.38	0.66	17.4	22
Conex		1.38	0.6	–	35
SVM	Heterocyclic aramid	1.42	3.1	122	3.5
Rusar C		1.46	5.2	153	2.6
Rusar HT		1.47	5	172	2.6
Aramid III		1.43	4.5	130	4.1

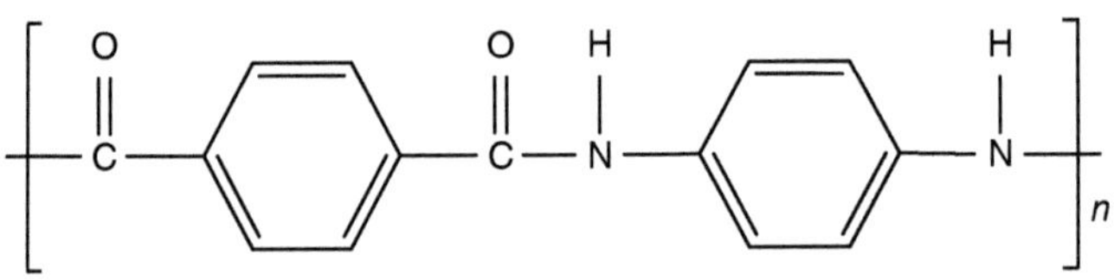

Figure 4.9 Molecular structure of para-aramid fibers.

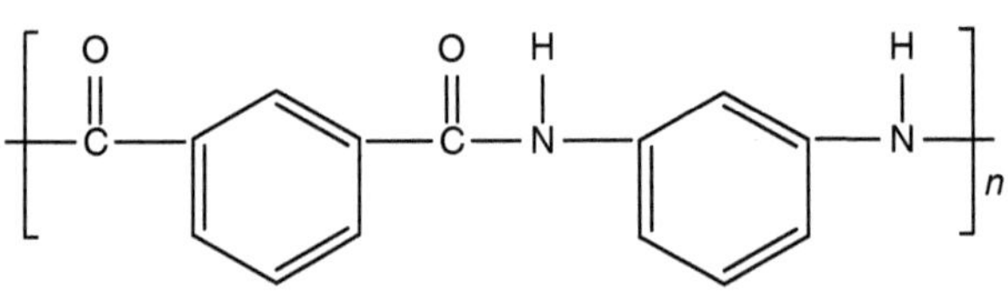

Figure 4.10 Molecular structure of meta-aramid fibers.

fibers. Nomex from DuPont (USA), Teijinconex® from Teijin Corporation (Japan), and Aramid 1313 from China are all meta-aramid fibers, which present excellent heat resistance, flame retardancy, and insulation properties. What's more, the meta-aramid fibers present low modulus and high elongation, and their fabric is more comfortable than the fabric made of para-aramid fibers. Therefore, meta-aramid fibers are mainly used in high-temperature protective fabrics, electrical insulation films, and high-temperature conveyor belts.

4.2.1.1.3 Heterocyclic Aramid Fibers

Heterocyclic aramid fibers refer to the polyaramide fibers whose molecular chains contain N, O, and S atoms, and so on. [94, 95]. Armos and SVM from Russian and Aramid III from China (Figure 4.11) all belong to heterocycle aramid fibers. Aramid III fibers are a new type of heterocyclic aramid fibers developed in China, and the aggregation structure is the same as PPTA fibers. Aramid III fibers have a certain amount of heterocyclic structure containing N, O, and S elements on the molecular chain, and the tensile strength and modulus are higher than that of Aramid II fibers. In addition, the introduction of N, O, and S elements into the main chain for Aramid III fibers reduces the steric hindrance effect caused by the benzene ring and increases the polarity of the molecular chain, the surface energy and the number of chemical active points. F-3B aramid fibers, developed by China BlueStar Chengrand Research Institute of Chemical Industry, present high strength/modulus, impact resistance, wear resistance, fatigue resistance, dimensional stability, and low density due to the molecular main chain containing phenyl benzimidazole. F-3B aramid fibers are widely used in aircraft components (fuselage, main wing, tail, and so on), airborne and shipborne radomes and their wave penetration structures, and bulletproof materials (advanced body armors, bulletproof helmets, and bulletproof armors).

4.2.1.2 Preparation of Aramid Fibers

Aromatic polyamides are the raw materials for the preparation of aramid fibers. Aromatic polyamides have a high theoretical melting point, higher than their thermal decomposition temperature (560 °C). Therefore, the conventional melt-spinning method cannot be used to prepare aramid fibers. Aromatic polyamides belong to lyotropic liquid crystal polymers; the molecular weight reaches a certain value, and the solubility reaches a critical point in suitable solvents to form liquid crystal. Therefore, liquid crystal spinning is the preparation method of aramid fibers with high orientation using aromatic polyamide with low traction force [96, 97].

The meta-aramid fibers can be dissolved in amide solvents (dimethyl formamide, dimethyl acetamide, and *n*-methyl pyrrolidone) and dimethyl sulfoxides. In addition, the addition of lithium chloride, calcium chloride, and other salts can

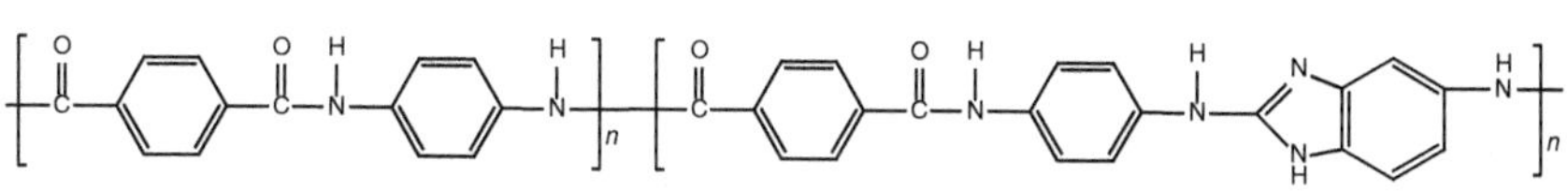

Figure 4.11 Molecular structure of heterocyclic aramid fibers.

effectively increase the solubility of meta-aramid fibers and improve the processing technology. Para-aramid fibers are only dissolved in strong acids, like sulfuric acid, hydrofluoric acid, and mesylate. Therefore, according to different aramid fibers, the corresponding spinning process is mainly divided into dry spinning, wet spinning, and dry spraying-wet spinning [98, 99].

4.2.1.2.1 Dry Spinning

The meta-aramid fibers are usually prepared by a dry spinning process. During the dry spinning process, meta-aromatic polyamides are firstly dissolved in dimethylformamide/lithium chloride mixed solution and subsequently extruded through the spinning nozzle on the top of the heating cylinder to form the primary silks. Then, the primary silks are dried by the hot air layer to evaporate the dimethylformamide. Next, the meta-aramid fibers initial products are obtained by cooling and solidification of the dried primary silks. Then, initial meta-aramid products are stretched properly in dilute solvent bath to improve their physical properties. Finally, the meta-aramid fibers are obtained by cleaning, drying, and heat treatment.

Hearst Corporation (Germany) and Tejin Corporation (Japan) have conducted related research on wet spinning and dry spinning. The spinning solution of meta-aromatic polyamides is sprayed from the nozzle into the first coagulation bath with organic solvent, then into the second coagulation bath with $CaCl_2$ aqueous solution to promote the formation of aramid virgin fibers. Finally, aramid virgin fibers are washed, hot stretched, dried, and stretched to obtain meta-aramid fibers.

4.2.1.2.2 Wet Spinning

The meta-aramid fibers are also prepared by wet spinning. During wet spinning process, meta-aromatic polyamides are extruded from spinners to form a spinning solution, which is then transported into the coagulation bath composed of high-concentration aqueous inorganic salts to form meta-aramid fibers. The keys of wet spinning process are the choice of solvent for meta-aromatic polyamides, the composition of coagulation bath,the temperature range, and the duration of the spinning solution in coagulation bath.

Teijin Corporation (Japan) adopted wet spinning to prepare aramid 1313 fibers. The temperature of aramid 1313 solution was controlled at about 22 °C and extruded through a spinnery nozzle to form the spinning solution. Then, the spinning solution was passed through a coagulation bath with *N*,*N*-dimethylacetamide/calcium chloride to obtain aramid virgin fibers, washed, and dried to obtain aramid 1313 fibers.

4.2.1.2.3 Dry Spraying-Wet-Spinning

Para-aramid fibers are usually obtained by dry spraying-wet-spinning process (Figure 4.12). During the spinning process, para-aramid polyamides are dissolved in concentrated sulfuric acid to form anisotropic solutions, which are extruded from the spinning nozzle to form the primary filament fibers. Next, the primary filament fibers pass through the hot air layer, cold water bath, and coagulation bath to remove the concentrated sulfuric acid and other acidic substances to obtain the primary products of para-aramid fibers. Subsequently, the primary products are wound

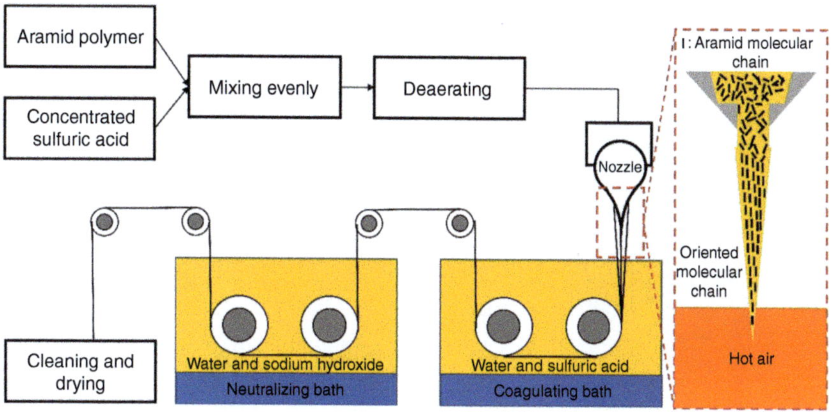

Figure 4.12 Schematic diagram of dry spraying-wet spinning process.

on a drum for drying to obtain para-aramid fibers. In conclusion, the forming of para-aramid fibers consists of three steps: the spinning stock solution is formed by shear forces as it passes through the spinning nozzle; the primary filament fibers are formed again under a certain stretching effect as they pass through the hot air layer; when the primary filament fibers pass through the coagulation bath, about 80 wt% sulfuric acid is removed, causing them to shrink laterally and form again.

During the dry spraying-wet spinning process, the aramid fibers, after solidification bath, should be fully solvated to give them excellent tensile properties. During the conventional dry spraying-wet spinning process, the diffusion speed of calcium chloride from the inside of the aramid fibers to the outside is low due to the high content of calcium chloride in the solidification bath. The high temperature of the solidification bath can increase the diffusion rate of calcium chloride, but affect the mechanical properties of aramid fibers. The Industrial Technology Research Institute of Taiwan reported a new kind of coagulation bath with low-temperature salt-free organic solvent for the preparation of aramid fibers with excellent properties.

4.2.1.3 Structure of Aramid Fibers

Aramid fibers exhibit internal skin-core structure, crystalline structure, primary fiber structure, and pleated structure [100–102]. Figure 4.13 shows the schematic diagram of the internal structures for aramid fibers.

4.2.1.3.1 Skin-Core Structure

The crystallinity of the cortex for aramid fiber is smaller than that of the core layer, and the thickness of the cortex is about 0.1–1 μm, which depends on the type of aramid fibers and the spinning process. The core layer consists of the single crystals of aromatic polyamide arranged along the axial direction. The orientation of the single crystals in the core layer is not as high as the orientation of the cortex. When the primary filament fibers pass through the coagulation bath during the spinning

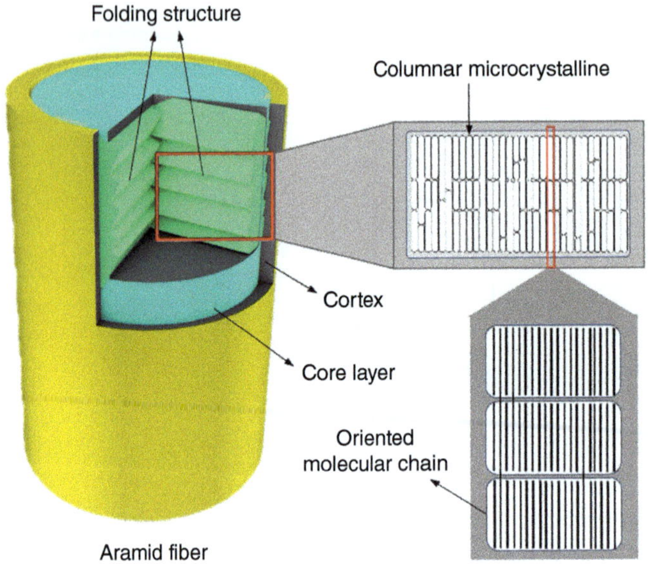

Figure 4.13 Schematic diagram of the internal structures for aramid fibers.

process, the cortex is formed first, and then the core layer is relaxed. The force is reduced with the attenuation of the spinning stress.

4.2.1.3.2 Crystalline Structure

Aramid fibers present high crystallinity above 76%. The molecular chains with amide bond and benzene ring are rigid, showing high orientation along the axial direction. In addition, aramid fibers possess an obvious subcrystalline structure, and there are localized single-phase and longer amorphous aromatic polyamide molecular chain segments in the crystalline state.

4.2.1.3.3 Fibril Structure

Aramid fibers possess a large number of fibril structures (a length of 3–5 cm and a diameter of about 600 nm) along the axial arrangement direction. From the perspective of the cross-section for aramid fibers, the microstructure of aramid fibers is not uniform, and the cortex presents higher density and orientation than that of the core layer. When aramid fibers are subjected to friction, the fibril structure is easy to separate, resulting in a gap of 30 ~ 40 nm, mainly ascribed to small transverse force of aramid fibers.

4.2.1.3.4 Fold Structure

During the preparation of aramid fibers, the transverse and longitudinal shear forces on the primary filament fibers decrease as they pass through the solidification bath. And the shear forces reduce from the surface layer to the interior, which relaxes the core layer and leads to the periodic transverse folding structure of the aramid fibers along the axis. The length of the fold is about 250–400 nm, and the two adjacent folds are directly at an angle of about 170°.

4.2.1.4 Properties of Aramid Fibers

4.2.1.4.1 *Mechanical Properties*

The mechanical properties of aramid fibers are anisotropic due to the strong covalent bond in the axial direction and the weak hydrogen bond in the transverse direction [103–105]. Moreover, the conjugated structures of the benzene ring in aramid fibers make the molecular chains difficult to rotate and fold. The molecular chains present the extended state along the axial direction of the aramid fibers. Therefore, aramid fibers present ultra-high modulus, and the axial tensile modulus can reach 80% of the theoretical axial tensile modulus. In addition, aramid fibers have excellent tensile strength (about 2.9–3.5 GPa).

4.2.1.4.2 *Dielectric Properties*

The molecular chains of aramid fibers are symmetrical in structure and composed of alternating amide bonds and benzene rings. The molecular chains with low polarity have very high orientation and ordered lateral arrangement. The carbonyls and nitrogen atoms of the molecular chains form the large π-conjugation with the benzene ring, which has low polarizability [81]. Therefore, aramid fibers present low ε (3–4) and $\tan\delta$ (0.001) and still maintain excellent dielectric stability under high-low temperature and high-frequency conditions [106, 107].

4.2.1.4.3 *Thermal Stability*

The dissociation energies of C—C and C—N bonds in the aromatic polyamide of aramid fibers are about 20% higher than that of aliphatic polyamide, and the thermal decomposition temperature of aramid fibers can reach over 450 °C. Particularly, the conjugated action of amide and benzene ring in para-aramid fibers make the molecular chains stable, and the corresponding decomposition temperature is up to 550 °C [108–110]. The hydrogen bonding between the molecular chains and their rigidity endows aramid fibers with high T_g. Para-aramid fibers present a low thermal expansion coefficient due to the nearly perfect axial orientation of the molecular chains; the longitudinal thermal expansion coefficient is -2×10^{-6} °C, and the transverse coefficient thermal expansion is 5.9×10^{-5} °C. Therefore, para-aramid fibers can hardly shrink when exposed to heat air or are impregnated with hot water.

4.2.1.4.4 *Chemical Stability*

Aramid fibers present excellent chemical stability and are resistant to dissolution in varieties of organic solvents. Still, aramid fibers are easily dissolved in some strong acids and sodium hypochlorite aqueous solutions [111, 112]. Aramid fibers even degrade in the above solutions, especially under the action of long-term high temperatures. Para-aramid fibers only dissolved in concentrated sulfuric acid, hydrofluoric acid, and mesylate. Meta-aramid fibers are dissolved in *N*-methyl-2-pyrrolidone and dimethylacetamide.

4.2.1.4.5 *Flame Retardant*

Aramid fibers possess excellent flame retardant and immediately stop burning after removing the fire source, and the limit oxygen index (LOI) can reach 28–29

[113, 114]. Aramid fibers in combustion are gradually brittle, carbonized until decomposition, and hardly produce any melt drops. In addition, the emissions of oxidation gas and toxic gas are less than other organic fibers during combustion.

4.2.1.4.6 UV Resistance

Aramid fibers present poor UV resistance because the benzene rings of molecular chains are sensitive to UV light [115, 116]. Under a certain wavelength of UV radiation, the molecular chains of aramid fibers are easily broken. Significantly, the surface color of aramid fibers can change from yellow to brown under long-term exposure to 300–450 nm wavelength of UV radiation, and the mechanical properties are greatly reduced. Therefore, aramid fiber products are rarely used in outdoor environments directly.

4.2.2 Ultrahigh Molecular Weight Polyethylene Fibers

UHMWPE fibers, also known as high strength and high modulus polyethylene fibers, have a relative molecular weight of more than 1 million [117, 118]. UHMWPE fibers possess outstanding impact and cutting resistance, chemical corrosion and UV resistance, and excellent resistance to low temperatures [119–121]. In addition, UHMWPE fibers have received extensive attention in polymer matrix wave-transparent composites because of their low ε and tanδ [122].

In the 1970s, Ward et al. [123] fused and crystallized polyethylene with 100 000 molecular weights by melt spinning method, further traction, and stretching to obtain medium-strength polyethylene fibers with a modulus of 70 GPa. Subsequently, the manufacturing technology of the Ward team [124] for polyethylene fibers was transferred by the British Technology Group to SNIA Viscosa Ltd. (Italy) and Celanese Corporation (the United States). In the 1990s, SNIA Viscosa Ltd. and Celanese Corporation launched UHMWPE fibers with trademarks Tenfor® and Certran TM®, respectively.

In the 1970s, DSM Corporation (Netherlands) reported the first patent for the preparation of UHMWPE fibers by using gel spinning-super stretching technology. In 1984, DSM, in collaboration with Toyobo Co., Ltd. (Japan), realized the dry spinning technology to prepare large quantities of UHMWPE fibers (Dyneema®). In the 1980s, Mitsui (Japan) developed the technology to prepare UHMWPE fibers by gel extrusion-super stretching technology using paraffin as a solvent, and the UHMWPE fibers (brand name Tekmilon®) were commercially produced in 1988. With the demand for low-cost UHMWPE fibers, Teijin Corporation (Japan) developed a new production process for UHMWPE fibers using film-cutting method, and the UHMWPE fibers (Endumax®) were commercially produced in 2012. China started the relative research on UHMWPE fibers in the early 1980s. In 1984, Donghua University cooperated with Beijing Tongyizhong Special Fiber Technology Development Co., Ltd., Hunan Zhongtai Special Equipment Co., Ltd., and Ningbo Dacheng New Material Co., Ltd. to fabricate industrialized UHMWPE fibers by wet spinning process.

4.2.2.1 Synthesis of UHMWPE

UHMWPE fibers have high requirements on the molecular weight, molecular weight distribution, and particle size of raw materials. The synthesis of UHMWPE resins with appropriate molecular weight is of great significance for fabricating high-performance UHMWPE fibers [125]. Compared with common polyethylene polymerization, the polymerization of UHMWPE resins has strict polymerization temperature and different catalyst concentration, which needs a small amount of hydrogen during polymerization. The production process of UHMWPE resins is basically similar to that of high-density polyethylene prepared by low-pressure slurry method [126]. The only difference is that the UHMWPE resins have no granulation procedure in the post-stage process.

The production process of UHMWPE resins is mainly divided into continuous polymerization process and intermittent polymerization process.

4.2.2.1.1 Continuous Polymerization Process

The device of continuous polymerization process adopts distributed control system and catalyst metering supply to make the quality of UHMWPE products stable. Continuous polymerization processes present low energy consumption, suitable for large-scale production [127–129]. The continuous polymerization process is mainly slurry polymerization, which is divided into the following steps:

Catalyst Configuration Loaded catalyst ($MgCl_2$ loaded $TiCl_4$) and alkyl aluminum (CH_9CH_6Al) are evenly mixed with appropriate amount of hexane and fed into the polymerization kettle by metering pump after complexation [130, 131]. During the polymerization process, the complexed catalyst is continuously added into the polymerization kettle by a metering pump to ensure rapid reaction.

Polymerization The high-purity ethylene and complexed catalyst are continuously transported into the polymerization kettle, and the slurry polymerization is carried out at 60–90 °C, 0.4–0.9 MPa. The generated heat during the polymerization process is removed by cooling the polymerization kettle and refluxing the mother liquor. The molecular weight of UHMWPE resin can be controlled by the amount of catalyst, temperature, and polymerization pressure.

Separation and Drying UHMWPE slurry obtained from Step ② is transported into the flash reactor through pressure difference, and the unreacted ethylene in the slurry is removed to be reused in the polymerization kettle by flash compressor. The degasified slurry is transported to the centrifuge through the slurry pump to separate UHMWPE and reaction solution. The separated UHMWPE is transported to the stripping kettle through a screw conveyor, and the reaction solution is returned to the polymerization kettle through the mother liquor pump. Subsequently, steam is passed under the stripping kettle to distill the remaining solvent of UHMWPE. After stripping, the UHMWPE is sent to a spiral dehumidifier, and removed the water to obtain UHMWPE powder.

Screening, Mixing, and Packaging The dried UHMWPE powder is screened, mixed by a vibrating screen, and packaged by packaging machine.

Solvent Recovery The solvent is firstly washed with water to remove impurities and then separated into low molecular weight polymer and high purity solvent through the vapor extraction tower. The low molecular weight polymer with the molten state is discharged from the bottom of the tower. The high-purity solvent is discharged from the top of the tower and then passed through the distillation tower to remove water, followed by entering the solvent tank for reusing.

4.2.2.1.2 *Intermittent Polymerization Process*

The catalyst is supplied quantitatively, and the molecular weight of UHMWPE is flexibly controlled by adjusting the process parameters. The production efficiency of the intermittent polymerization process is relatively low, which is suitable for the production of UHMWPE with less demand. For the same device, the energy and material consumption of the intermittent polymerization process are much higher than that of the continuous polymerization process.

The intermittent polymerization process also uses slurry polymerization [132], and the corresponding process is basically similar to continuous polymerization.

Catalyst Configuration The amount of catalyst added into the polymerization kettle is determined according to the molecular weight of UHMWPE. A certain concentration of supported Ziegler–Natta catalyst and alkyl aluminum (CH_9CH_6Al) solution are prepared and directly pressed into the polymerization kettle after complex processes.

Polymerization A solvent accounting for about 65% of the total volume is added into the polymerization kettle, and gas replacement is performed 2–3 times to ensure the oxygen content in the polymerization kettle reaches the target. When the polymerization kettle is warmed up to 50–60 °C, the catalyst and high-purity ethylene are continuously added into the polymerization kettle. The slurry polymerization is carried out under the condition of 60–90 °C and pressure of 0.4 ~ 0.9 MPa, and the polymerization time is 4 hours. The generated heat during the polymerization is removed by the cooling of the polymerization kettle jacket. The molecular weight of the UHMWPE polymer is controlled by the amount of catalyst, polymerization temperature, and pressure.

Stripping, Drying, and Packing The polymerization slurry is pressed directly into the autoclave, which is heated up by autoclave. Therefore, the solvent of the polymerization slurry evaporates through the pipeline into the condenser and returns to the solvent settling tank. When the solvent of the stripping kettle is completely distilled out, the UHMWPE powder is cleaned to remove the catalyst residue. UHMWPE powder in the stripping kettle is sucked into the suction pipe by induced draft fan, and the vapor–solid separation is carried out to make the purified powder fall into the wet powder tank, finally to obtain the dried powder.

4.2.2.2 Preparation of UHMWPE Fibers

The conventional method of extrusion molding tends to make the UHMWPE molecular chain slip, and the corresponding melt is prone to rupture due to high viscosity and low friction coefficient between molecular chains. Since the 1970s, a variety of fiber forming methods, mainly including the plasticized melt spinning method, solid-state extrusion method, super stretching or local stretching method, surface crystal growth method, and gel spinning method have been developed. However, only the gel spinning method has successfully realized industrial production.

4.2.2.2.1 Plasticized Melt Spinning Method

The plasticized melt spinning method reduces the viscosity of UHMWPE melt by adding plasticizers or diluents to improve the processability. Plasticizers are usually solid paraffin waxes or good solvents of UHMWPE, and the boiling point is higher than that of UHMWPE. During the processing, the concentration of UHMWPE is controlled between 20% and 40%, and the melt spinning adopts the twin-screw extruder. The UHMWPE mixture melt is extruded and shaped, directly being multi-stage stretched in extractant, followed by cleaned and dried to obtain UHMWPE fibers. Zhang et al. [133] added nano-scale montmorillonite mineral (OMMT) into UHMWPE during melt spinning to improve the flowability without affecting the crystalline structure.

4.2.2.2.2 Solid-State Extrusion Method

The solid-state extrusion method places UHMWPE resin powder into a heated extrusion device. UHMWPE powder is completely melted, extruded from the conical spray hole by high pressure, and then UHMWPE fibers are obtained by high-fold stretching. The schematic diagram of the preparation for UHMWPE fibers by solid-state extrusion method is shown in Figure 4.14. During the process of the solid-state extrusion, UHMWPE macromolecular chains can fully extend orientation under the action of high shear and tensile tension. However, the solid-state extrusion method is limited by the process equipment and the performance of UHMWPE, so it is difficult to realize industrial production. Kurtz et al. [134]

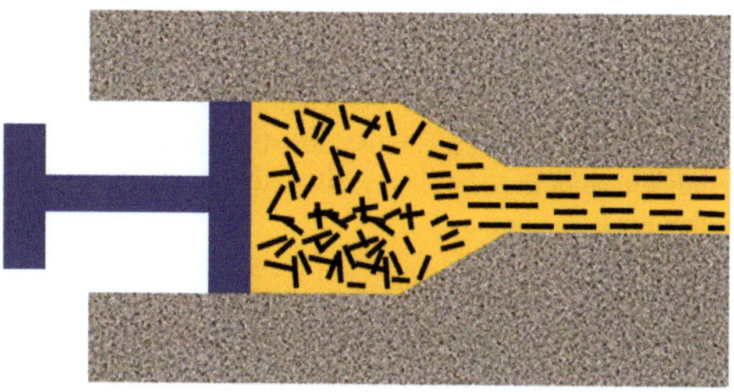

Figure 4.14 Schematic diagram of the preparation for UHMWPE fibers by solid-state extrusion method.

adopted a high-pressure solid-state extrusion method to obtain UHMWPE fibers with excellent oxidative stability. However, the obtained UHMWPE fibers possessed low strength and were not easily subjected to high-fold stretching.

4.2.2.2.3 Super Stretching or Local Stretching Method

Super stretching or local stretching is the hyperextend or partial stretching of the primary fibers for UHMWPE within the range of the crystallization dispersion temperature ($\alpha_c = 127\,°C$). In the range of crystallization dispersion temperature, the folded macromolecular chains of UHMWPE are fully extended and rearranged, and UHMWPE fibers with high strength and high modulus are obtained by appropriate stretching. However, the corresponding molecular weight of UHMWPE by this method should not be too high.

Toyo Textile Company (Japan) dissolved UHMWPE powder in paraffin oil or xylene, which was heated to more than 100 °C and cooled to room temperature to precipitate gel pellets with a diameter of 0.1 mm. The gel pellets were collected and pressed to form a gel film (the molecular chains were arranged in a straight line according to one direction), and UHMWPE fibers were prepared by stretching. In addition, Toyo Textile Company (Japan) extruded the UHMWPE solution through a slit spinning head to form a thin film, which was quickly cooled and then extruded through a roller (some of the solvent was extruded) to form 0.32 mm thick sheets. Next, the sheets were cut into strips and stretched at 120 °C and 150 °C, respectively, to prepare UHMWPE fibers.

4.2.2.2.4 Surface Crystal Growth Method

The surface crystallization growth method adds the dilute solution of UHMWPE into the crystallization device. It rotates the circular roller in the center of the device to form a layer of UHMWPE lyophilized film on the surface, then putting the crystalline seeds into the dilute solution of UHMWPE. When the crystalline seeds are in contact with the lyophilized film, UHMWPE crystals are induced, and fibrous crystals grow continuously. The fibrous crystals are then led out of the duct and further hot-drawn to prepare UHMWPE fibers. It is worth noting that the direction of extraction for the UHMWPE fibrous crystal is opposite to the direction of rotation for the central roller so that the growth of the UHMWPE fibrous crystal is affected by the axial tensile force along the fibers. However, the slow crystalline growth rate and low yield of UHMWPE limit the broad application. Pennings et al. [135] from DSM Corporation (Netherlands) first prepared UHMWPE in a dilute solution to make it flow and crystallize, then to produce high modulus (closed to 22 GPa) UHMWPE fibers by traction. The advantage of the surface crystal growth method is that the molecular chain entanglement is less in the dilute solution.

4.2.2.2.5 Gel Spinning-Super-Stretching Method

The gel spinning-super-stretching method is to dissolve UHMWPE resin in an appropriate solvent to form a dilute solution, extruded and cooled through a spinneret to form gel protofilaments. Then, the solvent in the gel protofilaments is removed, followed by multi-stage super-stretching to obtain UHMWPE fibers.

The process of configuring UHMWPE resins into dilute solution is to untangle UHMWPE molecular chains. The function of forming gel protofilaments is to maintain the state of untangling for UHMWPE macromolecules and facilitate the full extension of the molecular chains. The super-stretching process is an important process to transform the folded chains sheet crystal structure to the elongated chain crystal structure to obtain UHMWPE fibers with high crystallinity and high orientation.

Gel spinning is divided into dry spinning and wet spinning. The difference between the two spinning methods is mainly the different solvents and the different methods of removing the solvents from the gel protofilaments. Dry spinning is to dissolve UHMWPE resin with highly volatile solvents such as decalin and to obtain dry gel protofilaments by removing the solvent through solvent volatilization. Wet spinning adopts low volatile solvent (mineral oil, paraffin, kerosene, and so on) to dissolve UHMWPE. It uses high-volatile extractants to replace the low-volatile solvent from the gel protofilaments through the multi-stage extraction process. Finally, the gel protofilaments can be obtained through multi-stage drying.

Dry Spinning In dry spinning, the solvents of decahydronaphthalene, xylene, and so on, present high volatility. UHMWPE spinning solution through the twin screw metering pump and spinning components can form the gel protofilaments, which are directly fed into the heating channel. Under the action of the hot air layer, the highly volatile solvents inside the gel protofilaments are quickly removed to prepare UHMWPE fibers through the super-stretching. Figure 4.15 shows the process flow chart of preparing UHMWPE fibers by dry spinning.

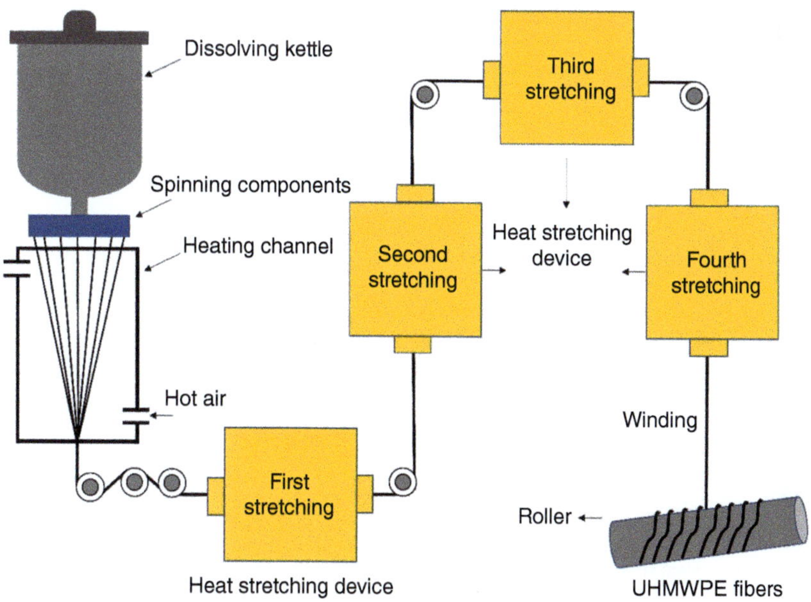

Figure 4.15 Process flow chart of preparing UHMWPE fibers by dry spinning.

Dyneema fibers developed by DSM Corporation (Netherlands) are the typical representative of UHMWPE fibers prepared by dry spinning [136–138]. Toyobo Co., Ltd. developed a new dry spinning route to prepare ultra-strong UHMWPE fibers by flow-induced phase separation with molecular-level analysis.

Wet Spinning Wet spinning involves adding UHMWPE resin powder into some low-volatile solvents, such as white oil, kerosene, and paraffin oil [139]. UHMWPE mixtures are swollen and dissolved in the twin-screw, followed by being extruded from the spinning jet. After a period of the air layer, the extruded UHMWPE enters the solidifying bath to be quenched into gel protofilaments. The low volatile solvent in the gel protofilaments is extracted by multi-stage extraction with a high volatile extractant, and then the extractant is removed by drying. The extracted gel protofilaments are stretched to obtain UHMWPE fibers. The process flow chart of the preparation for UHMWPE fibers by wet spinning is shown in Figure 4.16.

Spectra® fibers, developed by Honeywell Corporation (USA), are one kind of typical representative for UHMWPE fibers prepared by wet spinning [140, 141]. However, the UHMWPE fibers obtained by wet spinning are slightly inferior to the UHMWPE fibers obtained by dry spinning in terms of strength and stability. The wet spinning presents a long process flow, complicated extraction solution recovery, and treatment process. During the process of wet spinning, the volatilization of extractant is easy to cause environmental pollution.

4.2.2.3 Structure of UHMWPE Fibers

UHMWPE fibers are one kind of high-performance organic fibers with extended straight chain structures [142–144]. The chemical structure formula

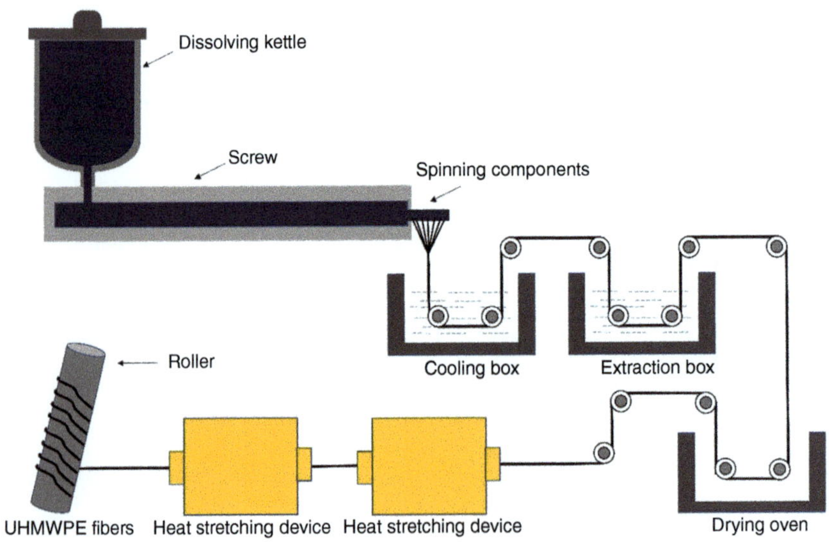

Figure 4.16 Process flow chart of the preparation for UHMWPE fibers by wet spinning.

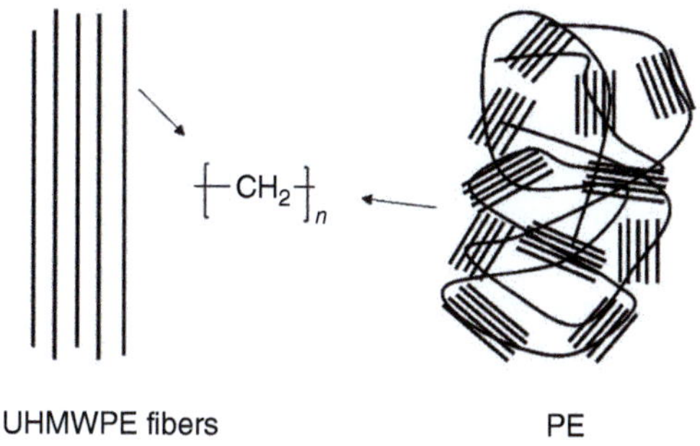

Figure 4.17 Schematic diagram of the structure for UHMWPE fibers and PE structure.

is $-(CH_2-CH_2)n-$, and the molecular weight is more than 1 million. From the perspective of molecular structure, the macromolecular chains of UHMWPE fibers are connected by highly symmetrical methylene structure without side chain groups. During the gel spinning process, after tens of times of hot stretching, the macromolecular chains of UHMWPE fibers are fully stretched along the fiber axial direction, forming the parallel-aligned straightened chain structure. Therefore, UHMWPE fibers present high crystallinity and high orientation, far higher than those of conventional polyethylene (PE). Normally, the crystallinity of UHMWPE fibers is as high as 85%, and the orientation is close to 100%. Compared with the tangled and folded chains of ordinary PE, the UHMWPE fibers present elongated chain structures, and the macromolecular chains are highly oriented. The schematic diagram of the structure for UHMWPE fibers and PE structure is shown in Figure 4.17.

4.2.2.4 Properties of UHMWPE Fibers

4.2.2.4.1 Mechanical Properties

The molecular chains of UHMWPE fibers are closely arranged, and the high orientation and crystallinity endow UHMWPE fibers with excellent strength and modulus. The tensile strength and modulus are 3.0 and 172.0 GPa, respectively. In addition, UHMWPE fibers present high specific strength/modulus due to the very low density (0.97–0.98 g/cm^3). The abrasion resistance, flexural fatigue strength, and impact resistance of UHMWPE fibers are also higher than those of aramid fibers, but the elongation at break is relatively low, only 3–6%.

4.2.2.4.2 Dielectric Properties

The molecular chains of UHMWPE fibers are composed of $-CH_2-$ with nonpolar, and the side chains have no polar groups, which is in favor of the extremely low polarizability of the molecular chain. Therefore, UHMWPE fibers present low ε (3.0) and tanδ (0.0001) and are widely used in high-performance radomes.

4.2.2.4.3 Thermal Stability

The melting point of UHMWPE fibers is usually between 144 and 155 °C. With the increase in temperature, the thermal expansion coefficient of orthogonal cell for a-axis (about 2.7×10^{-4} °C) in UHMWPE fibers begins to be much higher than that of *b*-axis (about 6.1×10^{-6} °C). When the size of the *a*-axis cell parameter is three times that of the *b*-axis, the structure of UHMWPE fibers transitions from orthogonal to hexagonal-like crystals. If certain restriction is imposed on the UHMWPE fibers, the crystalline structure hinders the transition to the melt, resulting in increasing melting temperature.

4.2.2.4.4 Chemical Stability

The molecular chains of UHMWPE fibers are composed of methylene ($-CH_2-$), presenting a highly oriented crystalline structure without other functional groups. Therefore, UHMWPE fibers present excellent chemical corrosion resistance and are difficult to react chemically in strong acids and alkalis, with only mild swelling in a few organic solvents. In addition, the properties of UHMWPE fibers remain basically unchanged, even in strong acids and alkalis, for a long time.

4.2.2.4.5 Biocompatibility

UHMWPE fibers possess excellent biocompatibility and can reduce the possibility of tissue inflammation and irritation in the biological environment due to the chemically inert structure. Consequently, UHMWPE fibers have a broad application prospect in medical sutures, artificial ligaments, artificial blood vessels, and so on.

4.2.3 Poly-*p*-Phenylene Benzobisoxazole Fibers

In the 1970s, the United Nations Air Force Materials Laboratory commissioned the Stanford Research Institute (SRI) to carry out relative research on high-performance fibers. Wolfe et al. [145] screened nearly 100 model polymers for para-aromatic polymers containing 2,6-benzobisheterocycles on the main chain, namely poly-*p*-phenylene benzobisoxazole (PBO), which possessed high strength and modulus and excellent temperature resistance. However, due to the limitation of the polymerization process, the molecular weight of the PBO was low, resulting in poor properties. In the mid-1980s, Dow Chemical Co., Ltd. obtained PBO patents from SRI and improved the synthesis method of PBO monomer, which increased the yield of PBO monomer and laid a foundation for realizing the industrialization of PBO fibers. In 1990, Dow Chemical Co., Ltd. and Toyobo Co., Ltd. developed a new spinning technology to produce PBO fibers with much higher strength and modulus than that of Kevlar fibers. In 1995, Toyobo Co., Ltd. started commercial production of PBO fibers (Zylon).

China also obtained some achievements in the synthesis of PBO polymers and the preparation of PBO fibers. At the end of the 1980s, East China University of Science and Technology pioneered the synthesis process of 4,6-diamino-1,3 resorcinol and PBO polymers. Afterward, Shanghai Jiao Tong University and Zhejiang University successively carried out the research on PBO polymeric monomers. Sinopec Group and Donghua University developed a new PBO polymerization process. They designed a reactive extrusions-liquid crystal spinning process to first prepare PBO fibers with a tensile strength of up to 4.3 GPa and thermal decomposition temperature above 600 °C.

4.2.3.1 Synthesis of PBO Monomer and PBO Polymers

4.2.3.1.1 Synthesis of 4,6-Diamino-1,3 Resorcinol Hydrochloride

4,6-Diamino-1,3 resorcinol hydrochloride (DAR) is an important monomer for the synthesis of PBO polymers [146–148]. Initially, the synthesis of DAR is based on resorcinol as raw material. However, the nitration process produces a variety of nitrosylation byproducts, which increases the difficulty of purification and leads to high production costs and low yield. In order to solve the problem of byproducts in the nitrification process, Dow Chemical Co., Ltd. adopted trichlorobenzene as the raw material, which was converted into 2,3,4-trichloro-1,5-dinitrobenzene and 2-chloro-4,6-dinitrobenzene-1,3-diol in turn to obtain DAR·2HCl. The chlorine atom protected the benzene ring neighbors before and after nitration to avoid excessive byproducts, and the yield of DAR·2HCl was more than 80%. The synthetic route of trichlorobenzene as raw material for DAR·2HCl is shown in Figure 4.18.

Considering the highly toxic and expensive of trichlorobenzene, Dow Chemical Co., Ltd. developed a new synthetic route of DAR in 1993. Aniline was firstly converted to benzene diazonium chloride and then reacted with resorcinol to form intermediate products. Next, the intermediates were catalytically hydrogenated to obtain the DAR·2HCl (Figure 4.19).

4.2.3.1.2 Polymerization Process of PBO Polymers

The synthetic routes of PBO polymers with high molecular weight can be divided into five categories.

Polyphosphate System Synthesis Method Figure 4.20 shows the route of the polyphosphate system synthesis method, and the reactants are DAR and terephthalic

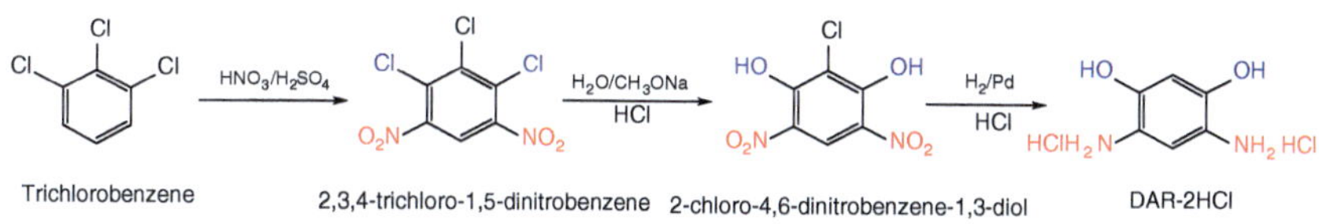

Figure 4.18 Synthetic route of trichlorobenzene as raw material for DAR·2HCl.

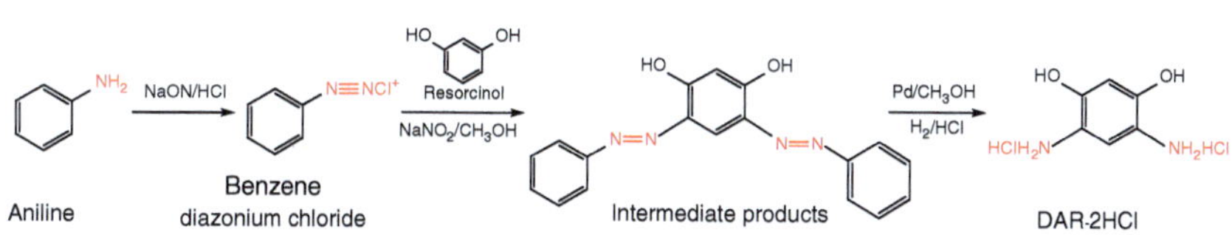

Figure 4.19 Synthetic route of aniline as raw material for DAR.

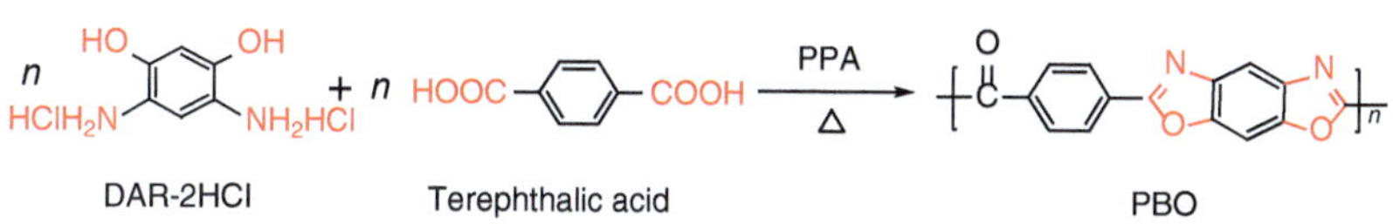

Figure 4.20 Synthetic route of the polyphosphate system synthesis method.

acid. Since the reaction system is a condensation reaction, phosphorus pentoxide needs to be added continuously during the reaction to keep the reaction going in the positive direction. The low solubility of terephthalic acid in the polyphosphoric acid system requires the addition of additional benzoic acid for molecular chain capping in the later stages of the reaction, thus controlling the molecular weight of the PBO polymers.

Methanesulfonic Acid System Synthesis Method The mechanism of the synthesis of the PBO polymers by methanesulfonic acid system is basically the same as that of polyphosphate system. The reaction solvent of methanesulfonic acid system is methanesulfonic acid, and the reactants are DAR, and terephthalic acid, or paraphthaloyl chloride. The methanesulfonic acid system presents the advantages of low solution viscosity and fast reaction speed.

Complex Salt Polymerization Method The mechanism of complex salt polymerization is that DAR hydrochloride reacts directly with terephthalic acid, and the self-condensation process occurs by heating. Figure 4.21 shows the synthetic route of the complex salt polymerization method. DAR can effectively avoid oxidation after complexing with hydrochloric acid and ensure that the content of functional group molar ratio for DAR and terephthalic acid does not change, which is in favor of uniform molecular weight distribution for PBO polymers. However, in the late stage of the reaction by complex salt polymerization, the overall system reacts incompletely due to the increase in viscosity of the reaction mixture, resulting in a low molecular weight of PBO polymers.

Trimethylsilylation Method The trimethylsilylation method is to react DAR with trimethylsilane to form intermediate products and further polycondensation with terephthaloyl chloride to obtain PBO polymers through high-temperature

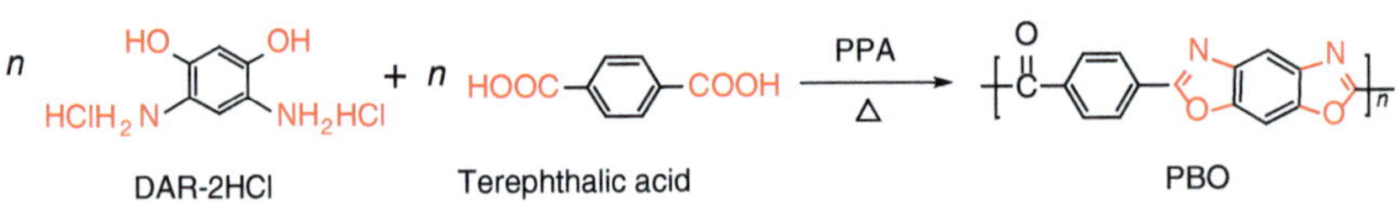

Figure 4.21 Synthetic route of complex salt polymerization method.

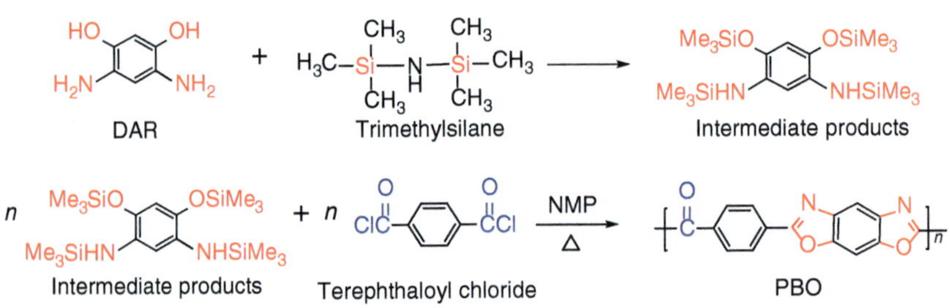

Figure 4.22 Synthetic route of trimethylsilylation method.

ring closure. The corresponding synthetic route of trimethylsilylation method is shown in Figure 4.22. The advantage of the trimethylsilylation method is that NMP presents good solubility with PBO polymers and reduces the reaction viscosity. Still, the disadvantage is that the obtained PBO polymers need to be dissolved into the solvent again for spinning into PBO fibers.

AB-Type Monomer Self-Condensation Method The mechanism of the AB monomer self-condensation method is to adopt DAR to react with methoxycarbonylated benzoyl to obtain transition intermediates, further hydrogen reduction to reduce the nitro groups to amino groups, and finally to obtain PBO polymers through high-temperature self-condensation. The synthetic route of the AB-type monomer self-condensation method is shown in Figure 4.23. Compared with the above four methods, the AB-type monomer self-condensation method presents the advantages of less water removal, less amount of phosphorus pentoxide, higher reaction efficiency, and shorter reaction time. However, the reaction steps of this method are too complicated to achieve large-scale production.

4.2.3.2 Preparation of PBO Fibers

PBO polymers are one kind of typical lyotropic liquid crystal polymers [149]. When the concentration reaches or exceeds the critical value concentration of the liquid crystal state, the PBO molecular chains in the liquid crystal state can be stretched and oriented in the direction of high traction force. In addition, the high concentration of PBO polymers can increase the viscosity of the system and difficult to spin into fibrous products.

At present, the preparation method of PBO fibers mainly adopts dry spray-wet spinning technology, and the main process parameters include PBO polymer viscosity, storage tank, spinning temperature and pressure, spinneret aperture size, traction stretching speed, and air bath length.

During the preparation of PBO fibers, the spinning temperature depends on the viscosity of the reacting PBO polymers. When the viscosity of the reaction system is high, the temperature in the storage tank is increased accordingly, and the reaction temperature is generally controlled between 180–200 °C. In addition, the spinning pressure is adjusted according to the content and viscosity of the PBO polymers in the spinning solution, and the pressure is controlled 5–20 MPa. The initial pressure serves to provide the initial velocity of the PBO polymers, which are pushed to the spinneret then spun into PBO fibers under high drafting forces. The stretching process favors aligning the liquid crystal molecular chains in the PBO polymers.

The soaking time and temperature of PBO fibers in a coagulation bath (acidic system, neutral system, or alkaline system) would affect the comprehensive properties. When the mixed solution of water and methanesulfonic acid is used as the

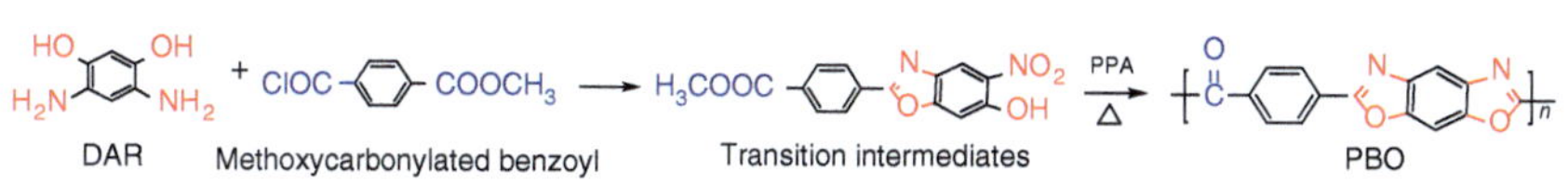

Figure 4.23 Synthetic route of AB-type monomer self-condensation method.

solution of the coagulation bath, the molecular chain arrangement of PBO fibers is relatively uniform. The reason is that the methanesulfonic acid/water, as a weak acid system, provides a slow diffusion environment that allows sufficient time for the molecular chains inside the PBO fibers to reorient.

The drying and heat treatment process is the last process for preparing PBO fibers. The drying heat treatment removes the water inside the PBO fibers and rearranges the internal molecular chains under tensile force. Before drying and heat treatment, most of the water inside the PBO fiber should be removed in advance. Otherwise, direct drying heat treatment of PBO fibers can result in rapid vaporization of internal moisture and create a large number of pores.

4.2.3.3 Structures of PBO Fibers

The molecular chains of PBO fibers contain benzene rings and oxazole rings, and the molecular unit linkage angle is 180°. The whole molecular chain is in a choke state, presenting strong rigidity. In addition, PBO fibers display an obvious core-skin structure, as shown in Figure 4.24. PBO fibers have a selective orientation in the cross-sectional direction, and the microfibers are aligned along the radius of the cross-section. The surface layer of PBO fibers is a cortical region without micropores, while the core layer is composed of many microfibers, and there are some micropores among the microfibers. The main reason is that when the initial PBO fibers enter the solidification bath, the surface layer quickly loses solvent and starts to solidify rapidly. The solvent of the core layer is removed slowly, resulting

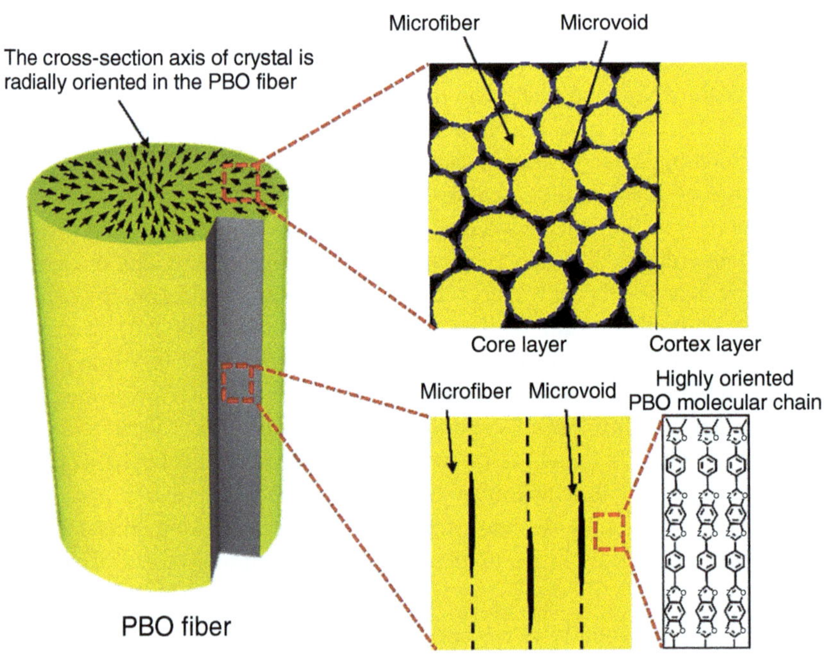

Figure 4.24 Structure diagram of PBO fibers.

in discontinuous surface and core layers of PBO fibers. In addition, PBO molecular chains are highly oriented along the fiber axis due to external tractive action.

4.2.3.4 Properties of PBO Fibers

PBO fibers, known as the twenty-first century super fibers, present outstanding mechanical properties and heat resistances and are currently the best comprehensive properties of organic fibers [150–152]. At present, there are two main kinds of commercial PBO fibers. One is directly spun protofilament, called PBO-AS fibers, and the other is the protofilament prepared by heat treatment, called PBO-HM fibers. Table 4.5 lists the main physical performance of PBO-AS and PBO-HM fibers.

4.2.3.4.1 Mechanical Properties

PBO fibers present excellent mechanical properties, and the tensile strength (5.8 GPa) and modulus (270 GPa) are about twice that of para-aramid fibers. The modulus of PBO fibers is considered to reach the limit of straight-chain polymer fibers. The shear modulus of PBO fibers at room temperature is about 1 GPa, much higher than that of common commercial organic fibers (about 0.5 GPa). However, the compression strength of PBO fibers (200–300 MPa) is only one-twentieth of the tensile strength and slightly lower than that of Kevlar fibers (400 MPa).

4.2.3.4.2 Dielectric Properties

The benzene and oxazole rings of the molecular chains in the PBO fibers form conjugate interaction with large free volume, and the side chains have no polar groups, which makes the polarizability extremely low. Therefore, PBO fibers present low ε (3) and tanδ (0.001), and are of wide interest in airborne and aircraft radomes.

4.2.3.4.3 Thermal Performance

PBO fibers are the highest-known organic fibers in terms of heat resistance due to the rigid aromatic and oxazole rings and highly ordered internal crystal structures. The thermal decomposition temperature of PBO fibers in air is about 650 °C, about 100 °C higher than that of aramid fibers, and far higher than that of UHMWPE fibers (300 °C). The thermal decomposition temperature of PBO fibers can reach 700 °C in an inert environment (nitrogen or argon). In addition, PBO fibers can be used for a long time at 300 °C, and even retain more than 95% of the initial weight after being

Table 4.5 Main physical performance of PBO-AS and PBO-HM fibers.

	Properties					
Fibers	**Density ($g\,cm^{-3}$)**	**Tensile strength (GPa)**	**Modulus (GPa)**	**Tensile elongation at break (%)**	**Thermal decomposition (°C)**	**LOI**
PBO-AS	1.54	5.8	180	3.5	650	68
PBO-HM	1.56	5.8	270	2.5	650	68

heated for 180 minutes at 400 °C. In addition, PBO fibers possess a low coefficient of thermal expansion (-6×10^{-6} °C) and excellent flame retardant, and the LOI is 68.

4.2.3.4.4 Chemical Stability

PBO fibers present good chemical stability. The mechanical strength basically does not decrease when immersed in most organic solvents and alkali solutions for a long time. However, PBO fibers are not resistant to strong acid solutions and can be completely dissolved in concentrated sulfuric acid, methanesulfonic acid, and polyphosphoric acid. Compared with that of aramid fibers, PBO fibers have better stability in hypochlorous acid and can maintain higher strength after a long time of soaking.

4.2.3.4.5 Hygroscopicity and Dimensional Stability

The moisture absorption rates of PBO AS and PBO-HM fibers are 2.0% and 0.6%, respectively, far lower than that of aramid fibers (about 4.5%). PBO fibers also present excellent dimensional stability, and the effect of temperature and moisture on the dimensional change of PBO fibers is minimal. Meantime, the plastic deformation of PBO fibers under 50% fracture load for 100 hours does not exceed 0.03%.

4.3 Summary

This chapter focuses on the structural composition, preparation methods, and physicochemical properties of the two main types of reinforced fibers (inorganic and organic fibers) applied in polymer matrix wave-transparent composites. The effects of the micro-structures on the macro-properties for different reinforced fibers and the advantages and disadvantages of different preparation methods are analyzed in detail. Organic fibers possess relatively better mechanical properties than those of inorganic fibers, and inorganic fibers present relatively lower manufacturing costs and higher heat resistance than those of organic fibers. At present, the preparation of reinforced fibers with high strength and heat resistance as well as low ε and tanδ is still a difficult task. In future research, it is of great theoretical significance and practical value to master the basic structures, preparation principles, and physicochemical properties of the reinforced fibers for fabricating polymer matrix wave-transparent composites with excellent mechanical properties, heat resistance, and wave-transparent performances.

References

1 Shi, X.L., Chen, W.Y., Zhang, T. et al. (2021). Fiber-based thermoelectrics for solid, portable, and wearable electronics. *Energy & Environmental Science* 14 (2): 729–764.

2 Zaarour, B., Zhu, L., Huang, C. et al. (2021). A review on piezoelectric fibers and nanowires for energy harvesting. *Journal of Industrial Textiles* 51 (2): 297–340.

3 Tang, L., Zhang, J.L., Tang, Y.S. et al. (2021). Polymer matrix wave-transparent composites: a review. *Journal of Materials Science and Technology* 75: 225–251.

4 Ben Amor, I., Ghallabi, Z., Kaddami, H. et al. (2010). Experimental study of relaxation process in unidirectional (epoxy/palm tree fiber) composite. *Journal of Molecular Liquids* 154 (2–3): 61–68.

5 Li, J., Lin, J., Wang, Y. et al. (2022). Cross-linked and rigid polyimide composite foams with prominent fire resistant, thermal insulating, and wave-transparent properties. *ACS Applied Polymer Materials* 4 (10): 6994–7003.

6 Ghazzawi, Y.M., Osorio, A.F., and Heitzmann, M.T. (2019). Fire performance of continuous glass fibre reinforced polycarbonate composites: the effect of fibre architecture on the fire properties of polycarbonate composites. *Journal of Composite Materials* 53 (12): 1705–1715.

7 Singh, J., Kumar, M., Kumar, S., and Mohapatra, S.K. (2017). Properties of glass-fiber hybrid composites: a review. *Polymer-Plastics Technology and Engineering* 56 (5): 455–469.

8 An, J., Meng, Z., and Su, Y. (2016). Research progress of mechanical property of long glass fiber reinforced polypropylene composites. *Engineering Plastics Application* 44 (6): 132–136.

9 Silva, R.V., Voltz, H., Itman, A. et al. (2021). Hybrid composites with glass fiber and natural fibers of sisal, coir, and luffa sponge. *Journal of Composite Materials* 55 (5): 717–728.

10 Sathishkumar, T.P., Satheeshkumar, S., and Naveen, J. (2014). Glass fiber-reinforced polymer composites – a review. *Journal of Reinforced Plastics and Composites* 33 (13): 1258–1275.

11 Merter, N.E., Baser, G., and Tanoglu, M. (2016). Effects of hybrid yarn preparation technique and fiber sizing on the mechanical properties of continuous glass fiber-reinforced polypropylene composites. *Journal of Composite Materials* 50 (12): 1697–1706.

12 Beales, K.J., Day, C.R., Duncan, W.J. et al. (1976). Preparation of sodium borosilicate glass-fiber for optical communication. *Proceedings of the Institution of Electrical Engineers-London* 123 (6): 591–596.

13 Li, H.B., Yan, M.L., Qi, D.T. et al. (2011). Corrosion of E-glass fiber in simulated oilfield environments. *Journal of Petroleum Science and Engineering* 78 (2): 371–375.

14 Naito, K. (2022). Static and fatigue tensile properties of carbon/glass hybrid fiber-reinforced epoxy composites. *Scientific Reports* 12 (1): 6298.

15 Shah, D.U., Schubel, P.J., and Clifford, M.J. (2013). Can flax replace E-glass in structural composites? A small wind turbine blade case study. *Composites Part B: Engineering* 52: 172–181.

16 Kawada, H. and Srivastava, V.K. (2001). The effect of an acidic stress environment on the stress-intensity factor for GRP laminates. *Composites Science and Technology* 61 (8): 1109–1114.

17 Wang, Q.X., Ding, Y.N., and Randl, N. (2021). Investigation on the alkali resistance of basalt fiber and its textile in different alkaline environments. *Construction and Building Materials* 272: 121670.

18 Tanks, J.D., Arao, Y., and Kubouchi, M. (2018). Diffusion kinetics, swelling, and degradation of corrosion-resistant C-glass/epoxy woven composites in harsh environments. *Composite Structures* 202: 686–694.

19 Steenkamer, D.A. and Sullivan, J.L. (1997). Recycled content in polymer matrix composites through the use of A-glass fibers. *Polymer Composites* 18 (3): 300–312.

20 Gao, G., Li, Y., Wang, Z. et al. (2012). State equation of a new type alkali glass fiber reinforced cloth. *Ordnance Material Science and Engineering* 35 (1): 5–7.

21 Zimin, D.E. and Khodakova, N.N. (2016). Chemical composition of rocks suitable for the production of basalt fibers resistant to corrosive media. *Glass and Ceramics* 73 (3–4): 82–87.

22 Liu, J.X., Jiang, M., Wang, Y. et al. (2013). Tensile behaviors of ECR-glass and high strength glass fibers after NaOH treatment. *Ceramics International* 39 (8): 9173–9178.

23 Alsaadi, M. (2019). Hybridization effects of S-glass fiber on charpy impact resistance of carbon/aramid fiber reinforced epoxy composite laminates. *Materials Research Express* 6 (12): 125342.

24 Cho, K., Wang, G.N., Raju, R. et al. (2019). Influence of surface treatment on the interfacial and mechanical properties of short S-glass fiber-reinforced dental composites. *ACS Applied Materials & Interfaces* 11 (35): 32328–32338.

25 Yeon, J., Chowdhury, S.C., and Gillespie, J.W. (2022). Mechanical properties and damage analysis of S-glass: a reactive molecular dynamics study. *Composites Part B-Engineering* 234: 109706.

26 Liu, Z., Zhu, Z., Xu, X., and Yan, R. (2011). The absorbing properties and optimization of the $Ba(Zn_{0.65}Co_{0.35})_2Fe_{16}O_{27}$ powders/M-glass fibers/epoxy composite panels. *Journal of Functional Materials* 42 (1): 124–127.

27 Ma, S., Zhu, Z., Xu, X., and Liu, Z. (2010). The absorption property of ferrite $Ba(Zn_{1-x}Co_x)_2Fe_{16}O_{27}$/epoxy-based composite panels. *Journal of Functional Materials* 41 (10): 1750–1752.

28 Kothe, J., Scherieble, G., and Ehrler, P. (1997). High-performance glass-fiber-reinforced thermoplastic composites made of textile prepregs. *Journal of Thermoplastic Composite Materials* 10 (4): 316–333.

29 Mavrich, A., Fondeur, F., Ishida, H. et al. (1994). The study of fiber-matrix interactions via FT-IR microscopy and NMR imaging. *Journal of Adhesion* 46 (1–4): 91–102.

30 Yao, Y., He, Y., and Chen, Z. (2016). The microwave dielectric properties of basalt fibers and CBF-EP composites. *Fiber Reinforced Plastics/Composites* 5: 65–67.

31 Yang, W., Xu, B., Qi, M. et al. (2020). Improving ablation properties of ceramifiable vitreous silica fabric reinforced boron phenolic resin composites via an incorporation of $MoSi_2$. *Plastics, Rubber and Composites* 49 (10): 456–469.

32 Takahashi, N., Hashimoto, S., Daiko, Y. et al. (2019). Fabrication and characterization of a novel lightweight adiabatic refractory composite consisting of alkaline earth silicate fibers and SiC particles. *Ceramics International* 45 (17): 23248–23255.

33 Parizi, M.J.G., Shahverdi, H., Pipelzadeh, E. et al. (2021). A finite element investigation into the cohesive properties of glass-fiber-reinforced polymers with nanostructured interphases. *Nanomaterials* 11 (10): 2487.

34 Shabarov, V.V., Snopatin, G.E., and Churbanov, M.F. (2010). Investigation of the flow of molten tellurite glass in a double-crucible die. *Theoretical Foundations of Chemical Engineering* 44 (2): 206–212.

35 Munoz, F., Pritula, O., Sedlacek, J., and Russel, C. (2006). The effect of the hydrostatic pressure on the drawing process of phosphate glass fibres. *Glass Technology-European Journal of Glass Science and Technology Part A* 47 (4): 97–102.

36 Zhu, C.K., Ahmed, I., Parsons, A. et al. (2019). The mechanical property, degradation and cytocompatibility analysis of novel phosphate glass fiber textiles. *Textile Research Journal* 89 (16): 3280–3290.

37 Saito, K., Sekiya, E.H., Ikushima, A.J. et al. (2006). Control of glass-forming process during fiber-drawing to reduce the rayleigh scattering loss. *Journal of the American Ceramic Society* 89 (1): 65–69.

38 Zheng, H.X., Hu, Y., and Mackenzie, J.D. (1991). Continuous drawing of Bi–Ca–Sr–Cu–O glass fibers from a preform. *Applied Physics Letters* 58 (15): 1679–1681.

39 Orcel, G. and Biswas, D. (1991). Influence of processing parameters on the strength of fluoride glass fibers. *Journal of the American Ceramic Society* 74 (6): 1373–1377.

40 Wang, Y., Li, J.F., Li, H. et al. (2017). Chitosan modified corn starch and its application as a glass fiber sizing agent. *Fibres & Textiles in Eastern Europe* 25 (3): 112–120.

41 Gorowara, R.L., Kosik, W.E., McKnight, S.H., and McCullough, R.L. (2001). Molecular characterization of glass fiber surface coatings for thermosetting polymer matrix/glass fiber composites. *Composites Part A: Applied Science and Manufacturing* 32: 323–329.

42 Liu, Z.Y., Zhang, L.T., Yu, E.L. et al. (2015). Modification of glass fiber surface and glass fiber reinforced polymer composites challenges and opportunities: from organic chemistry perspective. *Current Organic Chemistry* 19 (11): 991–1010.

43 Yang, L. and Thomason, J.L. (2013). Effect of silane coupling agent on mechanical performance of glass fibre. *Journal of Materials Science* 48 (5): 1947–1954.

44 Guo, X.Y., Lu, Y.G., Sun, Y. et al. (2018). Effect of sizing on interfacial adhesion property of glass fiber-reinforced polyurethane composites. *Journal of Reinforced Plastics and Composites* 37 (5): 321–330.

45 Murach, J., Jander, S., and Bruckner, R. (1997). Structure-sensitive investigations on alkali disilicate glass fibers with reference to metasilicate- and silica glass fibers. *Journal of Non-Crystalline Solids* 217 (1): 92–98.

46 Wang, D.P., Zhang, Z., and Wang, Q.Z. (2020). Study on fatigue failure of glass fibre-reinforced thermoplastic pipe. *Materials Research Express* 7 (4): 045203.

47 Wu, Z., Zhang, Z.Y., Yun, J.N. et al. (2013). Synthesis of alpha-Si_3N_4 crystallon by a solvothermal method at a low temperature of 180 °C. *Physica B: Condensed Matter* 428: 10–13.

48 Kuznetsov, A.S., Cuong, N.T., Tikhomirov, V.K. et al. (2012). Effect of heat-treatment on luminescence and structure of Ag nanoclusters doped oxyfluoride glasses and implication for fiber drawing. *Optical Materials* 34 (4): 616–621.

49 Doweidar, H. (2009). Optical properties and structure of R_2O–Ga_2O_3–SiO_2 and RO–Ga_2O_3–SiO_2 glasses. *Journal of Materials Science* 44 (11): 2899–2906.

50 Shekar, R.I., Rao, P.M.D., Siddaramaiah, K.M.N.S. et al. (2011). Microcrystalline and kinetic parameters of glass-PEEK hybrid composites. *Journal of Composite Materials* 45 (5): 549–556.

51 Hu, J., Zhang, J.X., Han, R.N., and Yang, Y. (2022). Correlation between glass fiber suspension characteristics and physical of glass fiber felt: role of beating time and speed. *Journal of Industrial Textiles* 51 (1): 1528S–1541S.

52 Asadi, H., Uhlemann, J., Stranghoener, N., and Ulbricht, M. (2021). Water influence on the uniaxial tensile behavior of polytetrafluoroethylene-coated glass fiber fabric. *Materials* 14 (4): 846.

53 Altay, P. and Ucar, N. (2022). Improvement of insulation properties of glass fiber fabric/epoxy composites modified by polymeric and inorganic fillers. *Polymer Composites* 43 (1): 225–238.

54 Liu, Y.J., Wang, C.X., Ma, Y.P. et al. (2021). Study of the dielectric properties and conductivity of double coated glass fibre felt composites. *Fibres & Textiles in Eastern Europe* 29 (4): 75–80.

55 Akonda, M.H., Kandola, B.K., Horrocks, A.R., and Myler, P. (2014). The effect of fibrous reinforcement on optical and impact performance of fibre-reinforced transparent glass composites. *Journal of Materials Science* 49 (4): 1903–1913.

56 Mathewson, C., Urbina, I., Badding, J.V. et al. (2021). Aluminosilicate glasses for zinc selenide tunable fiber laser cladding. *Journal of the American Ceramic Society* 104 (2): 691–696.

57 Czudkova, I., Penakova, H., and Helebrant, A. (2020). In-vitro testing of solubility of thermal insulation fibers. *Ceramics - Silikaty* 64 (3): 289–298.

58 Rahmani, K., Wheatley, G., Sadooghi, A. et al. (2021). The experimental analysis of creep and corrosion properties of polymeric tube reinforced by glass, carbon and Kevlar fibers. *Materials Research Express* 8 (6): 065307.

59 Wang, L., Yang, J., Cheng, W.H. et al. (2021). Progress on polymer composites with low dielectric constant and low dielectric loss for high-frequency signal transmission. *Frontiers in Materials* 8: 774843.

60 Nag, A., Rao, R.R., and Panda, P.K. (2021). High temperature ceramic radomes (HTCR) – a review. *Ceramics International* 47 (15): 20793–20806.

61 Liang, L. (2020). Study on preparation of quartz ceramics toughened with mineralizer. *Integrated Ferroelectrics* 207 (1): 197–207.

62 Zhang, H.W., Guo, S.J., Wu, J.J. et al. (2021). Effect of quartz crystal structure transformations on the removal of iron impurities. *Hydrometallurgy* 204: 105715.

63 Molaei, F. and Siavoshi, H. (2020). Molecular dynamics studies of thermal conductivity and mechanical properties of single crystalline alpha-quartz. *Solid State Communications* 320: 114020.

64 Saeed, A., Madkhli, A.Y., Dossari, M., and Abolaban, F. (2022). Electrical and dielectric properties of composites composed of natural quartz with aluminum. *Silicon* 14 (15): 9517–9531.

65 Kharitonov, D.V., Anashkina, A.A., Tychinskaya, M.S. et al. (2022). Investigation of the influence of the crystallization of quartz-glass pipe material on the quality of quartz ceramic workpieces. *Glass and Ceramics* 79 (3–4): 130–134.

66 Kolobov, A.Y. and Sycheva, G.A. (2021). Synthesis of nontransparent quartz glass for the production of refractory quartz ceramics. *Glass Physics and Chemistry* 47 (3): 209–218.

67 Li, P., Du, R., Liu, Y., and Zhao, G. (2021). Preparation of epoxy resin/quartz fiber wave-transparent composites. *Engineering Plastics Application* 49 (2): 29–33,9.

68 Li, J.W., Wang, Y.X., Zhao, W.Z. et al. (2022). High-performance quartz fiber/polysilazane and epoxy-modified cyanate ester microwave-transparent composites. *Advanced Composites and Hybrid Materials* 5 (3): 1830–1840.

69 Zu, Y., Zhang, F.F., Chen, D.D. et al. (2020). Wave-transparent composites based on phthalonitrile resins with commendable thermal properties and dielectric performance. *Polymer* 198: 122490.

70 Xu, P. and Yang, Y. (2020). Study on the effect of wire drawing rate on the preparation and irradiation resistance of quartz optical fibers. *Journal of Functional Materials* 51 (2): 02155–02159.

71 Matrenichev, V., Lessa Belone, M.C., Palola, S. et al. (2020). Resizing approach to increase the viability of recycled fibre-reinforced composites. *Materials (Basel, Switzerland)* 13 (24): 5773.

72 Zarrini, S. and Abrams, C.F. (2022). Modeling sizing emulsion droplet deposition onto silica using all-atom molecular dynamics simulations. *Composites Part B: Engineering* 235: 109712.

73 Xie, J., Zhang, L., Xing, H.Y. et al. (2020). Gas sensing of ordered and disordered structure SiO_2 and their adsorption behavior based on quartz crystal microbalance. *Sensors and Actuators B: Chemical* 305: 127479.

74 Tsuchiya, T. and Nakagawa, S. (2022). A new high-pressure structure of SiO_2 directly converted from alpha-quartz under nonhydrostatic compression. *Journal of Physics. Condensed Matter* 34 (30): 304003.

75 Skuja, L., Ollier, N., and Kajihara, K. (2020). Luminescence of non-bridging oxygen hole centers as a marker of particle irradiation of alpha-quartz. *Radiation Measurements* 135: 106373.

76 Ma, L., Fu, X.T., Zhang, C. et al. (2022). Effect of sizing agent on the mechanical, thermal, and electrical performance of basalt fiber/epoxy composites. *Polymers* 14 (17): 3533.

77 Yu, H.J., Tong, Z.W., Zhang, B.J. et al. (2021). Thermal radiation shielded, high strength, fire resistant fiber/nanorod/ aerogel composites fabricated by in-situ growth of TiO_2 nanorods for thermal insulation. *Chemical Engineering Journal* 418: 129342.

78 Bu, A.M., Zhang, Y.F., Xiang, Y. et al. (2020). Plasma electrolysis spraying Al_2O_3 coating onto quartz fiber fabric for enhanced thermal conductivity and stability. *Applied Sciences-Basel* 10 (2): 702.

79 Chen, S., Li, H., Chi, P. et al. (2021). High-temperature reusability of ceramics tile resistant to 1200 °C. *Journal of National University of Defense Technology* 43 (4): 69–75.

80 Xie, Y.D., Liu, S., Huang, K.W. et al. (2022). Ultra-broadband strong electromagnetic interference shielding with ferromagnetic graphene quartz fabric. *Advanced Materials* 34 (30): 2202982.

81 Yao, L., Li, W., Wang, N. et al. (2007). Tensile, impact and dielectric properties of three dimensional orthogonal aramid/glass fiber hybrid composites. *Journal of Materials Science* 42 (16): 6494–6500.

82 Gu, J., Guo, Y., Lv, Z. et al. (2015). Highly thermally conductive POSS-g-SiCp/UHMWPE composites with excellent dielectric properties and thermal stabilities. *Composites Part A: Applied Science and Manufacturing* 78: 95–101.

83 Zhang, R., Tang, P., Li, J. et al. (2014). Study on filler content dependence of the onset of positive temperature coefficient (PTC) effect of electrical resistivity for UHMWPE/LDPE/CF composites based on their DC and AC electrical behaviors. *Polymer* 55 (8): 2103–2112.

84 Zhang, C., Li, C., Wang, B. et al. (2013). Effects of atmospheric air plasma treatment on interfacial properties of PBO fiber reinforced composites. *Applied Surface Science* 276: 190–197.

85 Cosimbescu, L., Malhotra, D., Pallaka, M.R., and Swita, M.S. (2022). Kevlar-like aramid polymers from mixed PET waste. *ACS Omega* 7 (36): 32026–32037.

86 Dai, Y., Meng, C., Cheng, Z. et al. (2019). Nondestructive modification of aramid fiber based on selective reaction of external cross-linker to improve interfacial shear strength and compressive strength. *Composites Part A: Applied Science and Manufacturing* 119: 217–224.

87 Liu, Y., Ye, L., Shu, Y., and Zhao, X. (2018). In situ preparation of intrinsic flame retardant urea formaldehyde/aramid fiber composite foam: structure, property and reinforcing mechanism. *Composites Part A: Applied Science and Manufacturing* 115: 274–282.

88 Layman, P.L., Reisch, M.S., and Du, P. (1988). Akzo resolve long hassle over aramid fiber patents. *Chemical & Engineering News Archive* 66 (34): 10–13.

89 Zhang, H., Zou, X., Liang, J. et al. (2012). Development of electroless silver plating on para-aramid fibers and growth morphology of silver deposits. *Journal of Applied Polymer Science* 124 (4): 3363–3371.

90 Shebanov, S.M., Ivanov, V.B., Pavlikov, A.V. et al. (2021). Effect of treatment of the surface of para-aramid fiber with multiwall carbon nanotubes on its structure and properties infrared spectra. *Fibre Chemistry* 53 (4): 283–286.

91 Tian, W., Qiu, T., Shi, Y. et al. (2017). The facile preparation of aramid insulation paper from the bottom-up nanofiber synthesis. *Materials Letters* 202: 158–161.

92 Chantaratcharoen, A., Sirisinha, C., Amornsakchai, T. et al. (1999). Improvement of interfacial adhesion of poly(*m*-phenylene isophthalamide) short fiber–thermoplastic elastomer (SEBS) composites by *N*-alkylation on fiber surface. *Journal of Applied Polymer Science* 74 (10): 2414–2422.

93 Pramila, J., Melbiah, J.S.B., Rana, D. et al. (2018). Permeation characteristics of tailored poly (*m*-phenylene isophthalamide) ultrafiltration membranes and probing its efficacy on bovine serum albumin separation. *Polymer Testing* 67: 218–227.

94 Cao, K., Liu, Y., Yang, Y. et al. (2020). The preparation and characterization of a heterocyclic meta-aramid fiber with outstanding thermal stability. *High Performance Polymers* 33 (5): 554–567.

95 Wang, C., Su, D., Xie, Z. et al. (2021). Low-velocity impact response of 3D woven hybrid epoxy composites with carbon and heterocyclic aramid fibres. *Polymer Testing* 101: 107314.

96 Picken, S.J., Sikkema, D.J., Boerstoel, H. et al. (2011). Liquid crystal main-chain polymers for high-performance fibre applications. *Liquid Crystals* 38 (11–12): 1591–1605.

97 Teng, C., Li, H., Liu, J. et al. (2020). Effect of high molecular weight PPTA on liquid crystalline phase and spinning process of aramid fibers. *Polymers* 12 (5): 1206.

98 Li, T., Mao, Z., Du, J. et al. (2023). Shear-governed microstructural variation and evolution of PPTA in dry-jet-wet spinning process. *International Journal of Mechanical Sciences* 240: 107950.

99 Chaofeng, C., Qinglan, X., and Xinnian, F. (2022). Modification of heterocyclic para-aramid fibers with chlorine groups and the interfacial properties of the fiber/epoxy composites. *Journal of Macromolecular Science, Part B* 61 (9): 1156–1165.

100 Zhang, S.F. and Kang, C.L. (2013). Crystal structure analysis on aramid fiber/fibrids and paper by polarized light microscopy. *Key Engineering Materials* 531–532: 636–639.

101 Zavadskii, A.E., Zakharova, I.M., and Zhukova, Z.N. (1998). Features of the fine structure of aramid fibres. *Fibre Chemistry* 30 (1): 6–10.

102 Xie, F., Xing, L., Liu, L. et al. (2018). A facile method to prepare nanoscale polyacrylonitrile particles grafted aramid fibers for superior interfacial and mechanical properties of epoxy composites. *Polymer Composites* 39 (S4): E2436–E2444.

103 Mikheyev, P.V. and Berlin, A.A. (2019). Effect of splitting of polymer fibers on the strength of unidirectional composites. *Mechanics of Composite Materials* 55 (2): 267–274.

104 Cornelissen, B., Rietman, B., and Akkerman, R. (2013). Frictional behaviour of high-performance fibrous tows: friction experiments. *Composites Part A: Applied Science and Manufacturing* 44: 95–104.

105 Liu, S., Wu, D., Zhao, J. et al. (2022). The material removal mechanism in orthogonal cutting of woven AFRP. *Journal of Materials Science* 57 (34): 16301–16316.

106 Huang, S., Lin, P., Huang, H. et al. (2020). Tailored polyphenylene sulfite composite with desirable mechanical performance and low dielectric constant by constructing a controllable aramid fiber network. *Composites Part B: Engineering* 201: 108334.

107 Huang, S., Lin, P., Yao, S. et al. (2021). High-performance para-aramid paper strengthened by ultrafine fiber pulp of polyphenylene sulfide. *Composites Science and Technology* 216: 109073.

108 Yin, F., Tang, C., Wang, Q. et al. (2018). Molecular dynamics simulations on the thermal decomposition of meta-aramid fibers. *Polymers* 10 (7): 691.

109 Kuroki, T., Tanaka, Y., Hokudoh, T., and Yabuki, K. (1997). Heat resistance properties of poly(*p*-phenylene-2,6-benzobisoxazole) fiber. *Journal of Applied Polymer Science* 65 (5): 1031–1036.

110 Maity, J., Jacob, C., Das, C.K. et al. (2007). Fluorinated aramid fiber reinforced polypropylene composites and their characterization. *Polymer Composites* 28 (4): 462–469.

111 Xu, K., Ou, Y., Li, Y. et al. (2020). Preparation of robust aramid composite papers exhibiting water resistance by partial dissolution/regeneration welding. *Materials & Design* 187: 108404.

112 Wang, X., Li, P., Xiang, D. et al. (2022). Influence of high-temperature, high-pressure, and acidic conditions on the structure and properties of high-performance organic fibers. *Materials Testing* 64 (5): 623–635.

113 Jia, X., Li, G., Yu, Y. et al. (2009). Ablation and thermal properties of ethylene–propylene–diene elastomer composites reinforced with polysulfonamide short fibers. *Journal of Applied Polymer Science* 113 (1): 283–289.

114 Wu, K., Wang, X., Xu, Y., and Guo, W. (2020). Flame retardant efficiency of modified para-aramid fiber synergizing with ammonium polyphosphate on PP/EPDM. *Polymer Degradation and Stability* 172: 109065.

115 Chen, F., Zhai, L., Yang, H. et al. (2021). Unparalleled armour for aramid fiber with excellent UV resistance in extreme environment. *Advanced Science* 8 (12): 2004171.

116 Cheng, Z., Hong, D., Dai, Y. et al. (2018). Highly improved UV resistance and composite interfacial properties of aramid fiber via iron(III) coordination. *Applied Surface Science* 434: 473–480.

117 Gilman, A.B., Piskarev, M.S., Kuznetsov, A.A., and Ozerin, A.N. (2017). Modification of ultrahigh-molecular-weight polyethylene by low-temperature plasma (review). *High Energy Chemistry* 51 (2): 136–144.

118 Gogoleva, O.V., Petrova, P.N., Popov, S.N., and Okhlopkova, A.A. (2015). Wear-resistant composite materials based on ultrahigh molecular weight polyethylene and basalt fibers. *Journal of Friction and Wear* 36 (4): 301–305.

119 Du, J., Wang, Z., Yu, J. et al. (2018). Ultrahigh-strength ultrahigh molecular weight polyethylene (UHMWPE)-based fiber electrode for high performance flexible supercapacitors. *Advanced Functional Materials* 28 (20): 1707351.

120 Sokolov, A.V., Korolev, R.V., Murikhin, K.S. et al. (2020). Destruction products from gel spinning of ultrahigh-molecular-weight polyethylene fibers. *Fibre Chemistry* 51 (6): 409–411.

121 Zhang, Y., Yu, J., Chen, L. et al. (2009). Surface modification of ultrahigh-molecular-weight polyethylene fibers with coupling agent during extraction process. *Journal of Macromolecular Science, Part B* 48 (2): 391–404.

122 Zhang, X., Wang, Y., and Cheng, S. (2013). Properties of UHMWPE fiber-reinforced composites. *Polymer Bulletin* 70 (3): 821–835.

123 Wilding, M.A. and Ward, I.M. (1978). Tensile creep and recovery in ultra-high modulus linear polyethylenes. *Polymer* 19 (8): 969–976.

124 Hallam, M.A., Cansfield, D.L.M., Ward, I.M., and Pollard, G. (1986). A study of the effect of molecular weight on the tensile strength of ultra-high modulus polyethylenes. *Journal of Materials Science* 21 (12): 4199–4205.

125 Yuan, H., Long, C., Yu, J. et al. (2022). Synthesis of granular hydroxy-functionalized ultra-high-molecular-weight polyethylene and its fiber properties. *Advanced Fiber Materials* 4 (4): 786–794.

126 Anindyajati, A., Boughton, P., and Ruys, A.J. (2019). Mechanical and cyto-compatibility evaluation of UHMWPE/PCL/Bioglass® fibrous composite for acetabular labrum implant. *Materials* 12 (6): 916.

127 Joo, Y.L., Zhou, H., Lee, S.G. et al. (2005). Solid-state compaction and drawing of nascent reactor powders of ultra-high-molecular-weight polyethylene. *Journal of Applied Polymer Science* 98 (2): 718–730.

128 Leal, A.A., Veeramachaneni, J.C., Reifler, F.A. et al. (2016). Novel approach for the development of ultra-light, fully-thermoplastic composites. *Materials & Design* 93: 334–342.

129 Hong, J., Pan, Z., Tian, L. et al. (2015). Continuous fabrication of conductive UHMWPE yarns based on in-situ polymerization with different doping acids. *Synthetic Metals* 209: 512–520.

130 Li, W., Guan, C., Xu, J. et al. (2014). Disentangled UHMWPE/POSS nanocomposites prepared by ethylene in situ polymerization. *Polymer* 55 (7): 1792–1798.

131 Zhou, J., Zhang, X., Zhao, S. et al. (2022). Influence of modified Ziegler–Natta catalyst on the entanglement behavior and properties of ultrahigh-molecular-weight polyethylene (UHMWPE). *Industrial & Engineering Chemistry Research* 61 (48): 17512–17523.

132 Jamjah, R., Zohuri, G.H., Javaheri, M. et al. (2008). Synthesizing UHMWPE using Ziegler-Natta catalyst system of $MgCl_2$(ethoxide type)/$TiCl_4$/tri-isobutylaluminum. *Macromolecular Symposia* 274 (1): 148–153.

133 Zhang, Q., Wang, Q., and Chen, Y. (2013). Structure evolution of ultra high molecular weight polyethylene/montmorillonite nanocomposite fibers prepared by melt spinning. *Journal of Applied Polymer Science* 130 (6): 3930–3936.

134 Kurtz, S.M., Muratoglu, O.K., Evans, M., and Edidin, A.A. (1999). Advances in the processing, sterilization, and crosslinking of ultra-high molecular weight polyethylene for total joint arthroplasty. *Biomaterials* 20 (18): 1659–1688.

135 Penning, J.P., Vries, A.A., and Pennings, A.J. (1993). The effect of fibre diameter on the drawing behaviour of gel-spun ultra-high molecular weight polyethylene fibres. *Polymer Bulletin* 31 (2): 243–248.

136 Bajpai, P., Kumar, A., Kumar, A., and Dwivedi, P.K. (2023). Optimization of surface treatments for improving adhesion of UHMWPE nonwoven composite with dissimilar materials. *Fibers and Polymers* 24 (4): 1431–1440.

137 Fallah, A.S., Micallef, K., Langdon, G.S. et al. (2014). Dynamic response of Dyneema® HB26 plates to localised blast loading. *International Journal of Impact Engineering* 73: 91–100.

138 Luz, F.S., Garcia, F.D., Oliveira, M.S. et al. (2020). Composites with natural fibers and conventional materials applied in a hard armor: a comparison. *Polymers* 12 (9): 1920.

139 On, S.Y., Park, S., and Kim, S.S. (2019). Preparation and characterization of hybrid structured MWCNT/UHMWPE fiber sensors for strain sensing and load bearing of composite structures. *Advanced Materials Technologies* 4 (12): 1900807.

140 Boiko, Y.M., Marikhin, V.A., Myasnikova, L.P. et al. (2016). Statistical analysis of the strength of ultra-oriented ultra-high-molecular-weight polyethylene film filaments in the framework of the Weibull model. *Physics of the Solid State* 58 (10): 2141–2144.

141 Staniszewski, J.M., Boyd, S.E., and Bogetti, T.A. (2022). A multi-scale modeling approach for UHMWPE composite laminates with application to low-velocity impact loading. *International Journal of Impact Engineering* 159: 104031.

142 Pakhomov, P.M., Sokolov, A.V., Khizhnyak, S.D., and Galitsyn, V.P. (2021). Influence of porosity on the mechanical characteristics of UHMWPE fibers obtained by the gel spinning molding method. *Fibre Chemistry* 52 (5): 356–361.

143 Chen, Y., Zou, H., Cao, Y., and Liang, M. (2014). Melt miscibility of HDPE/UHMWPE, LDPE/UHMWPE, and LLDPE/UHMWPE blends detected by dynamic rheometer. *Polymer Science, Series A* 56 (5): 630–639.

144 Breuer, O., Tzur, A., Narkis, M., and Siegmann, A. (1999). HIPS/γ-irradiated UHMWPE/carbon black blends: structuring and enhancement of mechanical properties. *Journal of Applied Polymer Science* 74 (7): 1731–1744.

145 Wolfe, J.F. and Arnold, F.E. (1981). Rigid-rod polymers. 1. Synthesis and thermal properties of para-aromatic polymers with 2,6-benzobisoxazole units in the main chain. *Macromolecules* 14 (4): 909–915.

146 Hu, Z., Li, J., Tang, P. et al. (2012). One-pot preparation and continuous spinning of carbon nanotube/poly(*p*-phenylene benzobisoxazole) copolymer fibers. *Journal of Materials Chemistry* 22 (37): 19863–19871.

147 Hu, Z., Li, N., Li, J. et al. (2015). Facile preparation of poly(*p*-phenylene benzobisoxazole)/graphene composite films via one-pot in situ polymerization. *Polymer* 71: 8–14.

148 Zhang, T., Hu, D., Jin, J. et al. (2010). XPS study of PBO fiber surface modified by incorporation of hydroxyl polar groups in main chains. *Applied Surface Science* 256 (7): 2073–2075.

149 Alexandridis, P., Olsson, U., and Lindman, B. (1997). Structural polymorphism of amphiphilic copolymers: six lyotropic liquid crystalline and two solution phases in a poly(oxybutylene)-*b*-poly(oxyethylene)–water–xylene system. *Langmuir* 13 (1): 23–34.

150 Wang, Y., Wang, D., Song, Y. et al. (2017). Investigation of the mechanical properties of the modified poly(*p*-phenylene benzobisoxazole) fibers based on

2-(4-aminophenyl)-1*H*-benzimidazol-5-amine. *High Performance Polymers* 30 (5): 511–518.

151 Matsuo, M., Ishikawa, H., Xi, Y., and Bin, Y. (2007). High modulus and high strength fibers with high electric conductivity prepared by copper electroless plating on the surface of poly(*p*-phenylenebenzobisoxazole) (PBO). *Polymer Journal* 39 (4): 389–396.

152 Padhan, M., Marathe, U., and Bijwe, J. (2022). Exceptional performance of bi-directionally reinforced composite of PEEK manufactured by commingling technique using poly(*p*-phenylene-benzobisoxazole) (PBO) fibers. *Composites Science and Technology* 218: 109125.

5

Interfaces of Polymer Matrix Wave-Transparent Composites

5.1 Basic Concept of Interfaces

It is well known that there are three main factors affecting the properties of the polymer matrix wave-transparent composites: the properties of the polymer matrix, the properties of the reinforced fibers, and the interfacial bond between the polymer matrix and reinforced fibers [1–3]. The interface is a microscopic region connecting the interaction between polymer matrix and reinforced fibers with a certain thickness (nanometers) [4, 5]. Meanwhile, the physical and chemical properties of interfaces are significantly different from those of the polymer matrix and reinforced fibers [6–8]. Figure 5.1 shows the schematic diagram of the interfaces between polymer matrix and reinforced fibers in the polymer matrix wave-transparent composites.

As an important component of polymer matrix wave-transparent composites, the interface plays a vital role in the final comprehensive properties [9]. The interface function in polymer matrix wave-transparent composites mainly includes the following two aspects. First, the interface can effectively transfer the external load and give full play to the excellent mechanical properties of the reinforced fibers [10, 11]. However, the interface is also the main weakness in polymer matrix wave-transparent composites. When subjected to external stresses, the microscopic damage and cracks tend to develop at the interface, which leads to macroscopic damage of the polymer matrix wave-transparent composites. Therefore, increasing the interfacial bond strength between the polymer matrix and reinforced fibers is beneficial to limit the generation and propagation of cracks and further improve the mechanical properties of the polymer matrix wave-transparent composites [12–14]. In addition, the heat resistance and water absorption properties of the polymer matrix wave-transparent composites are closely related to their interfacial interaction. Second, the interface in the polymer matrix wave-transparent composites is susceptible to interfacial polarization [15]. Weak interfacial interaction is not conducive to charging migration at the interface to increased interfacial polarization, resulting in a decrease in the wave-transparent performances of the composites [16–18].

Polymer Matrix Wave-Transparent Composites: Materials, Properties, and Applications, First Edition.
Junwei Gu, Yusheng Tang, Jie Kong, and Jing Dang.

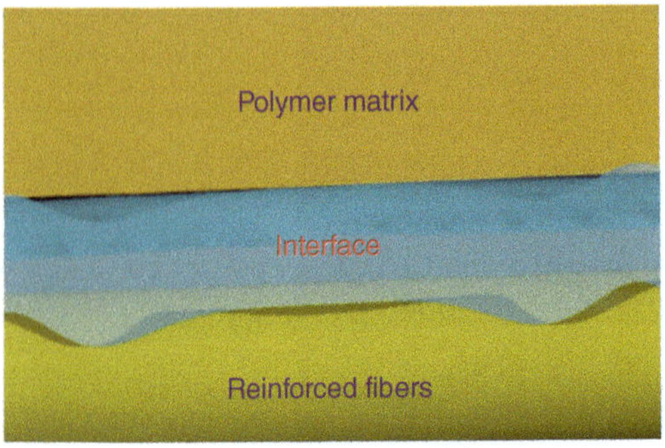

Figure 5.1 Schematic diagram of the interfaces for polymer matrix wave-transparent composites.

5.2 Formation of Interfaces

The composition, structures, and morphologies of the interfaces for polymer matrix wave-transparent composites are different from those of the polymer matrix and reinforced fibers [19, 20]. When the polymer matrix and reinforced fibers are in contact with each other, a chemical or physical interaction can form between polymer matrix and the sizing agents or chemical groups on the surface of the reinforced fibers, resulting in an interface phase different from polymer matrix and reinforced fibers [21–23]. Meanwhile, due to internal stresses generated during the curing of the polymer matrix or changes in the physical and chemical structures and stacking density of the polymer matrix close to reinforced fibers, the interface of the wave-transparent composites appears to have different characteristics of the polymer matrix and reinforced fibers [24–26].

The formation of the interfaces for polymer matrix wave-transparent composites can be divided into two stages [27]. The first stage is the contact and infiltration process between polymer matrix and reinforced fibers. Polymer matrix near the interfaces exhibits some structural differences due to the different adsorption abilities of the reinforced fibers to various groups and molecular chains of the polymer molecules [28, 29]. The second stage is the curing stage of the polymer matrix. The polymer in the interfaces is chemically cross-linked to form the fixed interfacial layer [30]. In the preparation of the polymer matrix wave-transparent composites, the ability of the polymer matrix to infiltrate reinforced fibers can be expressed in terms of contact angle (θ). When $\theta < 90°$, the polymer matrix can infiltrate reinforced fibers; when $\theta > 90°$, polymer matrix cannot infiltrate reinforced fibers; when $\theta = 0°$ or $180°$, it is completely infiltrated or not infiltrated at all [31, 32]. In general, surface functionalization of the reinforced fibers or the introduction of interfacial compatibilizers with specific structures in the polymer matrix wave-transparent composites can effectively improve the wetting of the polymer matrix and reinforced fibers [33].

5.3 Interfacial Interaction Mechanism of the Polymer Matrix Wave-Transparent Composites

In order to dissect the interfacial interaction mechanism of the polymer matrix wave-transparent composites, researchers have proposed a variety of interfacial interaction theories [34–36].

5.3.1 Mechanical Bonding Theory

Mechanical bonding theory is one of the most intuitive macroscopic theories of the polymer matrix wave-transparent composite interfaces [37–39]. For mechanical bonding theory, polymer matrix and reinforced fiber surfaces are mechanically engaged after the curing stage due to the roughness of the reinforced fiber surface and the capillary effect of fiber bundle, which can be favor of enhancing good interfacial bond strength between polymer matrix and reinforced fibers [40–42].

5.3.2 Adsorption Theory

For adsorption theory, the adhesive force at the interfaces of the polymer matrix wave-transparent composites is mainly generated by the mutual adsorption between molecules of the polymer matrix and reinforced fibers [43–45]. The interaction process between polymer matrix molecules and reinforced fiber surface molecules has two stages [46]. The first stage is the diffusion of the polymer matrix molecules toward the surface of the reinforced fibers using the Brownian motion and the mutual proximity of polar groups or chain links of the reinforced fibers [47, 48]. In this stage, the factors, such as elevating the temperature, applying contact pressure, and reducing the viscosity of the polymer matrix, are all conducive to enhancing the Brownian motion [49]. The second stage is the generation of interfacial adsorption, such as the Van der Waals forces, the hydrogen bonding interactions, or the chemical reaction between polymer matrix and reinforced fibers, to eventually form a stable interfacial layer [50–52].

5.3.3 Diffusion Theory

For diffusion theory, the bonding of the polymer matrix and reinforced fibers is caused by the mutual diffusion of molecules on the surfaces/interfaces, forming a stable interfacial layer [53–55]. The diffusion process is related to the molecular weight, flexibility, temperature of the polymer matrix molecular chain, and whether the compound system contains solvents or plasticizers [56, 57]. In general, the diffusion at the interface of the polymer matrix wave-transparent composites is similar to the process of substance dissolution, which is a "mixing" process. Especially when the solubility parameters of the polymer matrix and reinforced fibers are close to each other, the intermixing and diffusion can easily occur at their interfaces, which is beneficial to improving their interfacial bond strength [58–60].

5.3.4 Acid-Base Interaction Theory

Any substance that can accept foreign electron pairs (electron acceptor) is a generalized Lewis acid, and any substance that can provide electron pairs (electron donor) is a generalized Lewis base [61]. For acid-base interaction theory, there are a large number of electron acceptors and electron donors at the interfaces between polymer matrix and reinforced fibers [62, 63]. Interfacial bonding is formed through interaction between Lewis's acid–base pairs. The bonding energy (W_a) of the more polar reinforced fibers is composed of the ubiquitous dispersion adhesion work (W_{ad}^{d}) and the interaction adhesion work between Lewis's acid-base pairs (W_{ad}^{ab}) Eq. (5.1).

$$W_a = W_{ad}^{d} + W_{ad}^{ab} \tag{5.1}$$

If the acid–base interaction produces electron exchange, the covalent bond is formed, and the interface bonding strength is relatively large, usually stronger than the interface formed by the secondary valence bond forces, such as the hydrogen bonds, the dispersion forces, the dipoles, and the inductive forces.

5.3.5 Chemical Bonding Theory

Chemical bonding theory is one of the most widely used theory of interface formation [64]. When polymer matrix and reinforced fibers come into contact, a chemical reaction occurs between the functional groups on the surface of the reinforced fibers and polymer matrix to form a chemical bonding, resulting in an effective interfacial bond [65]. However, the chemical bonding strength depends on the number of reactive groups on surface of the reinforced fibers. Moreover, the contribution of chemical bonding between polymer matrix and the inert reinforced fibers to the interfacial adhesion is low [66].

However, each of the above theories has some basis, in fact, but also contradicts some experimental results [67]. Thus, researchers must select the most suitable interfacial interaction theory for polymer matrix wave-transparent composites according to the specific situation.

5.4 Characterization of Interfacial Performances

Investigating the physical and chemical properties, thickness, morphologies, and interfacial bond strength of interfaces and clarifying the interfacial mechanism for polymer matrix wave-transparent composites can further understand the interrelationship between the physical and chemical properties of interfaces and macroscopic properties of the polymer matrix wave-transparent composites [68–70]. The components, structures, and valence of the interfaces for polymer matrix wave-transparent

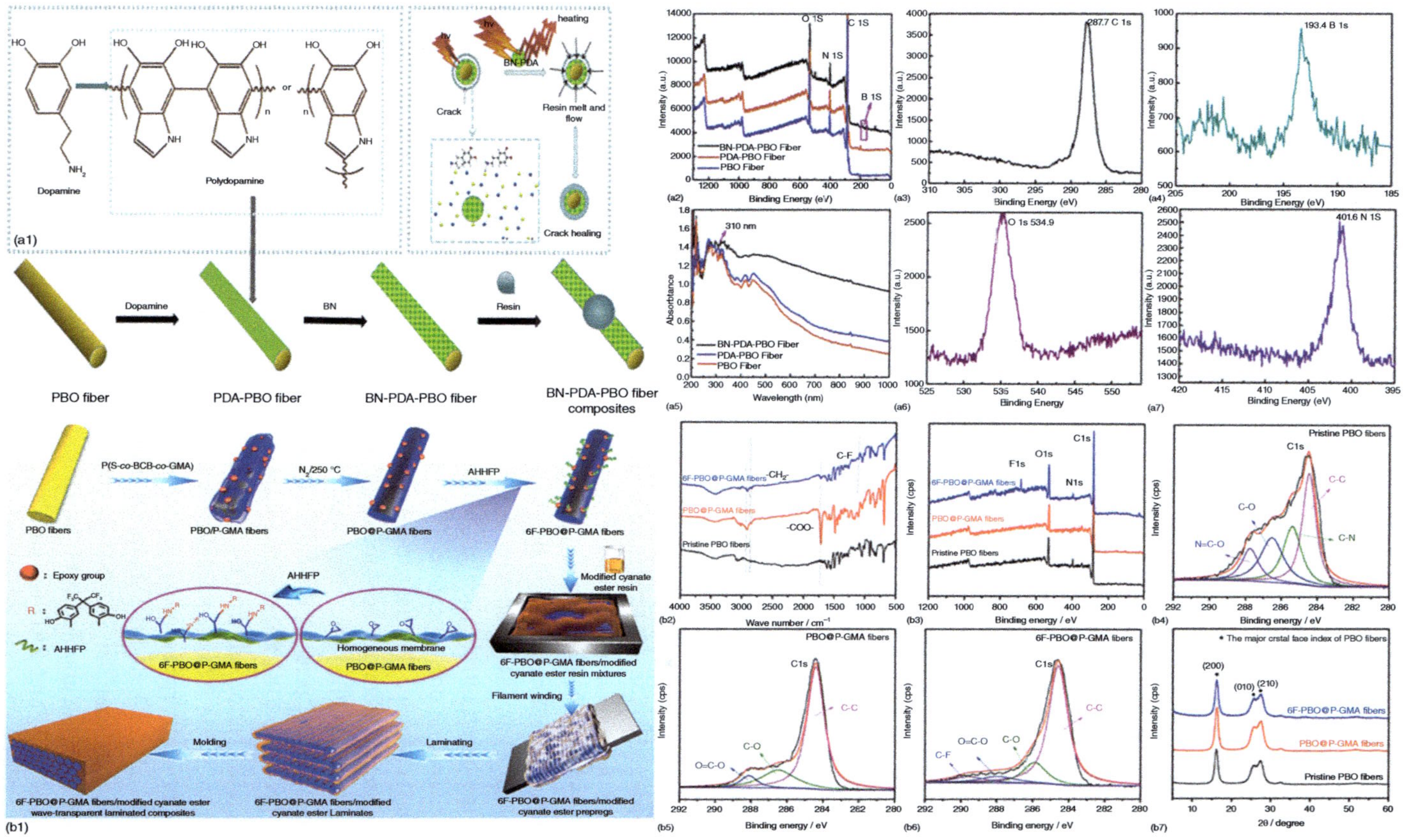

Figure 5.2 Schematic diagram and XPS spectra of the BN–PDA–PBO fibers (a) [84]; Schematic diagram of the preparation for 6F-PBO@P-GMA fibers, and their FT-IR, XPS and XRD spectra. Source: [85]/with permission of Elsevier.

composites can be characterized by many advanced testing methods [71]. Moreover, the interfacial bonding strength is also an important parameter to characterize the interfacial performances, which can be analyzed by the single fiber breakage tests, the single fiber pull-out tests, the fibers indentation/ejection tests, and the nanoindentation tests [72–74].

5.4.1 Characterization of Chemical Performances of Interfaces

Characterization of chemical performances at interfaces of the polymer matrix wave-transparent composites mainly refers to the surface chemical elements, surface functional groups, and surface energy of the reinforced fibers [75]. The surface chemistry performance of the reinforced fibers has a great influence on the interfacial performances of the polymer matrix wave-transparent composites [76]. In the surface reactive groups, surface energy of the reinforced fibers determines the wettability with polymer matrix, which can influence their interfacial bond strength [77].

5.4.1.1 Elemental Composition and Functional Groups on the Surface of the Reinforced Fibers

Currently, the main methods for characterizing the elemental compositions and functional groups on the surface of the reinforced fibers are Fourier transform infrared spectroscopy (FT-IR), X-ray photoelectron spectroscopy (XPS), and X-ray diffraction (XRD) [78–80]. Among them, FT-IR can characterize the change of chemical groups on the surface of the reinforced fibers before and after surface functionalization [81]. XPS can characterize the elemental compositions and bonding method on the surface of the reinforced fibers [82]. XRD can characterize whether the crystal structure of the reinforced fibers is changed before and after surface functionalization [83].

Shao et al. [84] used XPS to characterize the surface chemical performances of the boron nitride (BN) grafted poly (*p*-phenylene-2, 6-benzobisoxazole) fibers after coating polydopamine (PDA) to obtain BN–PDA–PBO fibers (Figure 5.2a). Results revealed that the surface of pristine PBO fibers was composed of C, O, and N elements. After surface functionalization, the content of C element on the surface of BN–PDA–PBO fibers increased, the content of O element decreased, and the characteristic peak of B element appeared, indicating the successful preparation of BN–PDA–PBO fibers. Chen et al. [86] used FT-IR to characterize the surface chemical performances of the aramid fibers coated with silica (SiO_2)/shape memory polyurethane (SMPU). Results showed that the modified aramid fibers showed the new absorption peaks of characteristic peaks of SMPU at 2958, 1729, 1610, and 1222 cm^{-1}, and the sharp absorption peak of Si–O–Si appeared at 1067 cm^{-1}, indicating that the SiO_2/SMPU coatings were successfully introduced onto the surface of the aramid fibers. In our previous works, Gu and coworkers [85] used FT-IR, XPS, and XRD to characterize the surface chemical performances of the 6F-PBO@P-GMA fibers after coated P(S-*co*-BCB-*co*-GMA) copolymer. Then, they grafted with 2,2-bis(3-amino-4-hydroxyphenyl)hexafluoro propane (AHHFP)

(Figure 5.2b). Compared with pristine PBO fibers, the appearance of new peaks (FT-IR) of 6F-PBO@P-GMA fibers at 1760, 2983/2850, and 1115 cm^{-1} could contribute to the $-CH_2-$ and –COO– in P(S-*co*-BCB-*co*-GMA) and C–F bonding in AHHFP, respectively. XPS curve of the 6F-PBO@P-GMA fibers also appeared at the F1s peak and C–F peak at 290.1 eV, demonstrating the successful grafting of AHHFP on the surface of the PBO fibers.

5.4.2 Surface Free Energy of the Reinforced Fibers

The surface free energy can visually reflect the surface activity of the reinforced fibers [87, 88]. Currently, the main methods for testing the surface free energy of the reinforced fibers include the contact angle method and the inverse gas chromatography (IGC) method [89–91].

Contact angle method is mainly based on the Wilhelmy principle, and the schematic diagram of testing principle for contact angle method can be seen in Figure 5.3 [92]. The single fiber is straightened and fixed to the parallel jig, keeping a certain distance between each single fiber. After that, the liquid pool is slowly raised until the reinforced fibers are in contact with the liquid, and the contact angle between the reinforced fibers and the testing liquid can be measured. The main factors affecting the testing results of the contact angle include the number of single fibers, the detection limit of testing instruments, the fibers length, the depth of fibers insertion into the liquid, the surface infiltration rate, and the infiltration rate of forward angle [93].

The surface energy of the reinforced fibers is calculated using the Wilhelmy method with the following equation [94]:

$$\gamma_1(1+\cos\theta) = 2\sqrt{\gamma_s^p\gamma_1^p} + 2\sqrt{\gamma_s^d\gamma_1^d} \tag{5.2}$$

$$\gamma_s = \gamma_s^d + \gamma_s^p \tag{5.3}$$

where, γ_1 is the surface energy of the testing liquid, mJ m^{-2}; γ_1^p and γ_1^d are respectively the polar and dispersive components of the surface energy of the testing liquid, mJ m^{-2}; γ_s is the surface energy of the fibers, mJ m^{-2}; γ_s^p and γ_s^d are respectively the polar and dispersive components of the surface energy of the fibers, mJ m^{-2}.

IGC method uses reinforced fibers as the stationary phase, a low molecular solvent with the known surface energy as the probe molecule [95]. Then, the vaporized

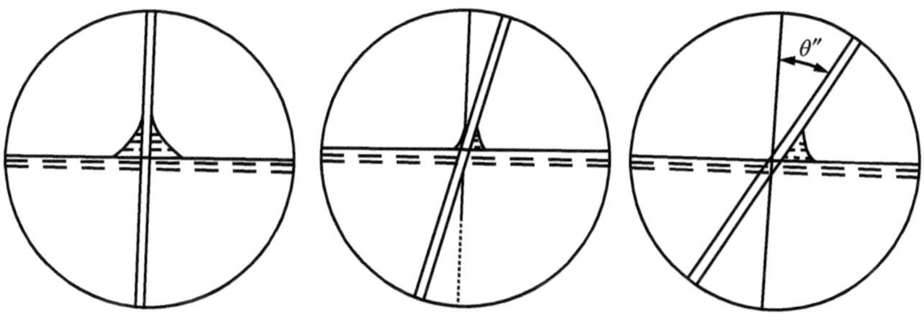

Figure 5.3 Schematic diagram of the test principle for contact angle method.

probe molecules are carried into the chromatographic columns with a carrier gas under certain conditions to investigate the surface performances of the reinforced fibers by testing the retention time of probe molecules through the chromatographic columns [96]. IGC method usually uses the dilute solvent probes and the net retention volume of solvent molecules (V_R^0) is calculated by testing the retention time of the low-molecular solvent probe through the column. The V_R^0 value is related to the force between solvent molecule and reinforced fibers and the relationship between their surface adsorption free energy (ΔG^0) is shown in Eq. (5.4) [97]:

$$\Delta G^0 = RT\ln V_R^0 + K \tag{5.4}$$

where, R is the ideal gas constant; K is related to the mass of the analyte in the column, the surface area of the reinforced fibers, and the adsorption state (the value of K is a constant for the same testing system). In addition, ΔG^0 is related to the surface area of the molecular probe and adhesion energy (W_A), as shown in Eq. (5.5) [97].

$$\Delta G^0 = \alpha N_A W_A \tag{5.5}$$

where, N_A is the Avogadro's constant; α is the surface area of the molecular probe. W_A is dominated by dispersion forces such as intermolecular Van der Waals forces, and the contribution of polar component can be neglected in the nonpolar solvents Eq. (5.6).

$$W_A = 2\sqrt{\gamma_s^d \gamma_l^d} \tag{5.6}$$

where, γ_s^d and γ_l^d are the surface energy of the fibers and probe molecular to further deduce the Eq. (5.7) [90].

$$RT\ln V_R^0 = 2\alpha N_A \sqrt{\gamma_s^d \gamma_l^d} + const \tag{5.7}$$

The slope of $RT\ln V_R^0$ versus $\alpha\left(\gamma_l^d\right)^{1/2}$ line for a series of n-alkanes at certain temperature conditions to obtain γ_s^d values.

The polar surface free energy is calculated using the Good-van Oss method by testing the adsorption free energy (ΔG^{sp}) of the two polar probe molecules (toluene and methylene chloride) to reinforced fibers Eq. (5.8) [98].

$$\Delta G^{sp} = 2\alpha N_A \left(\sqrt{\gamma_l^+ \gamma_s^-} + \sqrt{\gamma_l^- \gamma_s^+}\right) \tag{5.8}$$

where, γ_s^+ and γ_s^- are the contribution of electronic-feeding and electronic-accepting abilities of the reinforced fibers; γ_l^+ and γ_l^- are the contribution of electronic-feeding and electronic-accepting abilities of the probe molecules. The polar component γ_s^p of surface energy is calculated by acidic probe dichloromethane and basic probe toluene Eq. (5.9) [99].

$$\gamma_s^p = \sqrt{\gamma_s^+ \gamma_s^-} \tag{5.9}$$

The contact angle method and IGC method have different testing principles, resulting in some differences in the test data. Still, the overall trend of the above two methods for testing results is the same [100]. In comparison, the requirements of test condition of the contact angle method are more stringent, and the environmental factors or irregularities in the experimental operation during the measurement process are likely to cause the test errors. IGC method has advantages of high automation, simple testing process, and good data reproducibility, but its requirements for samples and equipment are relatively higher.

Currently, the surface tension and surface energy of the reinforced fibers are still usually tested by the contact angle method [101, 102]. Deng et al. [92] used the contact angle method to characterize the surface energy of the glass fibers and basalt fibers before and after removing sizing agent. Results showed that the surface energy of the glass fibers and basalt fibers was significantly reduced after removing the sizing agent. Song et al. [103] used the contact angle method (the equal altitude method) to characterize the surface energy of the UHMWPE fibers before and after the atmospheric pressure plasma treatment. Results revealed that the contact angle between modified UHMWPE fibers and water was 68.15°, significantly lower than that of 99.41° for pristine UHMWPE fibers.

In our previous work, Liu et al. [104] used the contact angle method to characterize the surface energy of the PBO fibers coated with poly(dopamine) hybrid zeolite imidazole framework (PBO@PDA*ZIF-8, Figure 5.4). Compared with pristine PBO fibers, the contact angles of the PBO@PDA*ZIF-8 fibers with water and glycol were respectively reduced by 39.2% and 57.6%, and the corresponding surface free energy

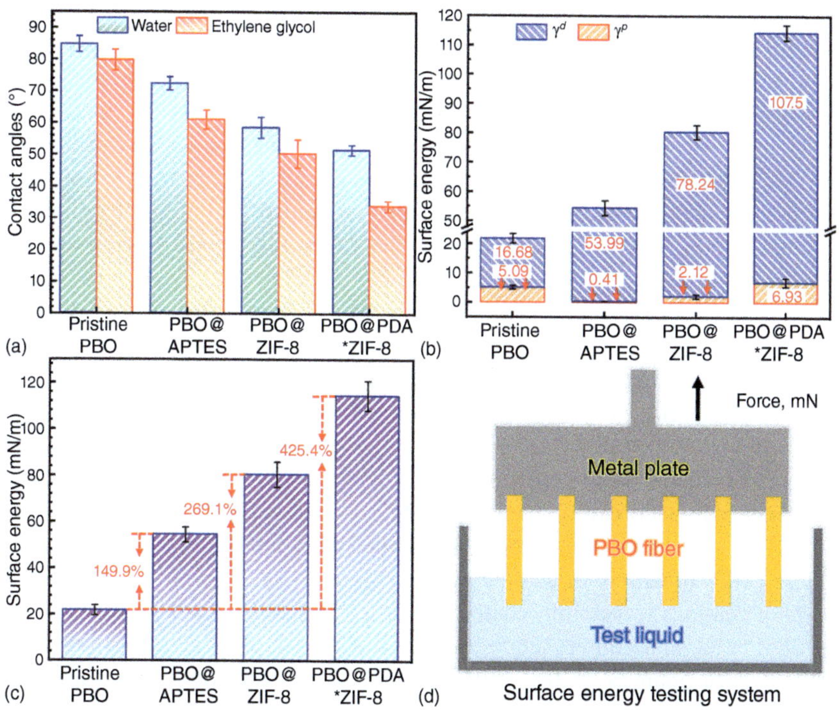

Figure 5.4 Dynamic contact angle and surface free energy of the PBO@PDA*ZIF-8 fibers. Source: [104]/with permission of Elsevier.

was increased by 425.4%. It could be mainly attributed that the large number of reactive groups (–OH, –NH– and –NH_2) on the surface of PBO@PDA*ZIF-8 fibers could enhance the interaction with the testing solvent in favor of increasing the surface free energy.

Moreover, some researchers replaced the testing liquid with the molten polymer matrix to further characterize the infiltration effect between polymer matrix and reinforced fibers by testing the contact angle at higher temperatures [105–107]. Zhu and Ma [108] used the high-temperature contact angle method to characterize the contact angle between basalt fibers (BF) modified by lanthanum ion (La^{3+}) – diethylenetetraacetic acid coating and bisphenol A dicyanate ester (BADCy) resin (Figure 5.5).

In general, the more active functional groups on the surface of the reinforced fibers are the smaller contact angle and the higher surface energy are, which are more favorable for the infiltration of the polymer matrix on the surface of the reinforced fibers, resulting in enhancing interfacial compatibility of the polymer matrix wave-transparent composites. Besides, the reactive functional groups on the surface of the reinforced fibers can also form hydrogen bonds or enhance intermolecular interaction with polymer matrix, which is also beneficial to enhancing the interfacial bonding of the polymer matrix wave-transparent composites [109, 110].

5.4.3 Characterization of Physical Performances of Interfaces

Physical performances of reinforced fiber surfaces also affect the interface of the polymer matrix wave-transparent composites [111, 112]. For example, the greater the surface roughness of the reinforced fibers is, the stronger the mechanical engagement between polymer matrix and reinforced fibers is, which can increase the interfacial interaction of the polymer matrix wave-transparent composites. The physical performances of interfaces include the surface morphology, surface roughness of the reinforced fibers, and the thickness of interfacial layer in the final composites [113–115].

5.4.3.1 Surface Morphology and Roughness of the Reinforced Fibers

At present, the surface morphology and roughness of the reinforced fibers are mainly characterized by scanning electron microscopy (SEM) and atomic force microscopy (AFM) [116]. Hua et al. [117] applied SEM to characterize the morphological changes on the surface of glass fibers before and after the coating of graphene oxide (GO)/carbon nanotube (CNT). Compared with the smooth surface of pristine PBO fibers, the surface of modified PBO fibers showed the wrinkled cladding (the thickness of coating layer was about 500 μm), indicating that the GO/CNT coatings successfully coated the surface of glass fibers. Jia et al. [118] used the SEM, XPS, and contact angle tests to characterize the changes in interfacial physical performances of the aramid fibers before and after the air-mediated barrier discharge plasma treatment. The surface roughness and surface energy of the modified aramid fibers increased significantly, and the O/C atomic ratio increased from 15.99% to 27.15% compared to pristine aramid fibers. Shao et al.

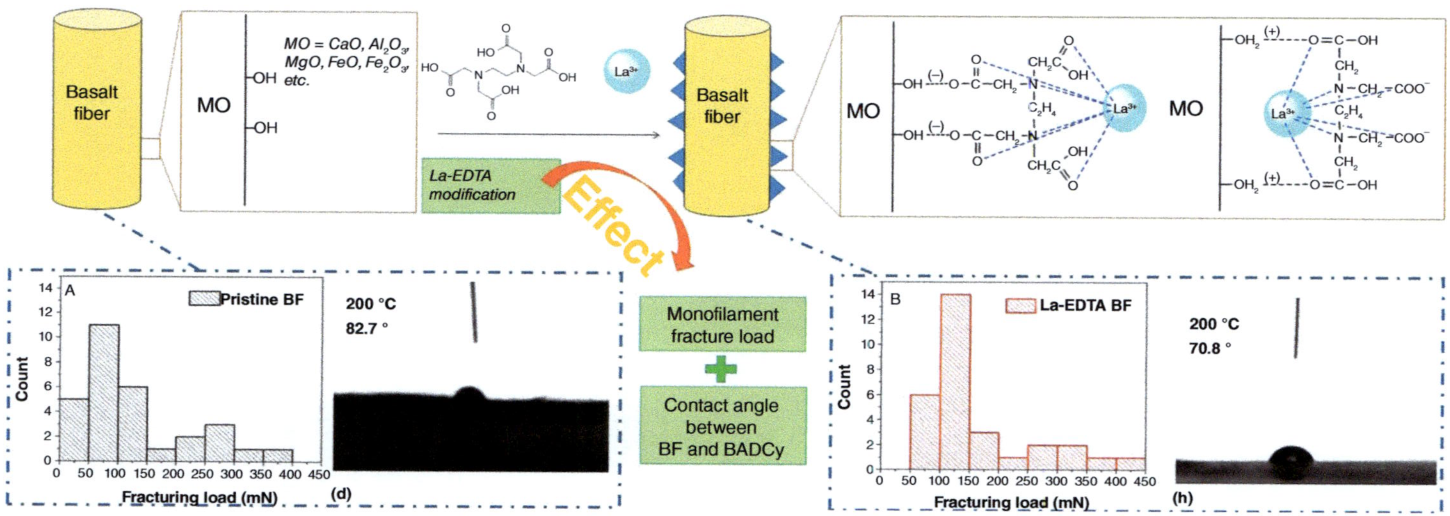

Figure 5.5 High-temperature contact angles and the modification mechanism between basalt fibers and BADCy resins before and after surface functionalization. Source: [108]. Zhu and Ma (2019)/Reproduced with permission from Elsevier.

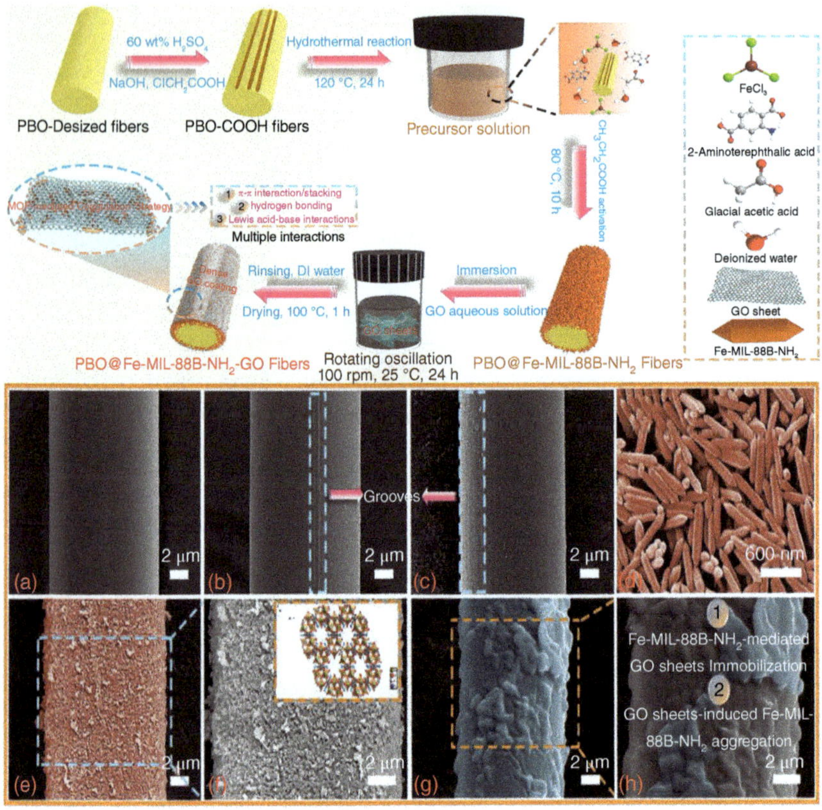

Figure 5.6 Schematic diagram of the preparation for PBO@Fe-MIL-88B-NH_2-GO fibers; SEM images of pristine PBO fibers (a), acidized PBO fibers (b), PBO-COOH fibers (c), Fe-MIL-88B-NH_2 nanocrystals (d), PBO@Fe-MIL-88B-NH_2 fibers (e–f), and PBO@Fe-MIL-88B-NH_2-GO fibers (g–h). Source: [119]. Shao et al. (2021)/Reproduced with permission from Elsevier.

[119] characterized the morphological changes of the PBO fibers surface during the coating of metal–organic framework (MIL-88B-NH_2) nanocrystals and GO by the high-resolution SEM (Figure 5.6). Compared with the smooth surface of pristine PBO fibers, the surface roughness of the modified PBO fibers increased after acid etching, and some grooves appeared on the modified PBO fibers surface. After the hydrothermal growth reaction, Fe-MIL-88B-NH_2 nanocrystals were uniformly grown on the surface of PBO fibers.

In our previous works, Tang et al. [120] characterized the changes in surface morphology and surface roughness of the random copolymer P(S-*co*-BCB-*co*-MMA) coating PBO fibers by SEM and AFM before and after UV etching (UV-PBO@P fibers, Figure 5.7). Pristine PBO fibers have a smooth surface (surface roughness of about 11 nm). After P(S-*co*-BCB-*co*-MMA) coating and UV etching (96 hours), the UV96-PBO@P fiber surface formed many holes extending to the internal PBO fibers surface, and the corresponding surface roughness increased to 46 nm.

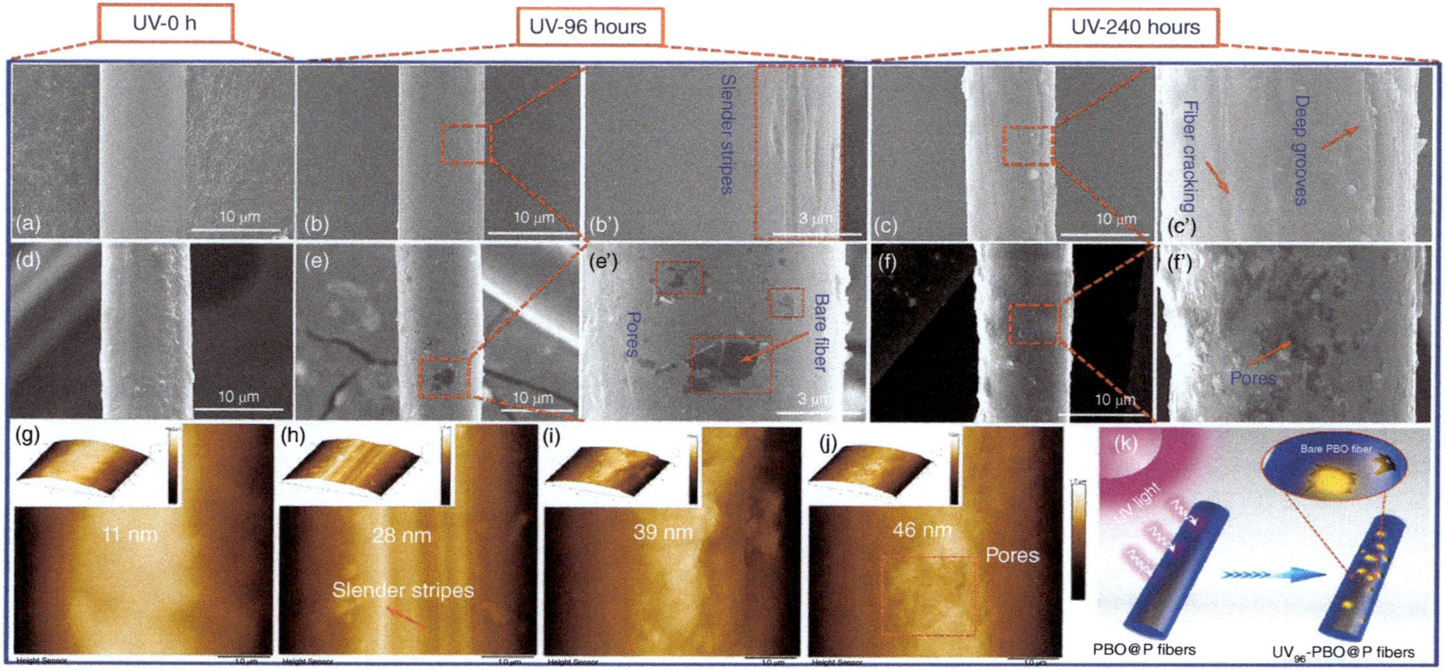

Figure 5.7 SEM images of pristine PBO (a), UV96-PBO (b, b'), UV240-PBO (c), PBO@P (d), UV96-PBO@P (e, e'), and UV240-PBO@P (f) fibers; AFM images of pristine PBO (g), UV96-PBO (h), PBO@P (i), and UV96-PBO@P (j) fibers; schematic diagram of UV etching mechanism for UV96-PBO@P fibers (k). Source: [120]. Tang et al. (2021)/Reproduced with permission from Elsevier.

5.4.3.2 Interface Layer of the Polymer Matrix Wave-Transparent Composites

The interfacial bonding and thickness of interfacial layer between polymer matrix and reinforced fibers in the polymer matrix wave-transparent composites are mainly observed and characterized through the ultrathin sectioning technique and high-resolution transmission electron microscopy (TEM) [121–123]. Anidha et al. [124] used TEM to characterize the interfacial bonding between the sugarcane bagasse/epoxy (BG/E) coating aramid fibers and epoxy resins. The BG/E interlayer appeared at the interface of the aramid fibers/epoxy resin composites, and the density flooding theory (DFT) simulation proved that the BG/E coatings could form the strong π–π interaction with the benzene ring in the aramid fibers, which was a benefit to enhancing the interfacial interaction between aramid fibers and epoxy resins. Chen et al. [124] utilized TEM to characterize the interface between the PBO fibers and epoxy resins after in-situ growth of zinc oxide nanowires (ZnO NWs). The introduction of Zn NMs bridged PBO fibers and epoxy resins matrix, enhancing the thickness of the interfacial layer of the PBO fibers/epoxy resins micro-composites. Meanwhile, the ZnO NWs in the interface were perpendicular to the axial direction of the PBO fibers, possessing a relatively better pinning effect.

Hu et al. [125] characterized the interfacial bonding between PBO fibers before and after surface growth of metal–organic framework (MOF) (UIO-66) nano-crystals (PUFs) and epoxy resins through the ultrathin sectioning technique and high-resolution TEM (Figure 5.8). The interfacial layer between pristine PBO fibers

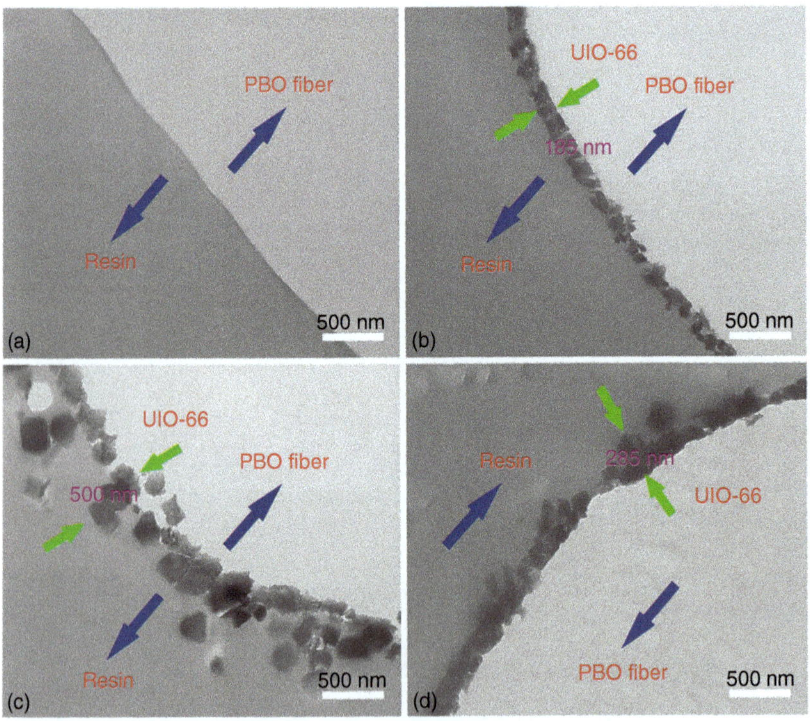

Figure 5.8 TEM images of interfaces between the PBO (a) or PUFs (b–d) fibers and epoxy resins. Source: [125]. Hu et al. (2018)/Reproduced from American Chemical Society.

and epoxy resins was the fuzzy boundary with a narrow thickness. Compared with pristine PBO fibers, an obvious interfacial layer could be observed between PUF fibers and epoxy resins with a thickness between 200 and 500 nm. Simultaneously, a large number of UIO-66 crystals appeared in the interface, which increased the mechanical engagement between PUF fibers and epoxy resins.

5.4.4 Characterization of Interfacial Bonding Strength

The test methods for interfacial bonding strength of the polymer matrix wave-transparent composites mainly include single fiber fracture test, single fiber pull-out test, fiber indentation/ejection test, and nano-indentation method [126–128].

5.4.4.1 Single Fiber Fracture Test

The single fiber fracture test is first applied to investigate the interfacial performances of the metal fibers/metal matrix composites [129, 130]. It is an early method to measure the interfacial bond strength of the polymer matrix wave-transparent composites, and the corresponding schematic test principle can be seen in Figure 5.9 [131]. The basic principle is to straighten a single fiber embedded in polymer matrix and then subsequently stretch the specimen axially along the single fiber to observe one or more breaks of the reinforced fibers [132, 133]. As the tensile strain increases, the reinforced fibers are broken, and the fracture occurs several times with the further increase of the strain. Then, the length of the remaining reinforced fiber gradually becomes shorter until the tensile stress is insufficient to make testing fibers fracture again. This situation is commonly referred to as saturation of the fiber fracture process. The length of the longest section of each fractured fiber is called the critical length L_c. The relationship between the L_c and the average fracture length L is shown in Eq. (5.10) [134, 135].

$$L_c = \frac{4}{3}L \tag{5.10}$$

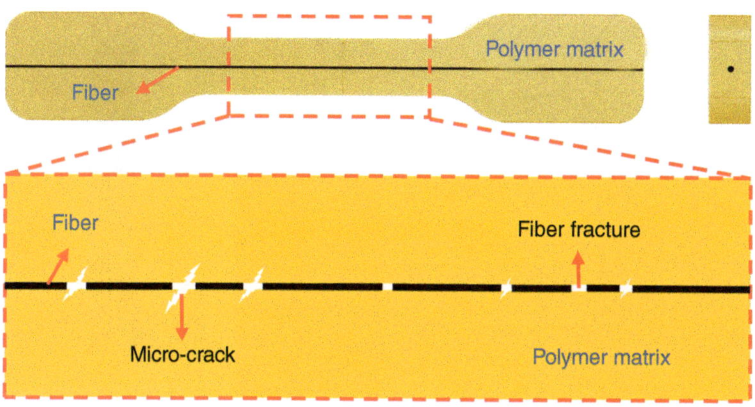

Figure 5.9 Schematic diagram of the single fiber fracture tests.

The critical shear strength (τ) can be further obtained by Eq. (5.11).

$$\tau = {}^{\sigma_f r}/_{2L_c} = \frac{3}{8} \times \frac{\sigma_f r}{L} \tag{5.11}$$

Organic fibers, such as aramid fibers, PBO fibers, and UHMWPE fibers, have extremely high fiber axial tensile strength, which is difficult to break during the stretching process [136]. Therefore, the single fiber fracture test is mostly used to investigate the interfacial bonding strength between polymer matrix and inorganic reinforced fibers.

5.4.4.2 Single Fiber Pull-Out Test

The single fiber pull-out test refers to dropping the polymer matrix liquid at the surface of the reinforced fibers and then blocking or pushing the polymer microspheres to apply a load (Figure 5.10) [137–139]. First, the load of single fiber pull-out test varies linearly up to the maximum value with simultaneously occurring the interfacial debonding between the polymer matrix and reinforced fibers, followed by the sharp drop to frictional load (F_s). Moreover, the maximum load (F) usually increases with the increase of embedding length between the polymer matrix and reinforced fibers. The fracture of the reinforced fiber will occur instead of interface damage when the embedding length reaches a threshold value (this embedding length is called the critical embedding length) [140, 141]. Assuming that the interfacial shear strength is uniformly distributed along the entire interface and the reinforced fibers

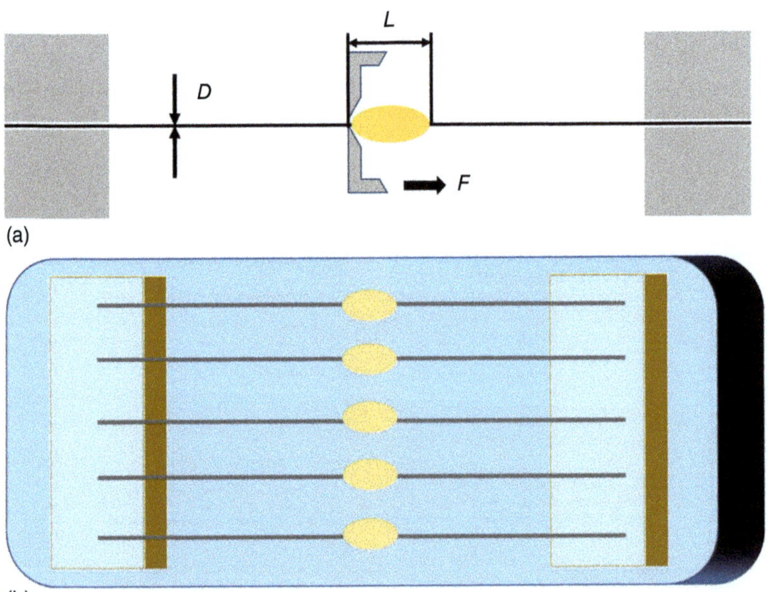

Figure 5.10 Schematic diagram of the single fiber pull-out tests (a) and sample preparation (b).

are cylindrical, the interfacial shear strength (τ) can be calculated by Eq. (5.12) [142].

$$\tau = \frac{F}{\pi DL} \tag{5.12}$$

where, F, D, and L are, respectively, the maximum load, fiber diameter, and encapsulation length.

Compared with the single fiber fracture test, the single fiber pull-out test can measure the instantaneous debonding force between polymer matrix and reinforced fibers, which applies to almost all polymer matrix wave-transparent composites [143, 144]. However, the single fiber pull-out test also has the following constraints. On the one hand, the debonding force is related to the embedding length, causing fracture of the reinforced fibers with the longer embedding length. On the other hand, the variability of parameters will result in a large dispersion of testing results.

The single fiber pull-out test is the most widely used method to characterize the interfacial bond strength of the polymer matrix wave-transparent composites [145–147]. Liu et al. [148] investigated the interfacial bond strength between basalt fibers/glass fibers and epoxy resins by single fiber pull-out test. The interfacial bond strength between basalt fibers and epoxy resins was 20.7 MPa, lower than that of glass fibers/epoxy resins (32.6 MPa), which can be attributed to the smooth surface and chemical inertness of the basalt fibers. Zhang et al. [149] characterized the interfacial bonding performances between nano-SiO_2 coating aramid fibers and epoxy resins through the single fiber pull-out test. Results revealed that the interfacial bonding strength between modified aramid fibers and epoxy resins increased significantly. When the pressure of supercritical carbon dioxide was 14 MPa, the single fiber pull-out strength of the modified aramid fiber/epoxy resins was the extreme value, higher than that of pristine aramid fiber/epoxy resin micro-composites. It was mainly attributed that the introduction of nano-SiO_2 on the surface of aramid fibers could increase the surface roughness in favor of enhancing the interfacial adhesion between aramid fibers and epoxy resins. Chen et al. [124] investigated the interfacial bond strength between modified PBO fibers by in-situ growth of ZnO NWs and polysesquisiloxane (POSS) and epoxy resins (Figure 5.11). Compared with pristine PBO fibers (40.4 MPa), the interfacial shear strength between PBO–ZnO–POSS fibers and epoxy resins increased to 74.1 MPa by 84.3%. It was mainly attributed to the presence of more active groups on the surface of PBO–ZnO–POSS fibers, which was more favorable for epoxy resin infiltration.

In our previous work, Gu et al. [150] characterized the interfacial bond between P(S-*co*-BCB-*co*-MMA) based random copolymer coating PBO (PBO@P) fibers and BADCy resins through the single fiber pull-out test and high-resolution TEM (Figure 5.12). Results showed that when the molecular weight of P(S-*co*-BCB-*co*-MMA) was about 30 000 (M_n = 32300), the surface roughness of the PBO@P-3 fibers increased from 11 (PBO fibers) to 39 nm. The single fiber pull-out strength of the PBO@P-3 fiber/BADCy resin micro-composites was 4.5 MPa, higher than that of pristine PBO fiber/BADCy resin (3.1 MPa).

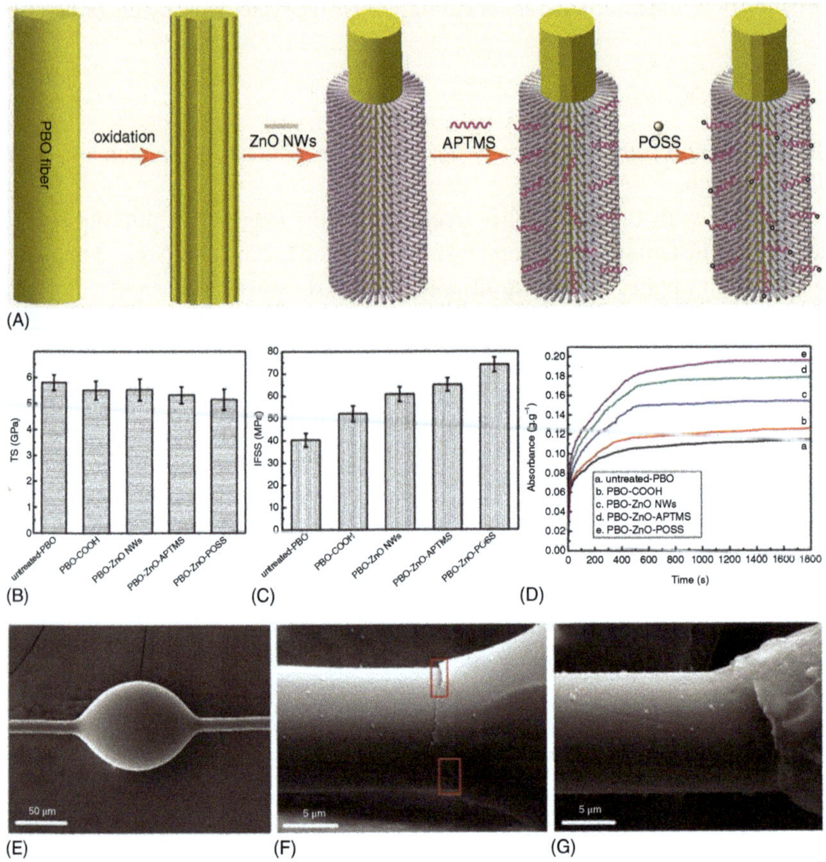

Figure 5.11 Schematic diagram of the preparation for PBO–ZnO–POSS fibers (A); tensile strength (B) and interfacial shear strength (C) of the different PBO fiber/epoxy composites; dynamic infiltration curves of the different PBO fibers (D); SEM images of the PBO fiber/epoxy composite after single fiber pull-out tests (E–G). Source: [124]. Chen et al. (2017)/Reproduced with permission from Elsevier.

5.4.4.3 Fiber Indentation/Ejection Test

Due to the large dispersion of single fiber pull-out test, the polymer matrix will climb up reinforced fibers because of surface tension, thus reducing the testing accuracy [151]. Fiber indentation/ejection test is developed to characterize interfacial binding (Figure 5.13) [152]. It can be seen that reinforced fibers undergo elastic deformation (Stage 1) and plastic deformation (Stage 2) with the increase of press-in depth [153, 154]. With further increase in the depth of indentation, the load does not change until the interface between polymer matrix and reinforced fibers break down, and the load value changes abruptly (Stage 3). After interfacial debonding, the force between polymer matrix and reinforced fibers is sliding friction (Stage 4) [155]. The value of IFSS between the polymer matrix and reinforced fibers can be calculated by Eq. (5.13).

$$\text{IFSS} = \frac{P}{\pi D t} \tag{5.13}$$

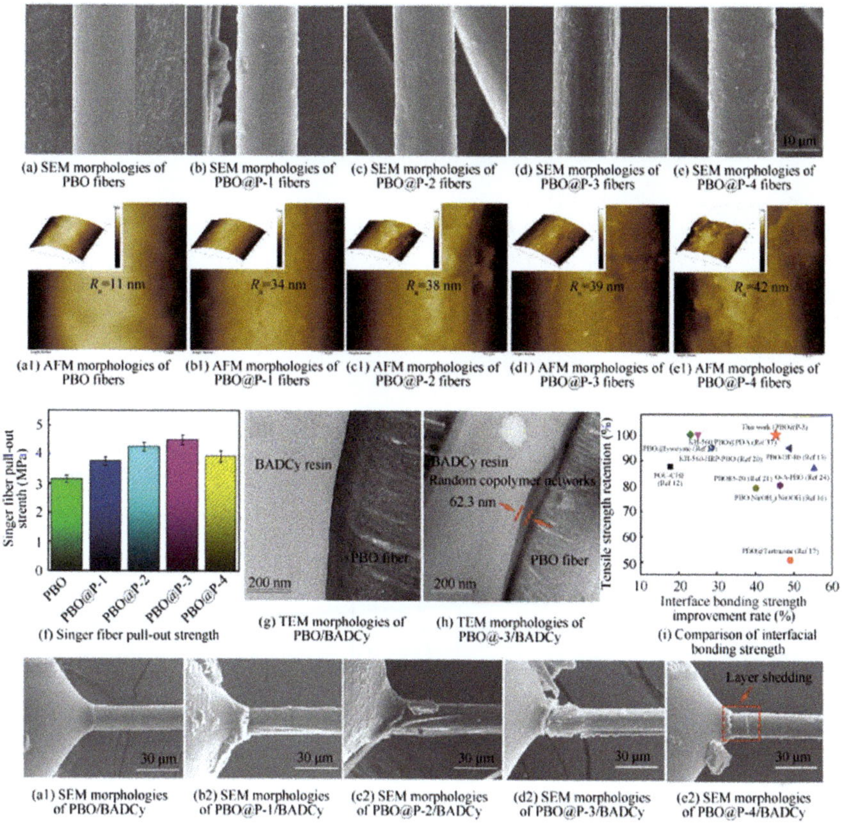

Figure 5.12 SEM images (a–e), single fiber pull-out strength (f), and interfacial layer thickness (g–h) of the different PBO fiber/BADCy micro-composites; comparison of interfacial performance with other research work (i). Source: [150]. Tang et al. (2021)/Reproduced with permission from Elsevier.

where D, t, and P are, respectively, the fiber diameter, the thickness of the specimen, and ultimate load.

Fiber indentation/ejection test is commonly used to characterize the interfacial bonding between carbon fibers and polymer matrix [156, 157]. For common inorganic reinforced fibers of the polymer matrix wave-transparent composites, the single fiber of the glass fibers and quartz fibers is extremely brittle, and the single fiber of the organic fibers, such as aramid fibers, UHMWPE fibers, and PBO fibers, cannot maintain the straightened state without pretension [158, 159]. Therefore, it is difficult to prepare the corresponding testing samples.

5.4.4.4 Nano-Indentation Method

In addition to the above three testing methods, the mechanical properties of the interface between polymer matrix and reinforced fibers can also be characterized by nano-indentation method [160, 161]. The load–displacement curves are obtained by testing selected points of the reinforced fibers, polymer matrix, and interfacial phase

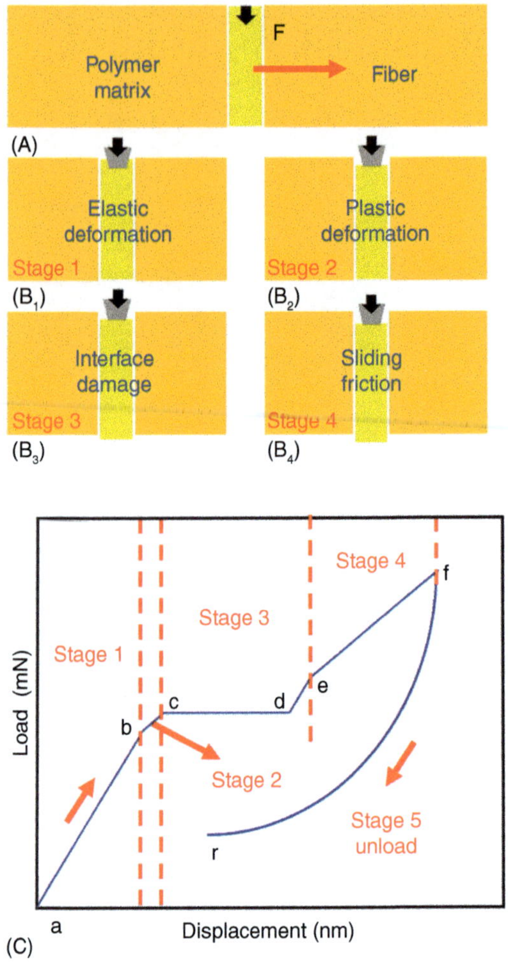

Figure 5.13 Schematic diagram of the fiber indentation/ejection test (A, B) and the load versus displacement curve (C).

under the same load and indentation depth [162]. The corresponding hardness and modulus of the interfaces for polymer matrix wave-transparent composites can be calculated from the load--displacement curves [163, 164]. The loading displacement curves of the nano-indentation present a clear distinction between polymer matrix, reinforced fibers, and the corresponding interfacial region [165]. The loading and unloading curves of the reinforced fibers almost overlap, but the loading and unloading curves of the polymer matrix form a large hysteresis circle. The mechanical response of interface between the polymer matrix and reinforced fibers shows a large dispersion [166].

Khanna et al. [167] investigated the elastic modulus and hardness at the interface between glass fibers and polyester using nano-indentation method. Results revealed that the elastic modulus of the glass fibers and polyester matrix were 7.45 and

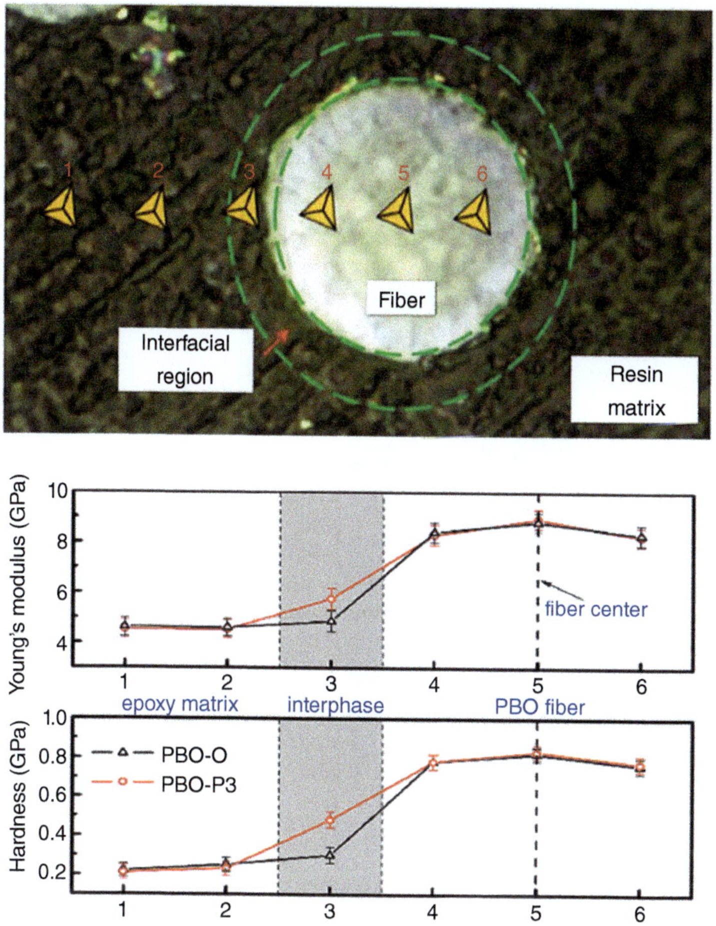

Figure 5.14 Characterization of mechanical properties for PBO fibers/epoxy resin by the nano-indentation method. Source: [168]. Song et al. (2021)/Reproduced from John Wiley & Sons, Inc.

1.8 GPa, respectively. Song et al. [168] used nano-indentation to study the elastic modulus and hardness of interface between PBO fibers and epoxy resins. Due to the poor interfacial compatibility between pristine PBO fibers and epoxy resin, the interfacial elastic modulus and hardness were close to those of BADCy matrix. The interfacial strength and modulus respectively increased to 5.8 and 0.48 GPa, resulting in increased interfacial compatibility of PBO fibers and epoxy resins (Figure 5.14).

In our previous works, Gu et al. [104] investigated the mechanical properties of interface between poly(dopamine)-hybrid zeolite imidazole (PDA*ZIF-8) uniformly coated PBO fibers and BADCy matrix (Figure 5.15). Results showed that the elastic modulus and hardness of the BADCy resins were respectively 4.2 and 0.34 GPa,

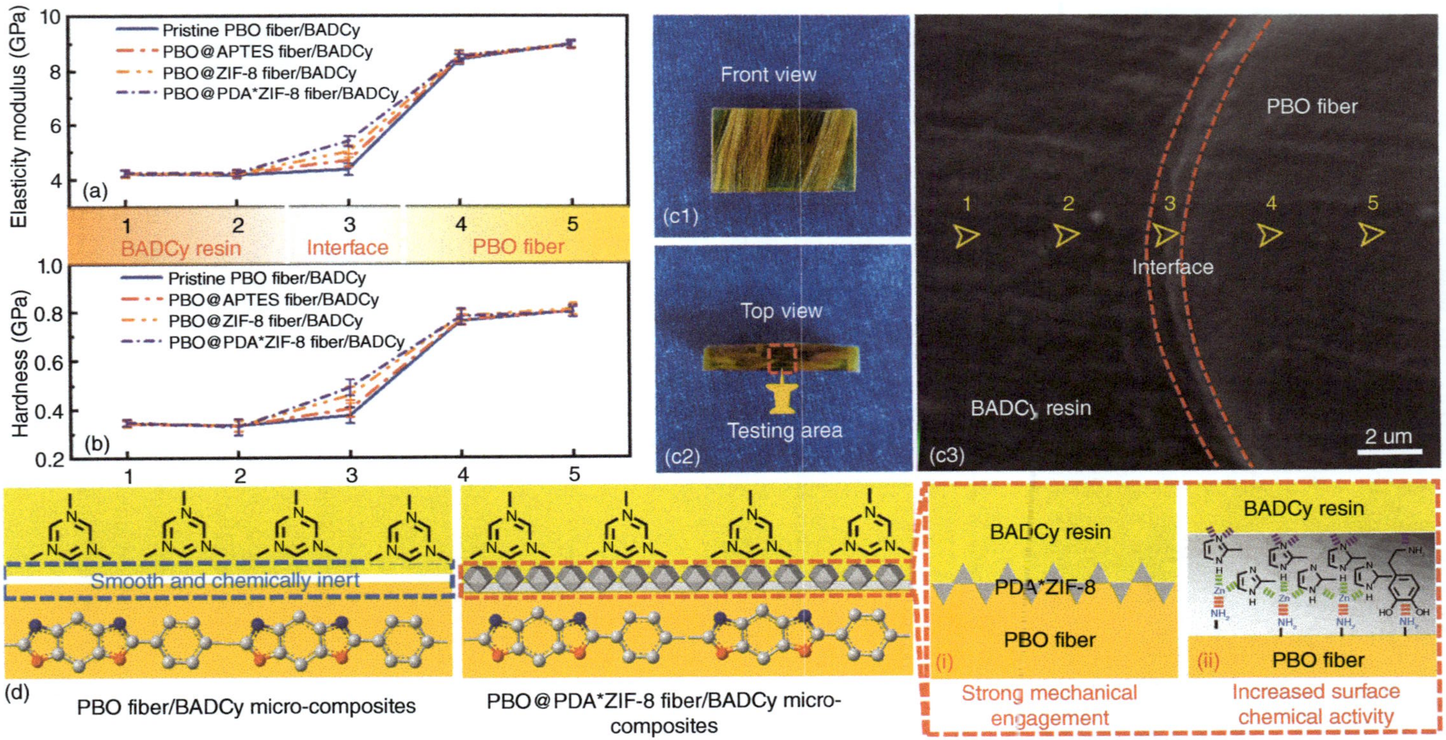

Figure 5.15 Nano-indentation test near the interface between PBO fibers and BADCy resins: elasticity modulus (a) and hardness (b) of indentation point, digital photos of the samples (c1–c2), and indentation point distribution (c3); schematic diagram of interfacial enhancement mechanism for PBO@PDA*ZIF-8 fiber/BADCy micro-composites (d). Source: [104]. Liu et al. (2022)/Reproduced with permission from Elsevier.

and that of PBO fibers were 8.9 and 0.80 GPa, respectively. Due to the extremely poor interfacial compatibility between pristine PBO fibers and BADCy resins, the interfacial elastic modulus (4.3 GPa) and hardness (0.38 GPa) were close to those of BADCy matrix. The elastic modulus and hardness at the interface of the modified PBO@PDA*ZIF-8 fiber/BADCy micro-composite increased significantly to 5.4 and 0.49 GPa, respectively. This was mainly attributed that the reactive proton groups ($-NH_2$) on the surface of PBO@PDA*ZIF-8 fiber could promote the cross-linked network formation of BADCy resins, which increased the elastic modulus and hardness at the interface of PBO@ZIF-8 fiber/BADCy micro-composites. Meanwhile, the PDA*ZIF-8 on the surface of PBO fibers could reinforce the BADCy resin, further increasing the elastic modulus and hardness of the PBO@ZIF-8 fiber/BADCy micro-composites.

5.5 Improvement of Interfacial Compatibility for Reinforced Fibers/Polymer Matrix

Physical and chemical performances of the polymer matrix and reinforced fibers for polymer matrix wave-transparent composites, such as thermal expansion coefficient, melting point, density, and surface energy, are very different, which makes it difficult to form an effective interfacial layer [169]. At present, the two main methods of surface functionalization of the reinforced fibers and synthesis of the interfacial compatibilizers are performed to improve the interfacial bond strength of the polymer matrix wave-transparent composites [170, 171].

5.5.1 Surface Functionalization of the Reinforced Fibers

Surface functionalization of the reinforced fibers improves the interfacial bonding strength with polymer matrix via enhancing the surface activity and roughness of the reinforced fibers to further achieve excellent chemical bonding and mechanical interlocking with polymer matrix [172, 173]. The surface functionalization of the reinforced fibers can be divided into the physical modification and chemical modification [174–176].

5.5.1.1 Physical Modification

Physical modification refers to activating the surface of the reinforced fibers by physical method to increase surface energy and contact area with polymer matrix, resulting in enhancing the interfacial bonding of the polymer matrix wave-transparent composites [177]. At present, the common physical modification methods for reinforced fibers mainly include plasma modification, acid etching modification, and high-energy ray modification [178].

5.5.1.1.1 Plasma Modification

Plasma modification refers to enhancing the interfacial bond strength of the polymer matrix wave-transparent composites by using high-energy plasma to bombard the surface of the reinforced fibers, increasing the surface activity [179, 180]. The effect of plasma modification is usually related to the process parameters, such as the type of plasma sources, ambient atmosphere, pressure, power, and treatment time [181]. Therefore, the optimal process parameters of the plasma modification need to be explored in the context of specific situation. In addition, the plasma modification mechanism is usually related to the type of the reinforced fibers.

For common inorganic fibers with high bond energy, such as glass fibers, quartz fibers, and basalt fibers, plasma treatment cannot break the chemical bonds, but slightly etch the surface of inorganic fibers to increase the mechanical engagement force with polymer matrix [182, 183]. Cech et al. [184] applied argon plasma for the surface treatment of the glass fibers. The interlaminar shear strength (ILSS, 29.8 MPa) of the modified glass fibers/unsaturated polyester composite was 65.6% higher than that of the pristine glass fibers/unsaturated polyester composite (18.0 MPa). Zhang et al. [185] utilized plasma-enhanced chemical vapor deposition to grow vertically aligned carbon nanotubes (VA-CNT) on the surface of the quartz fiber fabric. The water contact angle of the VA-CNT-modified quartz fibers fabric was reduced from 96.5° (pristine quartz fibers) to 30.6° (Figure 5.16a), which was beneficial to improving the interfacial compatibility with epoxy matrix. Kim et al. [186] used the low-temperature oxygen plasma to treat basalt fibers and then characterized the interlaminar fracture morphology of the modified basalt fibers/epoxy composites (Figure 5.16b). Compared with pristine basalt fibers, the interlaminar fracture toughness of the modified basalt fibers/epoxy composite was increased by 16.0%.

Common organic fibers, such as aramid fibers, PBO fibers, and UHMWPE fibers, can generate the polar groups to improve the interfacial compatibility of the polymer matrix wave-transparent composites [188–190]. Besides, the low-temperature plasma is usually used for treating surface modification of the organic fibers to avoid serious damage to their native structure [191, 192]. Fan et al. [187] utilized the low-temperature plasma for surface modification of the aramid fibers. The best modification of aramid fibers was achieved when the low-temperature plasma treatment power, time, and pressure were respectively 67.5 W, 11 min, and 2500 Pa. The active groups (the hydroxyl and carboxyl groups) and roughness of the modified aramid fibers were significantly improved (Figure 5.16c). Compared with pristine aramid fibers, the interfacial bonding strength between modified aramid fibers and epoxy resins increased by 35.5%, and there was no significant debonding at the interfaces. Liu et al. [193] modified PBO fibers using low-temperature air plasma. The ILSS of the modified PBO fibers/bismaleimide resins (BMI) composites increased from 43.9 (pristine PBO fibers/BMI) to 57.1 MPa when the air plasma power was 50 W/cm^3. Liu et al. [194] utilized the low-temperature plasma for surface modification of the UHMWPE fibers. When the pressure, treatment power, and treatment time of plasma modification process were respectively 50 Pa, 100 W,

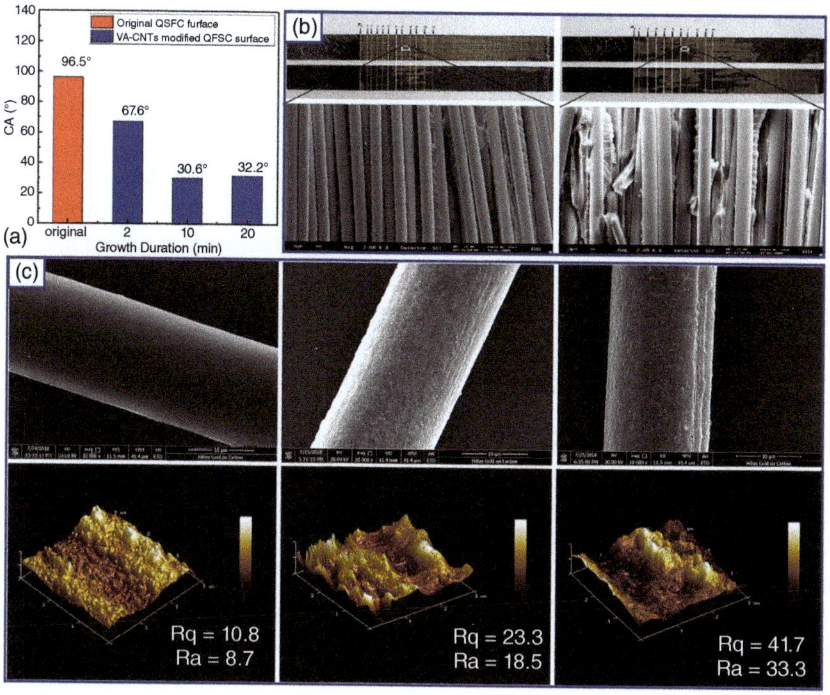

Figure 5.16 Contact angles of VA-CNT modified quartz fiber with water (a). Source: Adapted from [185]; SEM images of interlaminar fracture for basalt fibers/epoxy resins composites (b). Source: [186]. Kim et al. (2011)/Reproduced with permission from Elsevier; SEM and AFM images of low-temperature plasma modified aramid fibers (c). Source: [187]. Fan et al. (2018)/Reproduced with permission from Elsevier.

and 180 s, the surface roughness and water absorption of the modified UHMWPE fibers were significantly enhanced, which was beneficial to improving the adhesion between the surface of UHMWPE fibers and polymer matrix.

Plasma modification is usually a dry process, and the surface activity of the reinforced fibers can be increased without affecting the mechanical properties of the reinforced fibers. In addition, the plasma treatment is highly efficient and can adapt to continuous automated production [195, 196]. However, the modification effect of plasma treatment needs to be further improved.

5.5.1.1.2 Acid Etching Modification

Acid etching modification mainly utilizes the strong acids, such as the sulfuric acid, hydrochloric acid, and protonic acid, to treat the surface of the reinforced fibers, aiming to increase their surface roughness and the number of reactive groups, which is beneficial to improving the interfacial compatibility between polymer matrix and reinforced fibers [197, 198]. The factors affecting the effect of acid etching modification mainly include the strong acid type, the concentration of acid solution, and treatment time [199]. The acid solution concentration is too high or the treatment time is too long would damage the structure of the reinforced fibers, greatly reducing the mechanical properties of the reinforced fibers [200, 201]. In addition, similar to plasma modification, the mechanism of acid etching modification is related to the type of reinforced fibers.

The SiO_2 content of the quartz fiber is as high as 99.99% and does not contain other metal oxides, which makes the quartz fibers have excellent acid resistances [202, 203]. Therefore, there is fewer research on the treatment of quartz fibers by acid etching modification. Compared to quartz fibers, glass fibers and basalt fibers contain metal oxides, such as CaO, Al_2O_3, Na_2O, and MgO, which are susceptible to acid corrosion [204, 205].

Surface treatment of the glass and basalt fibers by strong acid can decompose the metal oxides to form micro-pores, grooves, and furrows on their surfaces, resulting in enhancing the mechanical engagement between reinforced fibers and polymer matrix [205, 206]. Moreover, acid etching can produce a large number of hydroxyl groups on the fibers' surface to further improve the surface reactivity. Tomao et al. [207] analyzed the effect of hydrochloric acid (HCl) concentration, treatment time, and temperature on the structures and properties of the glass fibers. The modified glass fibers obtained by treatment with 20% HCl at 100 °C for six hours had the highest hydroxyl content, provided that the native structure of glass fibers was not destroyed. Cho et al. [208] found that treatment of S-glass fibers with a mixture of hydrogen peroxide (H_2O_2) and sulfuric acid (H_2SO_4) (vol:vol = 1:4) for 24 hours to prepare the modified glass fibers (Gr-(iii)s glass fiber) with the highest surface activity. The ILSS values of the Gr-(iii)s glass fiber/polydimethacrylate composites increased from 7.38 (pristine S-glass fibers) to 10.30 MPa by 39.6% (Figure 5.17). Karunagaran and Rajadurai [209] immersed basalt fibers in H_2SO_4 for 24 hours to obtain modified basalt fibers. The flexural strength of the modified basalt fibers/epoxy resins composite was increased from 309 (pristine basalt fibers/epoxy resins) to 380 MPa.

Li et al. [210] analyzed the relationship between the flexural strength and tensile strength of the basalt fibers/epoxy resins composite and the HCl concentration, etching time, and temperature. When the concentration of HCl solution, etching time, and temperature were respectively 2 mol/l, two hours, and 80 °C, the flexural strength and tensile strength of the modified basalt fibers/epoxy resins composite, reached the maximum values, which were respectively 35.1% and 36.1% higher than that of unmodified basalt fibers/epoxy resins composites.

The mechanism of acid etching modified aramid fibers, PBO fibers, and UHMWPE fibers was usually to break the weaker chemical bonds in their molecular structures, such as amide bonds (aramid fibers), oxazole rings (PBO fibers), and $-CH_2-CH_2-$ (UHMWPE fibers) by strong acid etching [211]. Wang and coworkers [212] applied different concentrations of phosphoric acid to treat the surface of aramid fibers. Compared with pristine aramid fibers (52.9 MPa), the ILSS value of the modified aramid fibers/epoxy composite increased to 62.0 MPa, by 17.0%, after

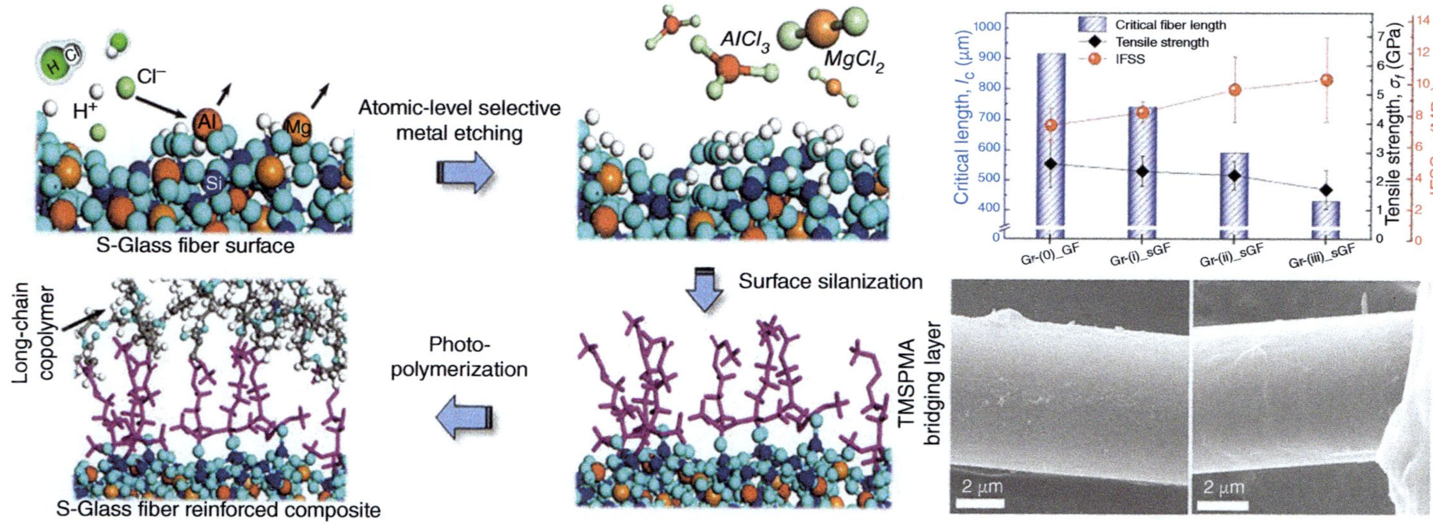

Figure 5.17 Schematic diagram of the acid etching modification for glass fibers [208].

the treatment of phosphoric acid (the concentration for 20%). Zhao [213] studied the effect of phosphoric acid concentration, treating temperature, and treating time on the surface activity of aramid fibers. When the concentration of phosphoric acid, the treating temperature, and the treating time were respectively 30%, 40 °C, and 5 min, the interfacial bond strength of the modified aramid fibers/epoxy resin composite was 42.1% higher than that of pristine aramid fibers/epoxy resins composites. The mechanical properties of modified aramid fibers did not decrease significantly. Wu and Shyng [214] used methane sulfonic acid (MSA) for etching the surface of PBO fibers. The surface energy of the modified PBO fibers was 62.1 mJ/m^2, higher than that of pristine PBO fibers (46.2 mJ/m^2), and the corresponding ILSS of the modified PBO fibers/epoxy resins composites was 40.2 MPa, also higher than pristine PBO fibers (34.7 MPa). Liu et al. [215] utilized polyphosphoric acid (PPA)/acetic acid for etching the surface of PBO fibers. The contact angle between modified PBO fibers and water decreased from 90.0° (pristine PBO fibers) to 42.8°, and the single fiber pull-out strength of the modified PBO fibers/epoxy resins micro-composites increased from 7.5 (pristine PBO fiber/epoxy resins micro-composites) to 9.7 MPa. In our previous work, Gu et al. [216] soaked high-modulus PBO (HMPBO) fibers with PPA and anhydrous ethanol (EtOH) (PPA:EtOH = 1:1, wt/wt) under ultrasonic treatment to prepare modified HMPBO fibers. Results revealed that the modified HMPBO fibers had better hydrophilicity (the contact angle with water was reduced from 75.1° (pristine HMPBO fibers) to 35.5°). Belgacemi [217] et al. studied the treatment effect of different concentrations of chromic acid solution on UHMWPE fibers. When the ratio of chromic acid solution, treating temperature, and treating time were respectively 1:4:30 ($K_2Cr_2O_7$:H_2O:H_2SO_4), 55°C, and 5°min, the IFSS value of the modified UHMWPE fibers/epoxy resins composite increased by 75.5%, compared to pristine UHMWPE fibers/epoxy resins composite.

Moreover, ultrasonic waves can accelerate the impregnation of the acid solution on the surface of the reinforced fibers, further increasing the interfacial bonding properties of the polymer matrix wave-transparent composites [218, 219]. Deng et al. [220] utilized an ultrasonic-assisted phosphoric acid solution to etch the surface of aramid fibers. When the phosphoric acid solution concentration, ultrasonic power, and treating time were respectively 30%, 500 W, and 5 min, the surface roughness and surface activity of the modified aramid fibers surface significantly increased. The contact angle between the modified aramid fibers and deionized water decreased from 102.9° (pristine aramid fibers) to 81.2°. Luo et al. [221] treated PBO fibers with PPA/EtOH solution under ultrasonic treatment. Then, they characterized the surface morphology, tensile strength, and interfacial bonding strength of the PBO fibers/epoxy resin composites. When the ultrasonic equipment power and reaction time were respectively 54 W and 4 min, the surface roughness and hydroxyl contents of PBO fibers significantly increased by the cavitation effect of ultrasound. The single fiber pull-out strength of the modified PBO fiber/epoxy resins micro-composites was 67.2% higher than that of unmodified PBO fibers/epoxy resins micro-composites. Jin et al. [222] applied ultrasonically assisted chromic acid etching of the UHMWPE fiber surface. When the temperature was below 60 °C, the chromic acid solution only etched the surface of UHMWPE fibers without destroying the molecular structure. When the temperature was further increased (>60 °C), the surface roughness and number of polar groups of the modified UHMWPE fibers increased, further enhancing interfacial bonding properties with polymer matrix.

The treatment effect of acid etching modification is better than that of plasma modification, attributing to simultaneously improving the chemical activity and roughness of the reinforced fiber surface [223]. However, the surficial structure of the reinforced fibers would be damaged after acid etching, causing a decrease in the mechanical strength [224]. Therefore, the processing parameters of acid etching should be strictly controlled.

5.5.1.1.3 High-Energy Ray Modification

High-energy ray modification mainly uses high-energy rays (most commonly γ-rays) to bombard the surface of the reinforced fibers, destroy their molecular chain structure, and produce reactive groups or free radicals, further enhancing the surface activity [225, 226]. The main component of the glass fibers, basalt fibers, and quartz fibers is SiO_2; high-energy radiation cannot be effective [227]. Therefore, the high-energy ray modification is often used for the surface functionalization of organic fibers, such as aramid fibers, PBO fibers, and UHMWPE fibers.

Zhang et al. [228] adopted γ-radiation to treat Kevlar fibers and then grafted 3-(2,3-epoxypropoxy)propyltrimethoxysilane (KH-560). Results revealed that the ILSS value of the modified Kevlar fibers/epoxy composite increased from 42.1 (pristine Kevlar fibers) to 55.6 MPa. Zhang et al. [229] used γ-ray radiation for grafting epichlorohydrin on surface of the PBO fibers. Results showed that the active groups and surface roughness of the modified PBO fiber synchronously increased. When the dosage of γ-ray radiation was 30 kGy, the surface free energy of the modified PBO fibers reached the maximum, 14.5% higher than that of pristine PBO fibers. Qiu et al. [230] used the ultraviolet radiation to graft acrylic acid (AA) for surface modification of UHMWPE fibers. When the AA monomer concentration was 50%, the tensile strength, flexural strength, and impact strength of UHMWPE fibers/epoxy composites were respectively increased by 68.4%, 35.6%, and 67.1% compared to unmodified UHMWPE fibers/epoxy composite.

Compared with plasma modification and the acid etching modification, the high-energy ray modification can be carried out at any temperature and has a relatively better modification effect. Simultaneously, the high-energy ray modification has the advantages of simple and convenient continuous processing and no environmental pollution [231]. However, the modification effect of high-energy rays would gradually attenuate with increasing exposure time of the reinforced fibers in the air.

It can be concluded that physical modification methods of the reinforced fibers are relatively simple to operate and convenient for continuous production. However, the modification effect of the physical modification needs to be further improved.

5.5.1.2 Chemical Modification

Chemical modification refers to treating reinforced fibers with chemical agents to introduce the active groups or coatings on their surfaces, thereby improving the surface wettability of the reinforced fibers and the bonding strength between polymer matrix and reinforced fibers [232, 233]. Chemical modification mainly includes the coupling agent treatment, the chemical grafting, and the coating modification.

5.5.1.2.1 Coupling Agent Treatment

Coupling agents are additives that can enhance the interfacial bonding properties between polymer matrix and reinforced fibers in the polymer matrix wave-transparent composites [234, 235]. The molecule of coupling agents contains two kinds of chemical groups with different properties. One group can be chemically bonded to the surface of the reinforced fibers, and the other group has a good compatibility with the polymer matrix [236].

There are various types of coupling agents, such as the silane coupling agents, the titanate coupling agents, and the aluminate coupling agents. Compared with the silane coupling agents, the titanium and aluminum elements in titanate coupling agents or aluminate coupling agents have more outer electrons, which are easy to be affected by an external electric field to decrease the insulation properties of the polymer matrix wave-transparent composites [237]. The decreasing insulation performances of the polymer matrix composites are not conducive to further improving the wave-transparent performances.

The X group of silane coupling agents are alkoxy, such as ethoxy or methoxy, which can produce Si–O–Si bonding or hydrogen bonding with the hydroxyl group on the surface of the reinforced fibers by dehydration condensation [236, 238]. The Y groups can be reacted with the polymer matrix, such as the –C=C–, $-NH_2$, and epoxy groups can be respectively involved in the curing process of unsaturated polyester, epoxy resins, and cyanate ester [239]. The action mechanism of the silanes coupling agents is shown in Figure 5.18. Commonly commercialized silane coupling agents can be divided into the vinyl, the amino hydrocarbons, the epoxy hydrocarbons, and the quaternary amine hydrocarbons, according to the different Y groups. In practical applications, silane coupling agents are usually selected by the type of the polymer matrix [240, 241].

The surface of the inorganic fibers, such as the glass fibers, basalt fibers, and quartz fibers, usually contains a certain amount of hydroxyl groups, which can usually be treated directly with commercial silane coupling agents [242, 243]. Park and Jin [244] used methacryloxypropyltrimethoxysilane to modify glass fibers. The ILSS value of the modified glass fibers/unsaturated polyester composite was increased from 16.2 (pristine glass fibers/unsaturated polyester composite) to 27.4 MPa. St John et al. [245] respectively utilized aminopropyltrimethoxysilane (APS) and KH-560 for surface modification of the glass fibers. Results revealed that the ILSS value of KH-560-glass fibers/phenolic resins composite was 78.4% higher than that of APS-glass fibers/phenolic resins composites. Furthermore, the surface functionalization methods of the quartz fibers are similar to those of glass fibers. Li et al. [246] used γ-aminopropyltriethoxysilane to modify the surface of

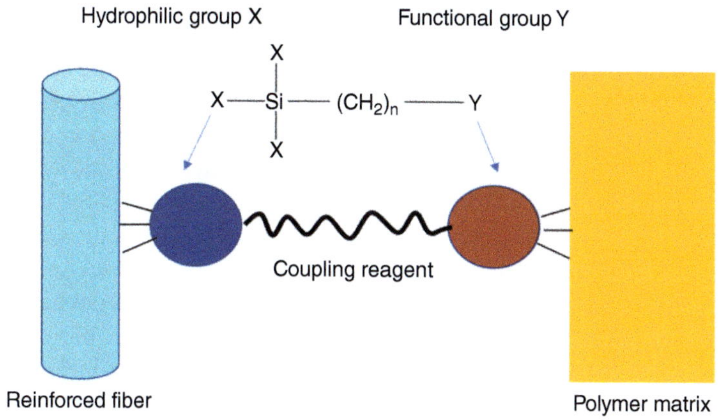

Figure 5.18 Schematic diagram of modification mechanism for silane coupling agents.

quartz fibers. The flexural strength and flexural modulus of the modified quartz fibers/epoxy composites increased to 376.4 MPa and 21.7 GPa, 41.1% and 22.6% higher than those of pristine quartz fibers/epoxy resins composites, respectively.

The surfaces of the organic fibers, such as aramid fibers, PBO fibers, and UHMWPE fibers, are extremely inert due to the formation of highly oriented skin-core structures after the spinning process [247]. For example, Liu et al. synthesized the novel coupling agent with the side chains and benzimidazole groups, then grafted onto the surface of pretreated aramid fibers. When the amount of coupling agents was 2.1 wt%, the ILSS and flexural strength of the modified aramid fibers/polythiazide resins composite were respectively 62.0% and 43.7% compared to pristine aramid fibers/polythiazide resins composite. Yu et al. [248] utilized γ-aminopropyltriethoxysilane to modify the surface of PBO fibers. Results revealed that the flexural strength of modified PBO fibers/polyimide resin composites was 12.1% higher than that of unmodified PBO fibers/polyimide resin composites. Zhang et al. [249] used γ-aminopropyltriethoxy silane to modify the surface of UHMWPE fibers. The ILSS value of modified UHMWPE fibers/epoxy resins composite was increased by 108.0% when the mass fraction of γ-aminopropyltriethoxysilane was 1.5 wt%. The corresponding tensile strength and modulus of modified UHMWPE fibers/epoxy resins composite were significantly improved compared with those of pristine UHMWPE fibers/epoxy resins composite.

Recently, design and synthesis of new large molecule coupling agents (molecular weight much larger than ordinary coupling agents, usually the graft or block polymers) is the new developing direction of coupling agents [250, 251]. Shen et al. [252] prepared the macromolecular coupling agents to modify glass fibers. Compared to pristine glass fibers/polystyrene composite, the IFSS value of the modified glass fibers/polystyrene composite was improved by 1.7 times. It was also superior to conventional coupling agent modification of the glass fibers. Gu et al. [253] prepared the novel acetylene-based capped polyetherimide macromolecular coupling agents (BDA-K). They investigated the effect on the toughening of the quartz fibers

(QF)/silicone-arylene (PSA) composites. The ILSS, flexural strength, and notched impact strength of the QF/PSA composites with the addition of macromolecular coupling agents respectively increased by 54.1%, 59.0%, and 23.8%; the ILSS and flexural strength retention were respectively 89.0% and 89.6% at 250°C. It was attributed to the ability of BDA-K to participate in the cross-linking curing reaction of PSA matrix, which produced effective chemical bonding at the interface of the QF/PSA composites. Simultaneously, the better heat resistance of imide rings in the molecular structure of BDA-K resulted in improving the heat resistances of the QF/PSA composites under high-temperature environment.

Coupling agent treatment modification is currently the most widely used chemical modification of the reinforced fibers [254]. Compared with the method of physical modification, the advantages of coupling agent treatment modification include the adequate reaction, good coating effect, and strong coating layer. The disadvantage is the environmental pollution problem from the adopted solvent [255]. The modified reinforced fibers would occur a moisture absorption phenomenon after the long period of storage, which is difficult for long-term storage of the modified reinforced fibers.

5.5.1.2.2 Chemical Grafting Modification

Chemical grafting modification improves the surface activity of the reinforced fibers by grafting some functional compounds on the surface of fibers [256]. However, it usually requires some reactive groups on the surface of the reinforced fibers as active sites for further grafting reaction. Inorganic fibers, such as glass fibers, quartz fibers, and basalt fibers, have low reactive hydroxyl groups on their surfaces, which usually cannot be used as active sites to initiate the secondary reaction [257, 258]. Among organic fibers, PBO fibers are extremely surface inert, and the benzene and oxazole rings in the molecular structures are stable, which does not benefit chemical grafting modification [259, 260]. Therefore, the common reinforced fibers modified by chemical grafting modification are mainly aramid fibers and UHMWPE fibers.

The hydrogen of the benzene ring (adjacent and opposite) in the aramid fibers has certain reactivity with certain electrophilic substituent groups [261, 262]. Xing et al. [263] utilized the Friedel–Crafts reaction to graft epichlorohydrin onto the benzene ring of the Kevlar fibers. When the reaction time was one hour, the single fiber pull-out strength of the modified Kevlar fiber/epoxy resins micro-composites increased from 8.0 MPa (Kevlar fiber/epoxy resins micro-composites) to 12.1 MPa. Benrashid et al. [264] treated aramid fibers with chlorosulfonic acid sulfonation and investigated the effects of different treating times and concentrations on the surface activity of aramid fibers. Results revealed that the best treatment effect of aramid fibers was achieved when the treating time and concentration of chlorosulfonic acid sulfonation were respectively 150 s and 0.2 wt%. The ILSS value of the modified aramid fibers/2,2-bis(4-(4-maleimidophenoxy)phenyl)propane resins composites increased from 19.2 (pristine aramid fibers reinforced composite) to 27.6 MPa.

The active hydrogen on the side chains of the UHMWPE fibers can be used as an active site for free radical polymerization reaction [265–267]. Ahmadi and coworkers [268] prepared the glycidyl methacrylate (GMA) as monomer and dibenzoyl peroxide as initiator to initiate radical polymerization on the surface of UHMWPE fibers to produce poly (GMA) layers, which contained a large number of epoxy groups on the surface. The surface energy of the modified UHMWPE fiber increased from 33.39 (UHMWPE fiber) to 50.74 mN/m, and the corresponding contact angle decreased from 95.0° (UHMWPE fiber) to 87.7°. In addition, the ILSS of the modified UHMWPE fibers/epoxy resins composite increased from 2.44 (UHMWPE fibers/epoxy resins composite) to 7.13 MPa.

The chemical grafting method has good affinity between the cladding layer constructed on the surface of the reinforced fibers and polymer matrix [217, 269]. The cladding layer can even form the chemically cross-linked interfacial layer with the polymer matrix. However, the process of chemical grafting is relatively cumbersome, and the adaptation is relatively narrow [270, 271].

5.5.1.2.3 Coating Modification

Coating modification refers to grafting or coating the surface of the reinforced fibers with a layer of active functional groups to improve the chemical inertness of the surface without damaging the mechanical strength of the reinforced fibers [272, 273]. The coating modification can be divided into the single functional coatings, biomass films, and the organic–inorganic hybrid coatings according to the type of coating constructed on the surface of the reinforced fibers.

① Single Functional Coatings

The surface activity of the reinforced fibers can be improved by constructing the highly adhesive functional coatings, which can avoid damage to the native structure of the reinforced fibers [274]. Common single-functional coatings include the polymer coatings and inorganic nanofilms.

Polymer coatings containing more reactive groups are usually introduced onto the surface of the reinforced fibers by in-situ polymerization or impregnation coating as an intermediate phase to enhance the interfacial bonding between polymer matrix and reinforced fibers [275, 276]. Cech et al. [277] used the plasma-enhanced chemical vapor deposition to construct a tetravinylsilane coating on the surface of the glass fibers. The ILSS value of the modified glass fibers/polyester composite was 14% higher than that of the pristine glass fibers/polyester composite. Bulatov et al. [278] constructed the polyimide coatings on the surface of the quartz fibers to improve the interfacial bond strength between modified quartz fibers and epoxy resins. Nasser et al. [279] reported that aramid nanofibers could be adsorbed onto the surface of aramid fibers to form nanofiber coatings, enhancing the interfacial performances between modified aramid fibers and epoxy resins by introducing mechanical interlocking. Compared with unmodified aramid fibers/epoxy resins composite, the interfacial shear strength of aramid nanofiber-coated aramid fibers/epoxy resins composite increased by 70.27% (Figure 5.19a).

Compared with polymer coatings, inorganic nanofilms have rich reactive groups and greater rigidity, which can improve the interfacial bond strength of the polymer matrix wave-transparent composites by using as the intermediate phase between

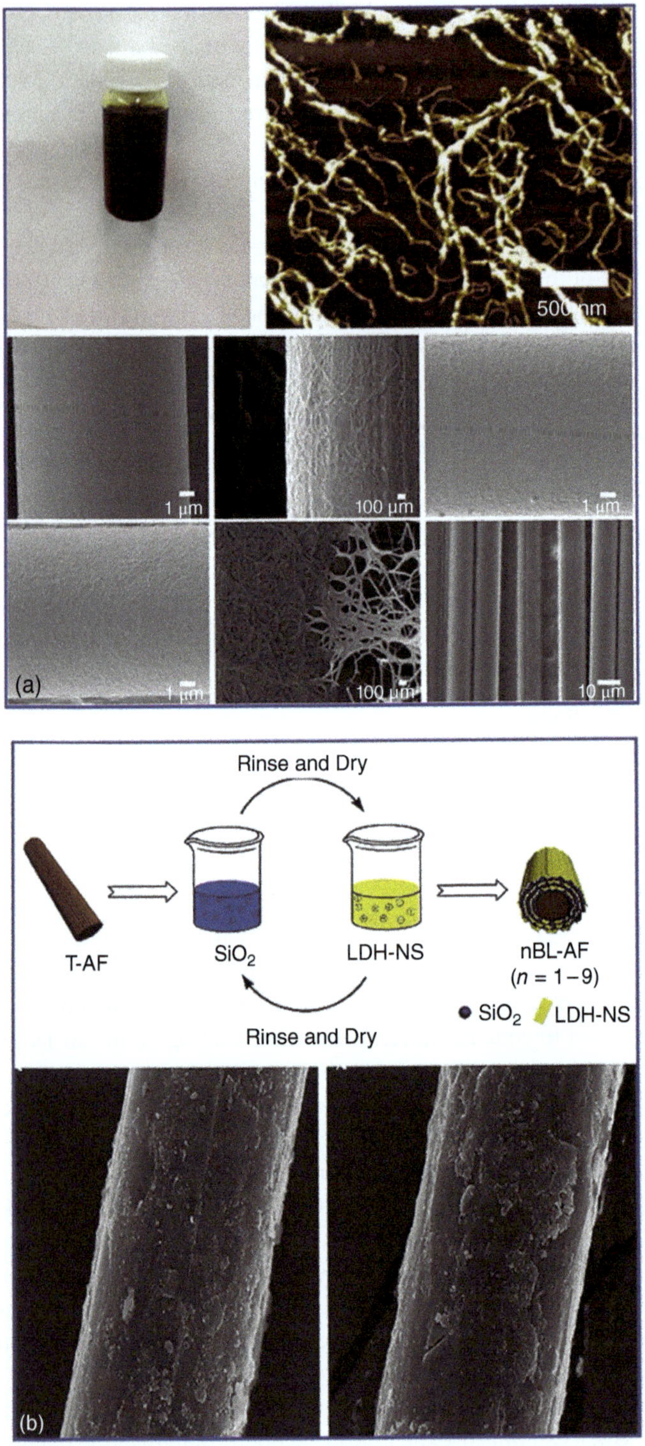

Figure 5.19 Aramid nanofiber adsorbed on the surface of aramid fibers (a); LBL self-assembly technology modified aramid fibers (b). Source: [279]. Nasser et al. (2019)/Reproduced with permission from Elsevier.

polymer matrix and reinforced fibers [280, 281]. Besides, some inorganic nanoparticles, such as the alloy particles, GO, and metal-organic framework materials (MOFs), possess photo-thermal conversion, thermal-electrical coupling, or induced adsorption, which can improve the interfacial bonding strength while imparting UV resistance or interfacial self-healing properties, etc. [282]. Liang and coworkers [283] constructed a new inorganic UV-absorbing particle $Ce_{0.8}Ca_{0.2}O_{1.8}$ coating (HSi-$Ce_{0.8}Ca_{0.2}O_{1.8}$@Kevlar, Figure 5.19b) on the surface of Kevlar fibers. Results showed that the surface energy of the HSi-$Ce_{0.8}Ca_{0.2}O_{1.8}$@Kevlar fibers increased by 40% compared with that of Kevlar fibers.

Yu et al. [284] prepared the graphene oxide-metal organic backbone (RGO-UIO-66) hybrid coating on the surface of the PBO fibers (PGU-CHF, Figure 5.20). The ILSS value of the PGU-CHF fibers/polyurethane (TPU) composite increased from 63.2 (PBO fibers/TPU composite) to 71.3 MPa. Shao et al. [119] immobilized iron-based metal framework material (MIL-88B-NH_2) nanocrystals and GO on the surface of the PBO fibers by hydrothermal reaction. The ILSS between the modified PBO fibers and polyurethane (TPU) was 72.24 MPa, 40.8% higher than the unmodified PBO fibers/TPU composites (51.31 MPa). Furthermore, the modified PBO fibers/TPU

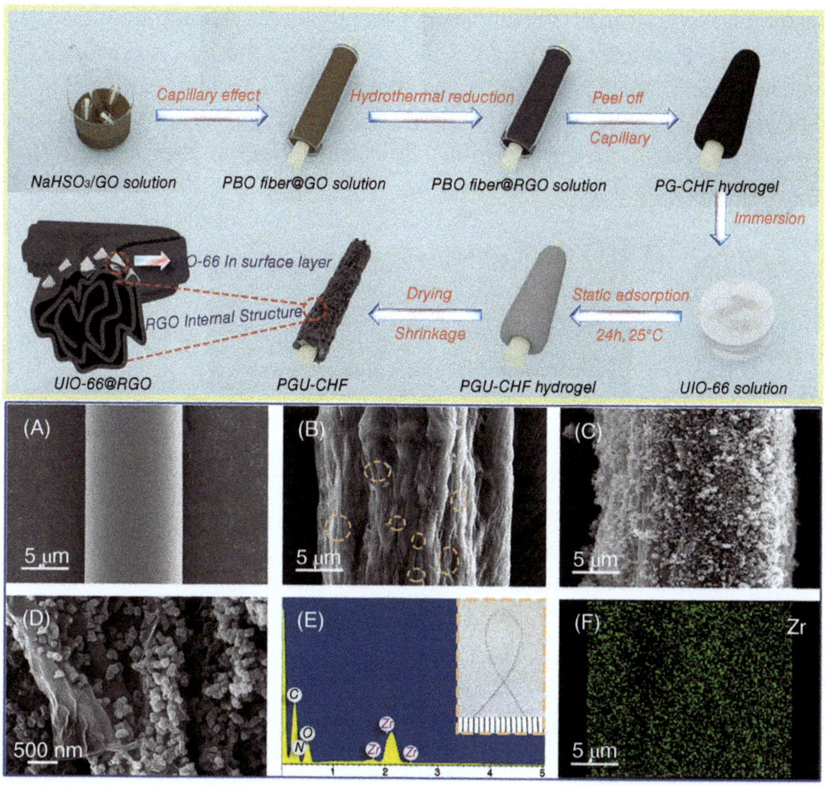

Figure 5.20 rGO-UIO-66 hybrid coatings modified PBO fibers. Source: [284]. Yu et al. (2019)/Reproduced from American Chemical Society.

laminated composites exhibited stable and efficient self-healing efficiency (~80%) for ILSS over five cycles.

② Biomass Membrane

The biomass membranes, with the characteristic of mild modification conditions, easy operation, and low damage to the mechanical properties of the reinforced fibers, can introduce reactive groups on the fibers' surface to further the interfacial compatibility between polymer matrix and reinforced fibers [285, 286]. Common biomass membranes mainly include the coating layers formed by self-polymerizing polyphenols such as PDA and polytannic acid and the biologically active layers such as catalase and lysozyme [287].

Dopamine (DA) with the catechol groups and amino groups can self-polymerize further to form PDA in a weak alkaline environment. The polymerization mechanism is shown in Figure 5.21 [288]. PDA coating layers have excellent adhesion to most substrates in humid environments and possess reactive groups such as the hydroxyl and amino groups, which can be used for further grafting reactions on the surface of the reinforced fibers [289].

Sa et al. [290] utilized the DA/KH-560 to modify the surface of Kevlar fibers (Kevlar-PDA/KH-560). Compared with the Kevlar fibers/rubber composite, the ILSS value of the Kevlar-PDA/KH-560 fibers/rubber composite was improved by 61.5%. Zeng et al. [291] modified the aramid fiber by the DA and amino-functionalized graphene, which significantly improved the interfacial bonding strength of the modified aramid fiber/epoxy resin composites (Figure 5.22a). In our previous work, Tang et al. [292] simultaneously modified the surface of Kevlar fibers by the DA and POSS to prepare the POSS-g-Kevlar@PDA (*f*-Kevlar) fibers. Compared with the

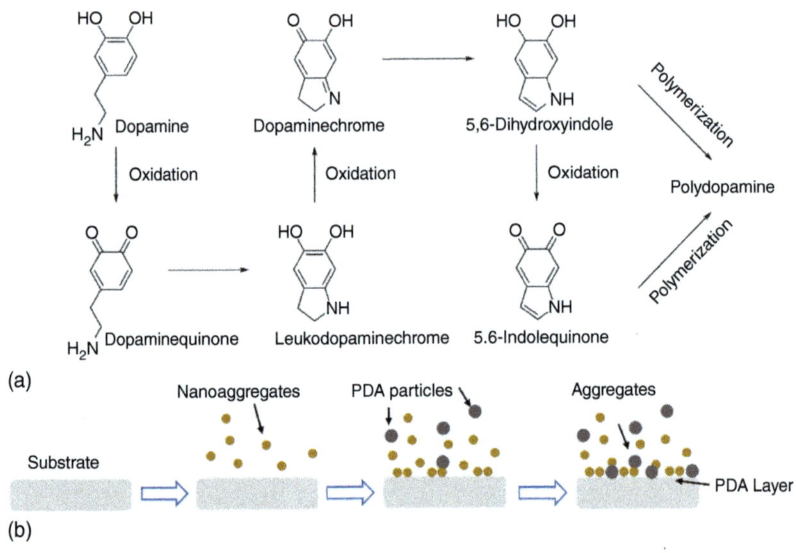

Figure 5.21 Schematic diagram of the mechanism for DA self-polymerization (a) and its adhesion (b).

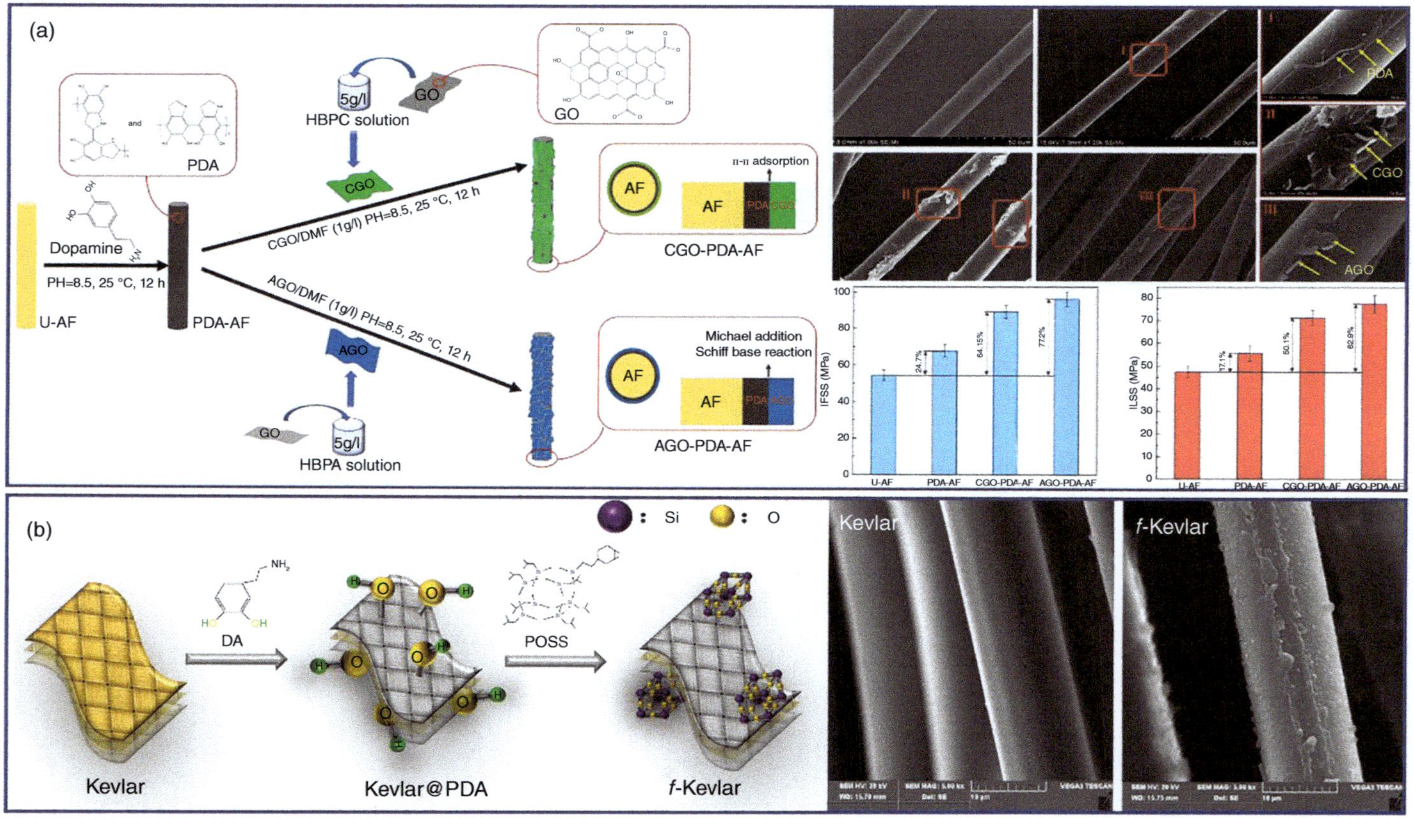

Figure 5.22 DA/amino-functionalized graphene modified aramid fibers (a). Source: Adapted from [291]; DA/epoxy POSS modified aramid fibers (b). Source: [292]. Tang et al. (2019)/Reproduced with permission from Elsevier.

pristine Kevlar fibers/BADCy, the ILSS value of *f*-Kevlar fibers/BADCy laminated composite was improved by 73.7%. The introduction of low-polarity POSS could further reduce the ε and $\tan\delta$ values of the *f*-Kevlar fibers/BADCy laminated composites (Figure 5.22b).

Similar to that of DA, tannic acid (TA), a plant-derived polyphenol, can quickly and stably adhere to the surface of various materials (polymers, metals and their oxides, inorganic nonmetallic materials, and so on) through oxidative self-polymerization [293, 294]. Meanwhile, the reaction of TA oxidation products with 3-aminopropyltriethoxysilane (APTES) generates TA-APTES films with a large number of reactive groups (hydroxyl and amino groups), which provides rich chemical sites for secondary functionalization [295, 296]. Wang et al. [297] deposited TA-APTES films on various substrate surfaces (polyvinylidene fluoride, polytetrafluoroethylene, polypropylene, stainless steel mesh, copper mesh, and so on). Results revealed that the modification effect of TA–APTES films was significantly better than that of PDA films due to the unique micro-nano structure of TA–APTES films and a large number of phenolic hydroxyl groups and amino groups. In our previous work, Liu et al. [298] deposited the hybrid polymer membrane (TA–APTES) on the surface of PBO fibers, synthesized by TA and APTES based on the Schiff's base and Michael addition reaction, and then grafted with POSS to obtain POSS-g-PBO@TA–APTES fibers (Figure 5.23). The ILSS and flexural strength of POSS-g-PBO@TA–APTES fibers/BADCy laminated composites were respectively enhanced from 36.7 and 587.4 MPa to 42.8 and 645.8 MPa, increased by 16.6% and 9.9% compared with those of PBO fibers/BADCy composites.

Lysozyme can rapidly self-assemble and aggregate to the two-dimensional nanofilms at the gas–liquid or liquid–solid interface under certain environmental conditions, with good colorless transparency, biocompatibility, and controlled

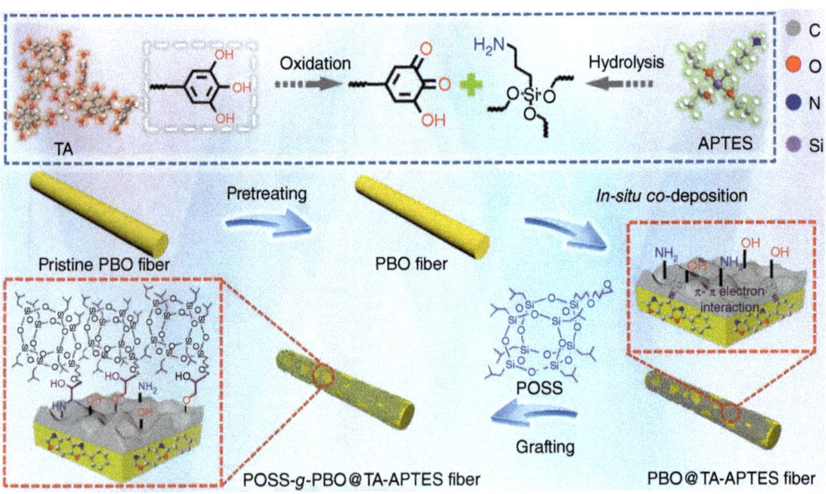

Figure 5.23 Schematic diagram of preparation for POSS-*g*-PBO@TA-APTES fiber. Source: [298]/Springer Nature.

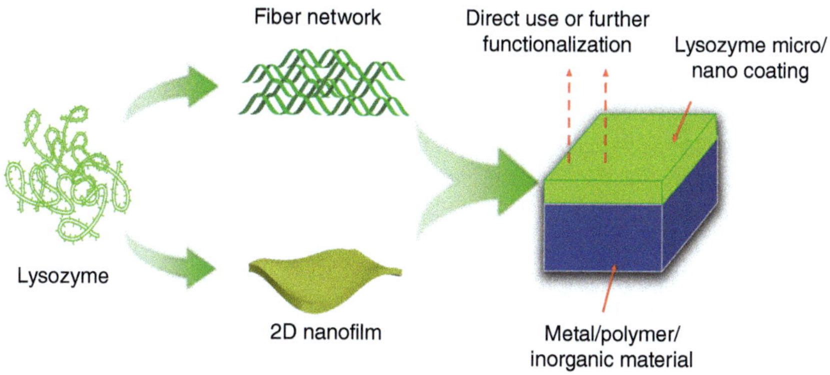

Figure 5.24 Biomimetic interfacial modification of lysozyme phase transition based on layered assembly. Source: Adapted from [300].

degradability, thus overcoming the shortcomings of PDA adhesion system [299]. Lysozyme phase transition assemblies can adhere rapidly and stably onto the surface of various materials (polymers, metals, metal oxides, inorganic nonmetallic materials, and so on) [300, 301], which also introduces chemical sites on the surface of substrates such as amino, sulfhydryl and carboxyl groups (Figure 5.24).

Zhong et al. [302] obtained functionalized titanium (Ti) metals by the mixtures of lysozyme and tris(2-carboxyethyl) phosphine solution (pH = 7.4) for two hours. Results revealed that the Ti metals were successfully coated with the two-dimensional nanofilm and introduced active groups for further modification. In our previous work, Gu et al. [303] prepared functionalized modified PBO fibers by coating PBO fibers with lysozyme and then grafting epoxy-based POSS (*f*-PBO fiber). The ILSS value of the *f*-PBO fibers/*m*-BADCy composite increased from 38.7 (pristine PBO fibers/*m*-BADCy) to 47.6 MPa by 23.0% (Figure 5.25).

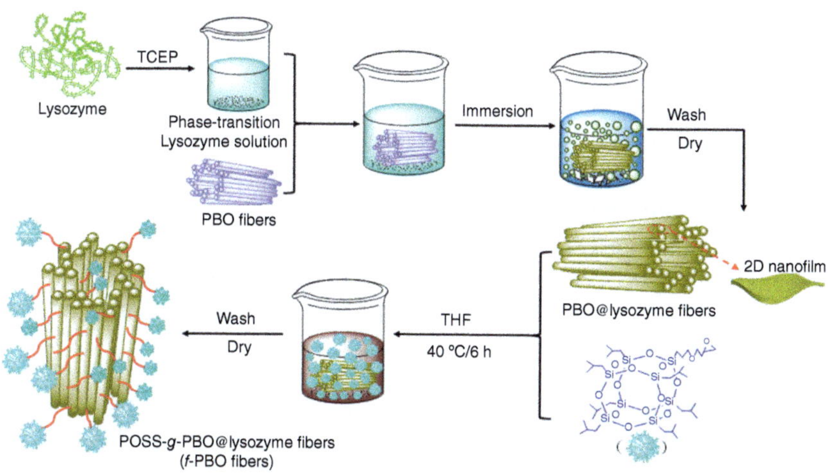

Figure 5.25 Synergistic modification of PBO fibers by lysozyme and POSS. Source: [303]/Royal Society of Chemistry.

Moreover, the horseradish peroxidase is also performed to functionalize the surface of the reinforced fibers. Liu et al. [304] modified aramid fibers by horseradish peroxidase. The polymer layer appeared on the surface of the modified aramid fibers, which simultaneously enhanced the roughness and chemical activity of the aramid fibers. The IFSS value of the aramid fibers/epoxy resin composites increased by 20.8% compared with that of pristine aramid fibers/epoxy resins. Wang et al. [305] treated PBO fibers with horseradish peroxidase. The surface hydrophilicity of the modified PBO fibers was increased, and the single fiber pull-out strength between the PBO fibers and the polymer matrix was increased by 21.7%. Yan et al. [306] successively used the horseradish peroxidase and KH-560 to functionalize the surface of PBO fibers and analyzed the chemical composition and wettability of the surface of PBO fibers before and after modification, as well as the bonding strength between fibers and cyanate resin. The surface free energy of the modified PBO fibers was increased by 77.5%, and the ILSS value of modified PBO fibers/cyanate esters composite was improved by 28.7% compared to unmodified PBO fibers/cyanate esters composites.

③ Organic–Inorganic Hybrid Coatings

Organic–inorganic hybrid coatings with "the soft and rigid" characteristics are constructed on the surface of the reinforced fibers to improve the strength and toughness of polymer matrix wave-transparent composites [307]. Shao et al. [308] constructed Al_2O_3-POSS organic-inorganic hybrid coating on the surface of quartz fibers via sol–gel method. The Al_2O_3–POSS hybrid coating formed a uniform protective layer on the surface of quartz fibers to increase the surface roughness. Li et al. [309] coated the ZnO silica particles (SiO_2@ZnO)/epoxy resins hybrid sol on the surfaces of the aramid fibers. The IFSS of the modified aramid fibers/epoxy resin composites was increased by 38.3% compared to the pristine aramid fibers/epoxy resins. Li et al. [310] constructed the carbon nanotube/dopamine organic-inorganic hybrid coatings on the surface of aramid fibers. Compared with the pristine aramid fiber, the single fiber pull-out strength of modified aramid fiber/epoxy resins micro-composites was improved by 41.2%.

Zhu et al. [311] constructed polyetherimide (PEI)/TiO_2 organic–inorganic hybrid coatings on the surface of PBO fibers. The single fiber pull-out of the modified PBO fiber/epoxy resins micro-composites was increased from 13.5 to 17.4 MPa compared to pristine PBO fibers/epoxy resins. Lv et al. [312] used the layered grafting method to graft melamine (MEL) and carbon nanotubes (CNTs) onto the surface of PBO (PBO–MEL–CNTs) fibers (Figure 5.26). The introduction of MEL–CNTs hybrid coatings could effectively improve the interfacial performances of PBO fibers/epoxy resin composites through the intermolecular Van der Waals forces and mechanical interlocking. Compared with pristine PBO fibers, the ILSS value of PBO–MEL–CNTs fibers/epoxy resins composite was improved by 180.4%. Chen et al. [313] modified the surface of PBO fibers by the ranched polyethyleneimine (*b*-PEI)/graphene oxide/PDA hybrid coatings (PBO@PDA-PEI-GO fibers). The surface roughness of the PBO@PDA-PEI-GO fibers was improved from 26.6 (pristine PBO fibers) to 69.4 nm. The single fiber pull-out strength of PBO@PDA-PEI-GO

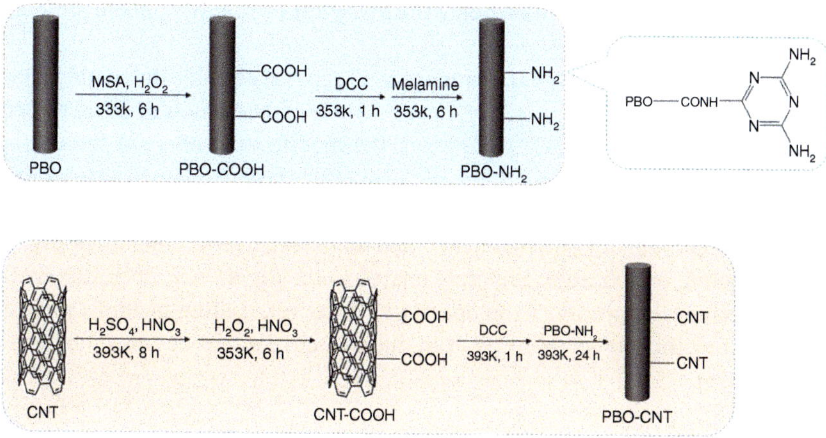

Figure 5.26 Schematic diagram of the preparation for PBO–MEL–CNTs fibers. Source: [312]/IOP Publishing Ltd.

fiber/epoxy resins micro-composite was 68.2 MPa, higher than that of PBO fibers/epoxy resins micro-composites (40.4 MPa). In our previous work, Gu and coworkers [314] utilized the *co*-precipitation method to synthesize calcium oxide doped cerium dioxide nanoparticles ($Ce_{0.8}Ca_{0.2}O_{1.8}$). Then they prepared a novel $Ce_{0.8}Ca_{0.2}O_{1.8}$/P(S-*co*-BCB-*co*-MMA) organic–inorganic hybrid coatings for surface functionalization of PBO fibers (PBO@P-$Ce_{0.8}Ca_{0.2}O_{1.8}$, Figure 5.27). When the mass fraction of $Ce_{0.8}Ca_{0.2}O_{1.8}$ nanoparticles was 0.6 wt%, the surface roughness of the PBO@P-$Ce_{0.8}Ca_{0.2}O_{1.8}$ fiber was increased from 11 (pristine PBO fiber) to 74 nm. The single fiber pull-out strength of the PBO@P-$Ce_{0.8}Ca_{0.2}O_{1.8}$ fiber/BADCy micro-composite was 4.6 MPa, higher than that of PBO fibers/BADCy

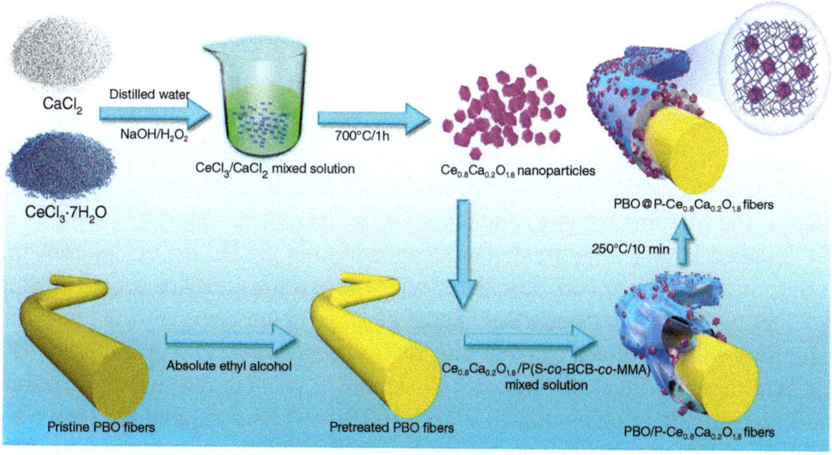

Figure 5.27 Mechanism diagram of the PBO fibers modified by $Ce_{0.8}Ca_{0.2}O_{1.8}$/P(S-*co*-BCB-*co*-MMA) hybrid coating. Source: [314]/with permission of Elsevier.

micro-composite (3.1 MPa). In addition, the PBO@P-$Ce_{0.8}Ca_{0.2}O_{1.8}$ fibers presented excellent UV resistance.

Compared with coupling agent treatment and chemical grafting, the coating modification can significantly improve the surface activity of the reinforced fibers without damaging their structures [173]. However, the modification effect of the coating modification is related to the thickness of the interface layer. Too small or too large a thickness of the interface layer is not conducive to improving the interface adhesion performance between the polymer matrix and reinforced fibers. Simultaneously, the effect of coating modification is closely related to the experience, operation methods, and technical maturity of the operators. Moreover, coating methods that can realize the modification of the reinforced fibers in large quantities must be further explored [315].

5.5.1.2.4 Synergistic Modification

Synergistic modification is the method of combining two or more modification methods, which is conducive to significantly improving the surface activity of the reinforced fibers and efficiently enhancing the interfacial bond strength with the polymer matrix [316].

Glass fibers, aramid fibers, and PBO fibers cannot be directly modified by chemical grafting due to the inert surfaces. Thus, the acid etching, coupling agent treatment, and coating modification are usually utilized to pretreat reinforced fibers to introduce active sites on the surface of the reinforced fibers before chemical grafting modification [317].

Sarr et al. [318] applied the 3-glycidoxypropyltrimethoxysilane for surficial preprocessing of glass fibers and then grafted cellulose nanofibers (CNF) to obtain GF-CNF fibers. The interfacial bond strength and flexural strength of the GF-CNF fibers/epoxy resins composites were respectively increased by 78% and 20% compared to the pristine glass fibers/epoxy resins composite. Huang et al. [319] used 3-(methacryloyloxy)propyltrimethoxy silane (MEMO) containing the carbon–carbon double bonds to pretreat glass fibers and then initiated the free radical polymerization by the benzoyl peroxide (BPO) to graft branched polyacrylic acid on the surface of glass fibers. When the amount of BPO and polymerization time were respectively 0.3%, and two hours, the grafting amount of acrylic acid on the surface of glass fiber was optimal. The surface energy of the modified glass fiber increased significantly, which was beneficial to enhancing the interfacial bond between the polyamide (PA-6) and glass fibers. In our previous work, Tang et al. [320] applied H_2SO_4 to pretreat glass fibers and then grafted the epichlorohydrin on the surface of glass fibers. Compared with the pristine glass fibers/epoxy resins, the ILSS and flexural strength of modified glass fibers/epoxy resins laminated composites were increased by 7.76% and 22.42%, respectively. Zhang and Teng [321] exploited the poly-1-3,4-dihydroxyphenylalanine to pretreat the surface of aramid fibers and then graft ZnO nanowires (Zn NWs). Results showed that the IFSS value of the modified aramid fibers/epoxy resin composites was 37.8% higher than that of pure aramid fibers/epoxy resin composite. Hu et al. [125] obtained PBO-COOH fibers by treating PBO fibers with sulfuric acid and then successfully grafted a MOF

(UIO-66) on the surface of PBO-COOH fibers to obtain PTF fibers. Compared with pristine PBO fibers, the single fiber pull-out strength of PTFs fiber/epoxy resin was increased by 49.0%.

In addition, plasma treatment and high-energy radiation modification are easy to damage the mechanical strength of the reinforced fibers, and the modification effect decays with the exposure time [322]. In response to the above problems, plasma treatment and high-energy radiation modification are usually used to slightly pretreat reinforced fibers under the small damage degree of the reinforced fibers, introducing an appropriate amount of active groups, which is beneficial for the secondary functionalization (such as the chemical grafting and coating modification) to further improve the effect of interface modification [323].

Jin et al. [324] modified UHMWPE fibers by oxygen plasma pretreatment and then coating the polypyrrole (PPy). The hydrogen bonding between UHMWPE fibers and PPy improved the interfacial bond strength between epoxy resins and UHMWPE fibers. The single fiber pull-out strength of modified UHMWPE fiber/epoxy resins micro-composites was increased from 1.06 (pristine PBO fiber) to 10.05 MPa. Li et al. [325] grafted GO onto the surface of UV-pretreated aramid fibers to significantly improve the chemical activity and roughness of aramid fibers' surface. In our previous work, Gu and coworkers [326] combined the oxygen plasma treatment and then grafting POSS to functionalize the surface of HMPBO fibers. Compared with pristine HMPBO fibers/BADCy, the ILSS of modified HMPBO fibers/BADCy composites was increased by 20.7%. Moreover, Tang et al. [120] modified PBO fibers by coating the random copolymer P(S-*co*-BCB-*co*-MMA) and UV etching treatment to prepare UV-PBO@P fibers (Figure 5.28). When the UV etching time was 96 hours, the ILSS and flexural strength of the UV-PBO@P fibers/BADCy laminated wave-transparent composites were 50.2 and 745.6 MPa, compared to the pristine PBO fibers/BADCy

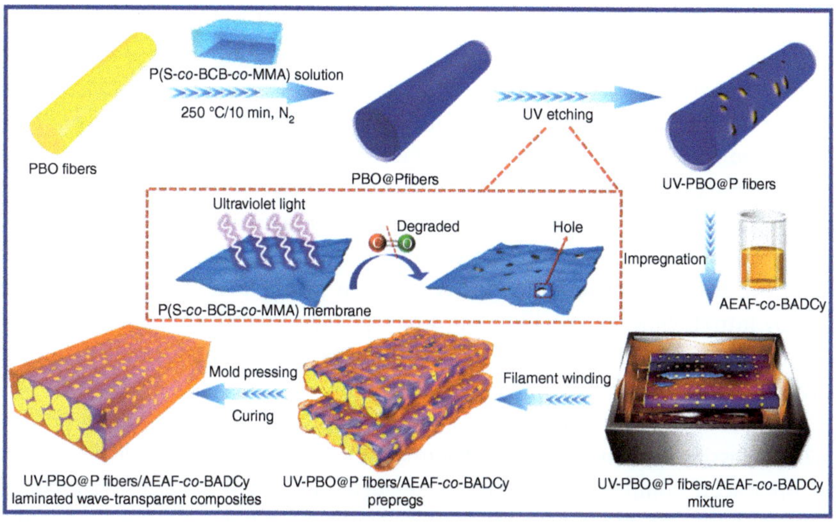

Figure 5.28 Schematic preparation diagram for UV-PBO@P fibers and its laminated wave-transparent composites. Source: [120]/with permission of Elsevier.

(the ILSS for 39.4 MPa and flexural strength for 613.3 MPa), increased by 27.4% and 21.6%, respectively.

In summary, the above-mentioned surface modification methods of the reinforced fibers can effectively improve the surface activity or roughness of the reinforced fibers [327]. However, these methods still have their disadvantages, such as the plasma, the acid etching, and the high-energy ray modification may destroy the micro-fiber structures on the surface of the reinforced fibers to impair the mechanical properties; the introduction of coupling agents usually needs the pretreatment and reduces the heat resistance of the final laminated wave-transparent composites; and the chemical grafting and coatings also introduce new interfaces into the polymer matrix wave-transparent composites, unfavorable for the wave-transparent performances [328, 329].

5.5.2 Interfacial Compatibilizers

Compared with the surface functionalization of the reinforced fibers, the introduction of interfacial compatibilizers into polymer matrix wave-transparent composites can improve the interfacial interaction between polymer matrix and reinforced fibers without damaging the mechanical strength of the reinforced fibers themselves. Meanwhile, this method has a great potential for large-scale industrial applications.

5.5.2.1 Definition of the Interfacial Compatibilizer

The interfacial compatibilizers are a series of compounds that can improve the compatibility and interaction force between polymer matrix and reinforced fibers, which acts as a "bridge" inner polymer matrix wave-transparent composites to form a more stable interface structure and effective stress transfer at the interfaces. In addition, the introduction of an appropriate amount of interfacial compatibilizers can not only reduce the interfacial tension and increase the bonding force of the interfacial layer, but also enable the polymer matrix wave-transparent composites to form phase structures with macroscopic uniformity and thermodynamic stability.

The main actions of introducing interface compatibilizers are to improve the interfacial interaction, thickness of interface layers, and mechanical properties and heat resistances of the polymer matrix wave-transparent composites [330]. Besides, the molecular structures of interfacial compatibilizer can be rationally designed to improve the interfacial compatibility between polymer matrix and reinforced fibers by introducing the specific groups or topologies, which can simultaneously enhance the mechanical properties and wave-transparent performances of the polymer matrix wave-transparent composites.

5.5.2.2 Classification and Action Mechanism of the Interfacial Compatibilizers

The interfacial compatibilizers can be divided into small molecule interfacial compatibilizers and macromolecular interfacial compatibilizers according to the molecular weight. Additionally, the interfacial compatibilizers can also be divided into reactive interfacial compatibilizers and nonreactive interfacial compatibilizers according to whether a chemical reaction between the interfacial compatibilizers and reinforced fibers or polymer matrix [331, 332].

Reactive interfacial compatibilizers can react with certain groups in the molecular structures of the polymer matrix or reinforced fibers. The stable chemical bonds between polymer matrix and reinforced fibers can be generated by the chemical reaction between the interfacial compatibilizers and polymer matrix or reinforced fibers to further enhance the interfacial compatibility of the polymer matrix wave-transparent composites [333]. Nonreactive interfacial compatibilizers without the reactive groups cannot react with the components while processing the polymer matrix wave-transparent composites. Nonreactive interfacial compatibilizers are usually the grafting copolymers and blocking copolymers prepared by the radical polymerization or ionic polymerization [334, 335]. The mechanism of nonreactive interfacial compatibilizers is to form the hydrogen bonds, van der Waals forces, and other secondary interactions or reduce the difference of solubility parameters between polymer matrix and reinforced fibers to improve further the interfacial interaction of the polymer matrix wave-transparent composites. Generally, the reactive interfacial compatibilizer can produce chemical bonding between polymer matrix and reinforced fibers and achieve simultaneous optimization of the comprehensive properties of the polymer composites [336]. However, the reactive interfacial compatibilizers are unsuitable for polymer matrix without containing reactive groups or reinforced fibers with strong surface inertness.

Moreover, the two chain segments of molecular structures for interfacial compatibilizers can be respectively designed in different types for the specific situation of the polymer matrix and reinforced fibers (e.g. reactive at one end; nonreactive at the other end) to combine the advantages of different types for interfacial compatibilizers [337, 338]. For example, for reinforced fibers with certain active groups on the surface and thermoplastic polymer composites system (glass fiber/polytetrafluoroethylene, and so on), the segment that interacts with the polymer matrix can be designed as a nonreactive interface compatibilizer. In contrast, reactive groups are introduced into the segments that interact with reinforced fibers to form chemical bonding to enhance interface compatibility between polymer matrix and reinforced fibers.

5.5.2.3 Design and Synthesis of Interface Compatibilizers

Interface compatibilizers are firstly applied to the wood–plastic composites (WPC) systems and blending-type polymer matrix composites with large polarity differences [339, 340]. Polymer matrix inner WPC systems are usually thermoplastic polymer, and the corresponding reinforced phase is an agricultural or forestry crop such as the wood and straw flour. Because the natural cellulose-based polymers contain a large number of hydroxyl groups with strong polarity, the common thermoplastic polymers, such as polyethylene (PE), polypropylene (PP), and polystyrene (PS), are usually nonpolar polymers, making them lower interface bonding strength between each other [341].

The interfacial compatibilizers with the olefin polymers or thermoplastic elastomers as main chains, followed by introducing the polar side groups, are beneficial to enhancing the interfacial interaction in WPC systems and polymer mixtures [342]. Catto and Santana [343] prepared a grafted interfacial compatibilizer by maleic anhydride grafting PP for the wood powder/PP composites. The results showed that the maleic anhydride grafting PP (CA) as an interfacial compatibilizer could improve the compatibility between the wood powder and plastic matrix and mechanical properties of the resultant composites. The tensile strength and Young's modulus of wood powder/CA/PP composites were respectively increased from 27.5 and 289 MPa (wood powder/PP) to 30 and 356 MPa. Yi et al. [344] produced an interfacial compatibilizer by grafting the GMA to polypropylene main chains for the PET/PP mixtures. Results revealed that the introduction of interfacial compatibilizers significantly improved the interfacial compatibility between PET/PP mixtures. In general, the interfacial compatibilizers of WPC systems and polymer mixtures usually attach polar units, such as the acrylate, the tert-butyl acrylate, and the maleic anhydride (MAH), to the main chains of the polymer matrix with the weak polarity through grafting reaction.

The interfacial compatibilizers for fiber-reinforced polymer matrix composites are usually required to takes into account both the molecular structures and the compatibility between polymer matrix and reinforced fibers, resulting in amphiphilic molecular structures or solubility parameters in between [345]. Zhang et al. [346] introduced a self-designed aminated PPS compatibilizer into carbon fibers reinforced polyphenylene sulfide (CF/PPS) composites. The amino groups in the interfacial compatibilizers could react with the carboxyl groups of the sizing agent on the CF fibers' surface. When the mass fraction of aminated PPS compatibilizer was 7 wt%, the flexural and tensile properties of CF/PPS composites reached the maximum. Still, the impact strength of CF/PPS composites was not significantly improved. Jia et al. [347] prepared the polyaryl ether ketone (PFEK-P) containing biphenyluorene and carboxyl side group structures based on the nucleophilic substitution ternary stepwise copolymerization of the phenolphthalein (PPL), bisphenol fluorene (BHPF), and 4,4′-difluorobenzophenone (DFK). Then, the polyaryletherketone interfacial compatibilizer containing the biphenylfluorene and ester pendant structures (PFEK-E) was prepared by the esterification reaction with PFEK-P and phenol under catalyzed by *p*-toluenesulfonic acid (PTSA). When the mass fraction of PFEK-E was 10 wt%, the flexural strength and flexural modulus of the CF/PFEK-PPENK laminated composites were respectively 1844 MPa and 148 GPa, and the corresponding short-beam shear strength was 84.3 MPa. In the high-temperature environment, the flexural strength and flexural modulus of the CF/PFEK-PPENK laminated composites were respectively 1250 MPa and 127 GPa, and the corresponding short-beam shear strength was 56.6 MPa, higher than that of CF/PPENK composites.

Recently, the design and synthesis of interfacial compatibilizers for polymer matrix wave-transparent composites has also received extensive attention. Chen and coworkers [348] prepared a series of different grades for interfacial compatibilizers by grafting different amounts of MAH onto the PP molecular chains. The tensile strength and flexural strength of the glass fibers/PP laminated composites with 3 wt% interfacial compatibilizers were respectively improved by 122% and 120%. Liu et al. [349] synthesized the hydroxyphosphated castor oil (PCO) blended with epoxy resins (E-51) as the interfacial compatibilizer. When the ratio of PCO to E-51 was 3:1, the single fiber pull-out strength of the PBO fiber/PCOO/epoxy resin micro-composite was 12.03 MPa, higher than that of unmodified PBO fibers/epoxy resin micro-composite (4.72 MPa), and the mechanical strength of PBO fibers was not damaged (Figure 5.29).

Wu et al. [350] synthesized a PBO fiber interfacial compatibilizer, 2,6-bis(2-hydroxy-4-aminophenyl) benzobisoxazole (HABO). Results revealed that the wettability, surface roughness, and compatibility of the modified PBO fibers with epoxy resins were significantly improved. The ILSS of the PBO fibers/epoxy resins composites was increased to 29.2 MPa, compared to unmodified PBO fibers/epoxy resins composites (20.8 MPa) increased by 40.4% (Figure 5.30).

Meanwhile, the special structures and groups can also be introduced into the molecular structures of interfacial compatibilizers, which can improve the interfacial bonding strength of the polymer matrix wave-transparent composites and endow them with the certain functional performances [351, 352].

In our previous work, a new method for synthesizing interfacial compatibilizers based on the design of PBO-like units, reactive groups, and molecular topology was proposed and established to address the critical problem of poor interfacial compatibility between PBO fibers and BADCy matrix. A series of interfacial compatibilizers with different molecular structures and topologies were synthesized, which could effectively solve the existing scientific problems of PBO fibers surface functionalization, such as damaging to PBO fibers, the low ILSS of laminated wave-transparent composites, or inability to achieve large-scale preparation. Firstly, the epoxy-terminated small molecule of PBO precursor (epoxy-*pre*PBO) was used as an interfacial compatibilizer to improve the interfacial compatibility between PBO fibers and BADCy resins by introducing PBO-like structures into the curing network of BADCy resins (Figure 5.31a) [353]. The epoxy-terminated PBO precursor modified BADCy resin (PBO-*co*-BADCy) had the optima comprehensive properties with the mass fraction of 7 wt% epoxy-*pre*PBO. Compared with the PBO fibers/BADCy, the ILSS and flexural strength of the PBO fibers/PBO-*co*-BADCy composites increased from 40.3 and 610.6 MPa to 48.4 and 658.5 MPa, respectively. The corresponding ε and $\tan\delta$ values were respectively 2.68 and 0.006, lower than the pristine PBO fibers/BADCy composites (ε for 3.13 and $\tan\delta$ for 0.010). Then, Liu et al. [354] also obtained the small molecule of fluorine-containing PBO precursors (epoxy-*pre*FPBO) to simultaneously improve the wave-transparent performances, mechanical properties, and interfacial compatibility between BADCy matrix and PBO fibers by the introduction of epoxy groups, fluorine-containing groups and PBO-like structures (Figure 5.31b). When the mass fraction of epoxy-*pre*FPBO was 7 wt%, the ε and $\tan\delta$ values of FPBO-*co*-BADCy resins at 1 MHz were respectively 2.51 and 0.0087, and the corresponding theoretical wave transmission rate ($|T|^2$) of 94.8%, higher than that of pure BADCy resins ($|T|^2$ for 92.6%). The flexural

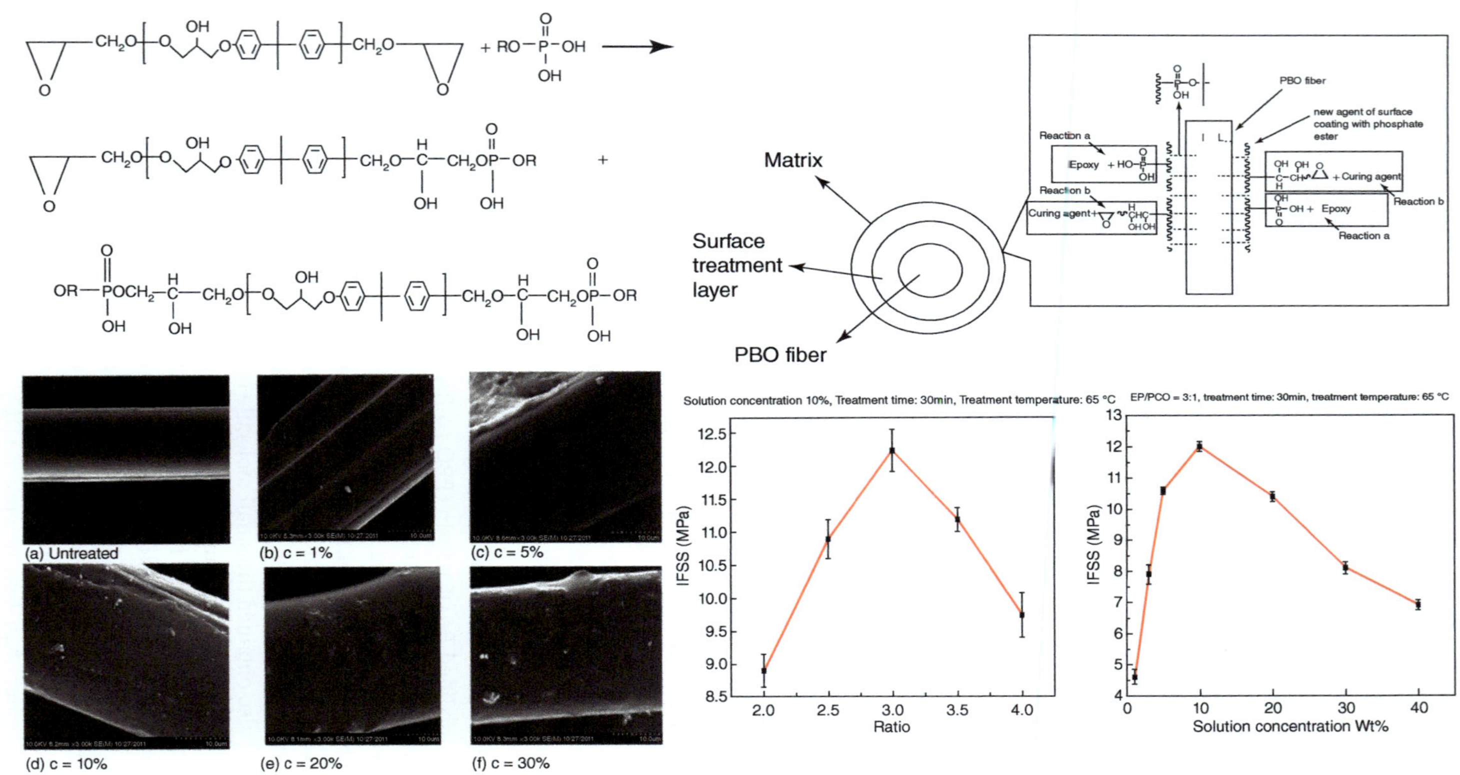

Figure 5.29 Schematic diagram of preparation for interfacial compatibilizer and the properties for PBO fibers/PCO/epoxy resin composites. Source: [349]. Liu et al. (2014)/Reproduced from John Wiley & Sons, Inc.

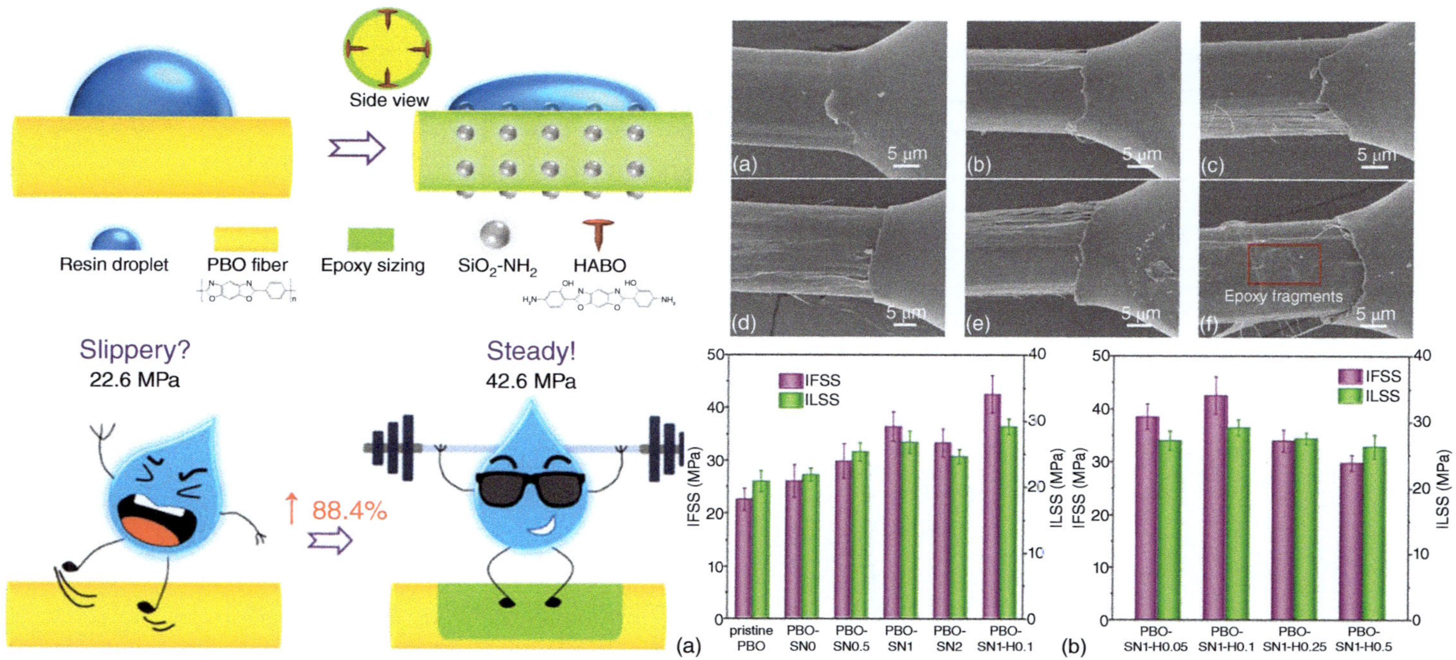

Figure 5.30 Schematic diagram of preparation for HABO interfacial compatibilizer. Source: [350]. Wu et al. (2020)/Reproduced from American Chemical Society.

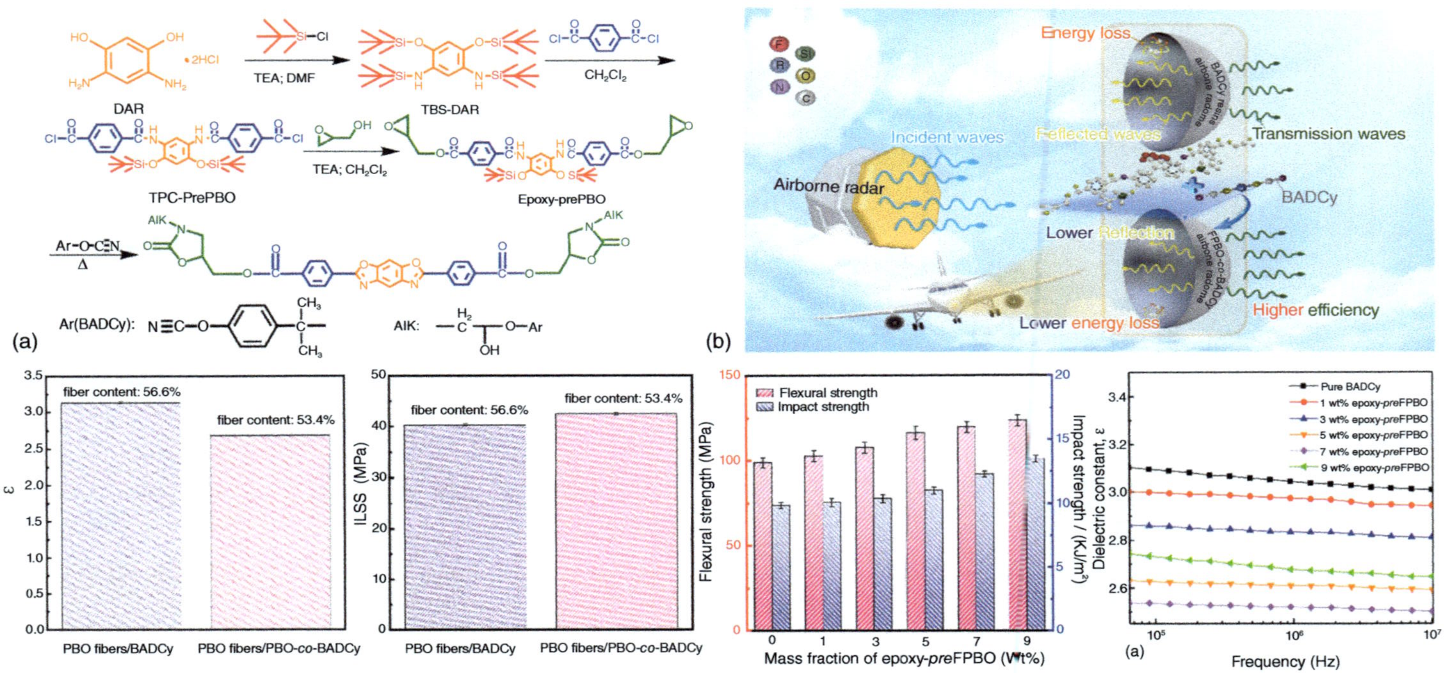

Figure 5.31 Schematic diagram of preparation for small molecule interfacial compatibilizers and properties for PBO fibers/BADCy laminated wave-transparent composites [353, 354]. Source: [354]. Zheng Liu et al. (2019)/Reproduced with permission from Elsevier.

and impact strengths of FPBO-*co*-BADCy resins were 119.9 MPa and 12.3 kJ/m^2, increased by 21.4% and 24.5% compared to the pure BADCy resins, respectively. Furthermore, FPBO-*co*-BADCy resins had relatively better interfacial bond strength with PBO fibers than pure BADCy resins. The single fiber pull-out strength of PBO fibers/FPBO-*co*-BADCy micro-composite increased from 2.9 (PBO fibers/BADCy) to 3.5 MPa.

To the best knowledge of our knowledge, thermoplastic polymer modified CE resins can greatly improve the overall properties of the CE resins owing to the formation of an interpenetrating polymer network (IPN) [355–357]. Liu et al. [358] prepared modified BADCy resins with a series of semi-interpenetrating polymer networks (semi-IPN) by dissolving thermoplastic polyimide (TPI) in BADCy. The modified BADCy resins demonstrated better mechanical properties (impact strength increased by 47–320%) and dielectric properties (ε stabled at 3.00–3.22) compared to the pure BADCy resin. In our previous work, Gu et al. [359] synthesized fluorine-containing linear PBO precursor (*pre*FLPBO) as the interfacial compatibilizer to prepare PBO fibers/FLPBO-*m*-BADCy laminated wave-transparent composites. The fluorine groups and PBO-like structures could, respectively, improve the dielectric properties of the BADCy matrix and the interfacial compatibility between the BADCy matrix and PBO fibers. Additionally, the partly interpenetrating network formed between *pre*FLPBO and BADCy could significantly improve the mechanical properties (Figure 5.32).

PBO fibers/FLPBO-*m*-BADCy laminated wave-transparent composites with 7 wt% *pre*FLPBO displayed the optimal comprehensive performances. The ε and $\tan\delta$ values of PBO fibers/FLPBO-*m*-BADCy were respectively 2.90 and 0.0042, and the corresponding $|T|^2$ at 1 MHz is 93.3%, higher than that of PBO fibers/BADCy laminated wave-transparent composites. The ILSS and flexural strength of PBO fibers/FLPBO-*m*-BADCy laminated wave-transparent composites were increased from 36.7 and 587.4 MPa to 44.8 and 628.0 MPa, respectively. In addition, the corresponding volume resistivity and breakdown voltage were $2.7 \times 10^{15}\ \Omega$ cm and 19.12 kV/mm, respectively.

Hyperbranched polymers display high solubility, low viscosity, abundant terminal groups, and so on [360]. The inside linear unit and special topological structures of the hyperbranched polymers can increase the free volume in favor of promoting the mechanical and dielectric properties of thermosetting resins. Liu et al. [361] used liquid crystal hyperbranched polysiloxane (LCPSi) with an amino terminator to modify EP/BADCy resins (eCE). The impact strength and flexural modulus of the LCPSi/eCE (with 1.5 wt% LCPSi) were 2.3 and 1.5 times higher than eCE, respectively. Besides, the ε and $\tan\delta$ values of LCPSi/eCE were lower than unmodified eCE. In our previous work, Liu et al. [362] introduced the hyperbranched fluorine and epoxy-containing PBO precursor (epoxy-*pre*FHBPBO) into BADCy resins. Compared with PBO fibers/BADCy, the ILSS and flexural strength of the PBO fibers/FHBPBO-*co*-BADCy laminated composites with 7 wt% epoxy-*pre*FHBPBO were respectively enhanced to 46.9 and 634.9 MPa, increased by 8.1% and 27.8%. The corresponding ε and $\tan\delta$ were 2.65 and 0.003, respectively, lower than those of PBO fibers/BADCy (3.06 and 0.006) composites. Meanwhile,

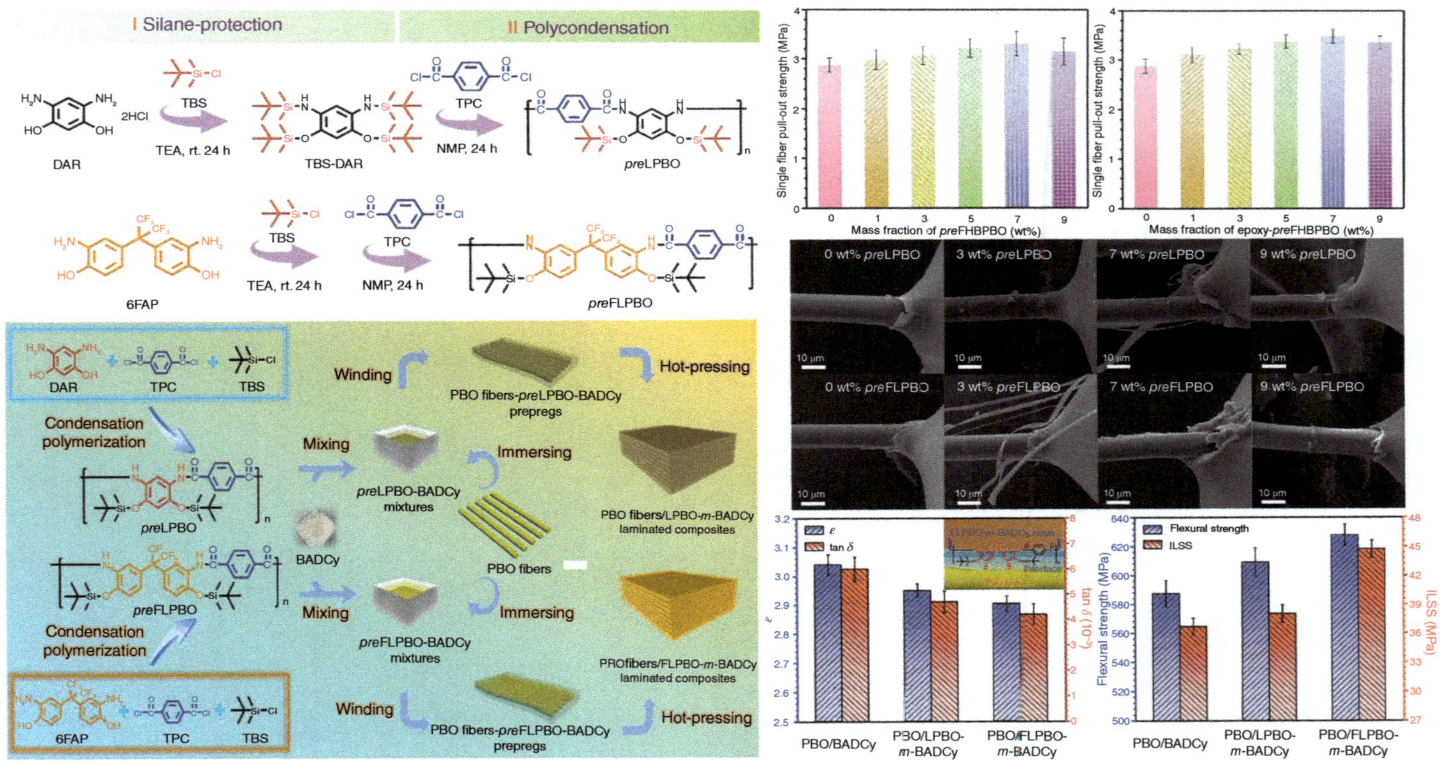

Figure 5.32 Schematic diagram of preparation for linear interfacial compatibilizers and properties of PBO fiber/BADCy laminated wave-transparent composites. Source: [359]. Liu et al. (2021)/Reproduced with permission from Elsevier.

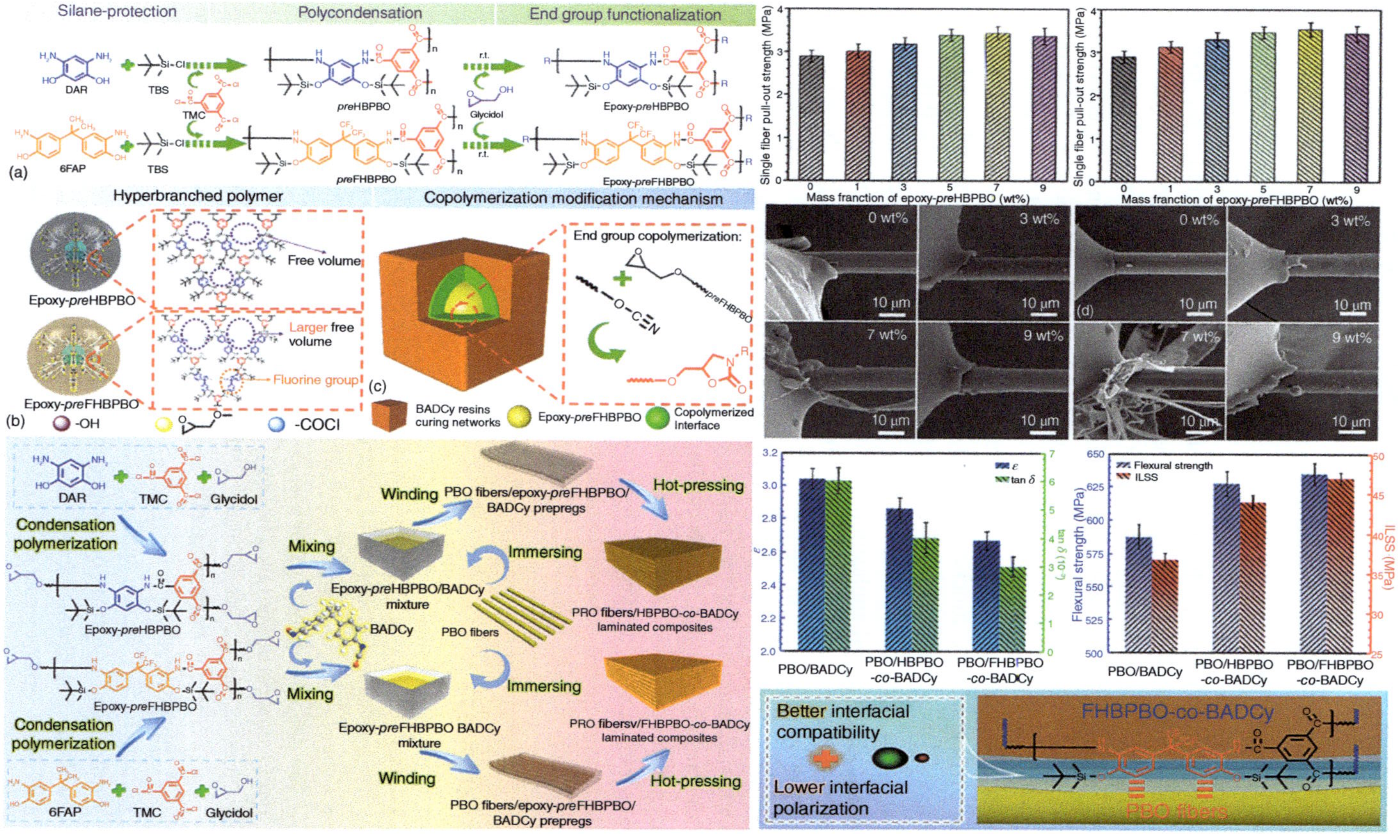

Figure 5.33 Schematic diagram of preparation for hyperbranched structure interfacial compatibilizers and properties PBO fiber/BADCy laminated wave-transparent composites. Source: [362]. Liu et al. (2021)/Reproduced with permission from Elsevier.

the corresponding volume resistivity and breakdown strength were 4.93×10^{15} Ω cm and 25.38 kV/mm, higher than PBO fibers/BADCy composites (2.2×10^{15} Ω cm and 17.69 kV/mm) (Figure 5.33).

5.6 Summary

This chapter describes the basic concept, interface formation, and interface mechanism of the interfaces for polymer matrix wave-transparent composites and then analyzes the interface design and optimal methods between polymer matrix and reinforced fibers. Simultaneously, the physical and chemical performances of the interfaces for polymer matrix wave-transparent composites, the series of characterization methods of interfacial bond strength, and the improvement effect and research progress of above methods are also introduced. Generally, the existing surface functionalization of the reinforced fibers and synthesis of interfacial compatibilizers still have some problems, such as damage to reinforced fibers at a certain extent, relatively poor modification effect, or difficulty in large-scale industrial application, resulting in poor interfacial bonding strength of the final polymer matrix wave-transparent composites. Analyzing the mechanism of interfacial interaction between polymer matrix and reinforced fibers of the polymer matrix wave-transparent composites can help researchers to select targeted functionalization methods for reinforced fibers or design interfacial compatibilizers with specific structures, which can be beneficial to simultaneously enhancing the mechanical properties and wave-transparent performances of the polymer matrix wave-transparent composites. It is an important development direction for the interface design and optimal regulation of wave-transparent/loading polymer-based composites in the future.

References

1 Cazan, C., Enesca, A., and Andronic, L. (2021). Synergic effect of TiO_2 filler on the mechanical properties of polymer nanocomposites. *Polymers* 13 (12): 2017.

2 Chen, J., Yan, L., Song, W., and Xu, D. (2018). Interfacial characteristics of carbon nanotube-polymer composites: a review. *Composites Part A: Applied Science and Manufacturing* 114: 149–169.

3 Chen, R., Yu, R., Pei, X. et al. (2021). Interface design of carbon filler/polymer composites for electromagnetic interference shielding. *New Journal of Chemistry* 45 (19): 8370–8385.

4 Chen, S., Wei, L., Cheng, B. et al. (2020). Influences of interface structure on tribological properties of engineering polymer blends: a review. *Journal of Polymer Engineering* 40 (8): 629–636.

5 Chen, X. and Vereecken, P.M. (2019). Solid and solid-like composite electrolyte for lithium ion batteries: engineering the ion conductivity at interfaces. *Advanced Materials Interfaces* 6 (1): 1800899.

6 Cheng, S., Guo, P., and Heng, L. (2020). Regulation of adhesion between polymer solid-liquid composite interface and liquid. *Acta Polymerica Sinica* 51 (5): 530–547.
7 Chung, D. (2019). A review of multifunctional polymer-matrix structural composites. *Composites Part B: Engineering* 160: 644–660.
8 Chung, D.D.L. (2019). Interface-derived solid-state viscoelasticity exhibited by nanostructured and microstructured materials containing carbons or ceramics. *Carbon* 144: 567–581.
9 Deng, Y., Cui, R., Wang, Z., and Zhuo, A. (2018). Research progress of polymer/biomass composites. *China Plastics* 32 (7): 10–20.
10 Di Boon, Y. and Joshi, S.C. (2020). A review of methods for improving interlaminar interfaces and fracture toughness of laminated composites. *Materials Today Communications* 22: 100830.
11 Du, B., Kong, X., Xiao, M. et al. (2018). Advances in thermal performance of polymer-based composites. *Transactions of China Electrotechnical Society* 33 (14): 3149–3159.
12 Fioravante, A., de Oliveira, P.C., Siqueira, E.J. et al. (2020). From brittle-to-ductile fracture of polymer composites: the incorporation of energy dissipation mechanisms by carbon nanotubes-based multilayered interface. *Journal of Applied Polymer Science* 137 (44): e49348.
13 Gao, F., Zhang, K., Guo, Y. et al. (2021). (Ba, Sr)TiO_3/polymer dielectric composites-progress and perspective. *Progress in Materials Science* 121: 100813.
14 Grundish, N.S., Goodenough, J.B., and Khani, H. (2021). Designing composite polymer electrolytes for all-solid-state lithium batteries. *Current Opinion in Electrochemistry* 30: 100828.
15 Guo, P., Wang, Z., Han, X., and Heng, L. (2021). Nepenthes pitcher inspired isotropic/anisotropic polymer solid-liquid composite interface: preparation, function, and application. *Materials Chemistry Frontiers* 5 (4): 1716–1742.
16 Guo, Y., Ruan, K., Shi, X. et al. (2020). Factors affecting thermal conductivities of the polymers and polymer composites: a review. *Composites Science and Technology* 193: 108134.
17 He, X. and Wang, Y. (2021). Recent advances in the rational design of thermal conductive polymer composites. *Industrial and Engineering Chemistry Research* 60 (3): 1137–1154.
18 Hejna, A., Przybysz-Romatowska, M., Kosmela, P. et al. (2020). Recent advances in compatibilization strategies of wood-polymer composites by isocyanates. *Wood Science and Technology* 54 (5): 1091–1119.
19 Huang, S., Fu, Q., Yan, L., and Kasal, B. (2021). Characterization of interfacial properties between fibre and polymer matrix in composite materials—a critical review. *Journal of Materials Research and Technology* 13: 1441–1484.
20 Huang, Y., Ellingford, C., Bowen, C. et al. (2020). Tailoring the electrical and thermal conductivity of multi-component and multi-phase polymer composites. *International Materials Reviews* 65 (3): 129–163.

21 Jia, H., Qiao, Y., Zhang, Y. et al. (2020). Interface modification strategy of basalt fiber reinforced resin matrix composites. *Progress in Chemistry* 32 (9): 1307–1315.

22 Khandelwal, S. and Rhee, K.Y. (2020). Recent advances in basalt-fiber-reinforced composites: tailoring the fiber-matrix interface. *Composites Part B: Engineering* 192: 108011.

23 Kim, H.S., Jang, J.U., Lee, H. et al. (2018). Thermal management in polymer composites: a review of physical and structural parameters. *Advanced Engineering Materials* 20 (10): 1800204.

24 Li, Y., Mei, Y., Wang, Y. et al. (2018). The state of art of nano-molding technique applying to the production of metal/polymer composites. *Materials Review* 32 (7A): 2295–2303.

25 Liu, S., Liu, W., Ba, D. et al. (2022). Filler-integrated composite polymer electrolyte for solid-state lithium batteries. *Advanced Materials* e2110423.

26 Luan, Y., Ren, D., and Xu, D. (2021). Research progress on effect of transcrystallinity on mechanical properties of natural fiber reinforced polymer composites. *Materials Review* 35 (3A): 5195–5198.

27 Luo, H., Zhou, X., Ellingford, C. et al. (2019). Interface design for high energy density polymer nanocomposites. *Chemical Society Reviews* 48 (16): 4424–4465.

28 Mashayekhi, F., Bardon, J., Berthe, V. et al. (2021). Fused filament fabrication of polymers and continuous fiber-reinforced polymer composites: advances in structure optimization and health monitoring. *Polymers* 13 (5): 789.

29 Mehra, N., Mu, L., Ji, T. et al. (2018). Thermal transport in polymeric materials and across composite interfaces. *Applied Materials Today* 12: 92–130.

30 Mohit, H. and Selvan, V.A.M. (2018). A comprehensive review on surface modification, structure interface and bonding mechanism of plant cellulose fiber reinforced polymer based composites. *Composite Interfaces* 25 (5–7): 629–667.

31 Mondal, R.K., Dubey, K.A., and Bhardwaj, Y.K. (2021). Role of the interface on electron transport in electro-conductive polymer-matrix composite: a review. *Polymer Composites* 42 (6): 2614–2628.

32 Mrajji, O., El Wazna, M., El Bouari, A., and Cherkaoui, O. (2022). The properties of feather fiber-reinforced polymer composites: a review. *Journal of Natural Fibers* 19 (13): 4868–4885.

33 Paul, R. and Dai, L. (2018). Interfacial aspects of carbon composites. *Composite Interfaces* 25 (5–7): 539–605.

34 Salahuddin, B., Faisal, S.N., Baigh, T.A. et al. (2021). Carbonaceous materials coated carbon fibre reinforced polymer matrix composites. *Polymers* 13 (16): 2771.

35 Sun, T., Wang, J., Zhang, Y. et al. (2018). Research progress of polymer-based thermoelectric composite materials. *Journal of Materials Science and Engineering* 36 (3): 511–518.

36 Sun, Z. (2019). Hyperbranched polymers in modifying natural plant fibers and their applications in polymer matrix composites: a review. *Journal of Agricultural and Food Chemistry* 67 (32): 8715–8724.

37 Colijn, I. and Schroen, K. (2021). Thermoplastic bio-nanocomposites: from measurement of fundamental properties to practical application. *Advances in Colloid and Interface Science* 292: 102419.

38 Grayson, B., Aljasar, A., Cheng, L. et al. (2022). Advances in shear strengthening of concrete bridge girders. *Journal of Bridge Engineering* 27 (6): 03122002.

39 Poggetti, I., Dyson, J., Sanchez, D.M. et al. (2018). Distributed internal strain measurement of the fluid-solid state coefficients of thermal expansion below the glass transition temperature during a composite manufacturing process. *Journal of Composite Materials* 52 (22): 3053–3084.

40 Wang, L., Yang, J., Cheng, W. et al. (2021). Progress on polymer composites with low dielectric constant and low dielectric loss for high-frequency signal transmission. *Frontiers in Materials* 8: 774843.

41 Wang, W., De Freitas, S.T., Poulis, J.A., and Zarouchas, D. (2021). A review of experimental and theoretical fracture characterization of bi-material bonded joints. *Composites Part B: Engineering* 206: 108537.

42 Wang, W. and Li, S. (2020). Research status and development of insulation breakdown in engineering solid dielectrics. *Chinese Science Bulletin* 65 (31): 3461–3474.

43 Bardestani, R., Patience, G.S., and Kaliaguine, S. (2019). Experimental methods in chemical engineering: specific surface area and pore size distribution measurements-BET, BJH, and DFT. *Canadian Journal of Chemical Engineering* 97 (11): 2781–2791.

44 Eder, M., Schaeffner, W., Burgert, I., and Fratzl, P. (2021). Wood and the activity of dead tissue. *Advanced Materials* 33 (28): 2001412.

45 Fan, Y., Fowler, G.D., and Zhao, M. (2020). The past, present and future of carbon black as a rubber reinforcing filler: a review. *Journal of Cleaner Production* 247: 119115.

46 Gao, X., Guo, C., Hao, J. et al. (2020). Adsorption of heavy metal ions by sodium alginate based adsorbent-a review and new perspectives. *International Journal of Biological Macromolecules* 164: 4423–4434.

47 Huang, J., Zhou, J., and Liu, M. (2022). Interphase in polymer nanocomposites. *JACS Au* 2 (2): 280–291.

48 Jia, Q., Song, H., Tang, S. et al. (2019). Synthesis of the functionalized porous materials and their applications in the specific recognition and separation. *Progress in Chemistry* 31 (8): 1148–1158.

49 Kamran, U., Rhee, K., Lee, S., and Park, S. (2022). Innovative progress in graphene derivative-based composite hybrid membranes for the removal of contaminants in wastewater: a review. *Chemosphere* 306: 135590.

50 Olbrich, W., Kadyk, T., Sauter, U., and Eikerling, M. (2022). Review wetting phenomena in catalyst layers of PEM fuel cells: novel approaches for modeling and materials research. *Journal of the Electrochemical Society* 169 (5): 054521.

51 Ren, Q., Wang, J., Yang, L. et al. (2020). Research progress of conductive polymer composites for resistive flexible strain sensors. *Materials Review* 34 (1A): 1080–1094.

52 Torres, L. Jr., and Bienek, D.R. (2020). Use of protein repellents to enhance the antimicrobial functionality of quaternary ammonium containing dental materials. *Journal of Functional Biomaterials* 11 (3): 54.

53 Chen, L., Zhou, J., Chu, H.W. et al. (2018). A review on water vapor pressure model for moisture permeable materials subjected to rapid heating. *Applied Mechanics Reviews* 70 (2): 020803.

54 Guan, H., Wang, Q., Wu, X. et al. (2021). Biomass derived porous carbon (BPC) and their composites as lightweight and efficient microwave absorption materials. *Composites Part B: Engineering* 207: 108562.

55 Li, X., Wang, X.Y., and Sun, J. (2021). Recent progress in the carbon-based frameworks for high specific capacity anodes/cathode in lithium/sodium ion batteries. *New Carbon Materials* 36 (1): 106–114.

56 Ma, C., Cui, W., Liu, X. et al. (2022). In situ preparation of gel polymer electrolyte for lithium batteries: progress and perspectives. *Infomat* 4 (2): 12232.

57 Qian, H. and Lu, Z. (2020). Interface properties in polymer/single-chain nanoparticle composite. *Acta Polymerica Sinica* 51 (1): 55–65.

58 Wang, J. and Liu, X. (2021). Forward osmosis technology for water treatment: recent advances and future perspectives. *Journal of Cleaner Production* 280: 124354.

59 Zhan, X., Sun, X., Xu, H., and Li, J. (2019). Membrane materials for desulfurization of gasoline via pervaporation. *Progress in Chemistry* 31 (5): 752–759.

60 Zhu, B., Jiang, X., He, S. et al. (2020). Rational design of poly(ethylene oxide) based membranes for sustainable CO_2 capture. *Journal of Materials Chemistry A* 8 (46): 24233–24252.

61 Oladunni, J., Zain, J.H., Hai, A. et al. (2018). A comprehensive review on recently developed carbon based nanocomposites for capacitive deionization: from theory to practice. *Separation and Purification Technology* 207: 291–320.

62 Venkatakrishnan, A. and Kuppa, V.K. (2018). Polymer adsorption on rough surfaces. *Current Opinion in Chemical Engineering* 19: 170–177.

63 Wang, H., Chen, H., Cui, M. et al. (2018). Progress in preparation of MOFs/polymer membrane for liquid separation. *New Chemical Materials* 46 (12): 11–15.

64 Lee, J., Lee, T., Char, K. et al. (2021). Issues and advances in scaling up sulfide-based all solid-state batteries. *Accounts of Chemical Research* 54 (17): 3390–3402.

65 Wang, F. and Sun, P. (2021). Solid-state NMR characterization of polymers. *Acta Polymerica Sinica* 52 (7): 840–856.

66 Liu, Y., Zhang, W., Zhao, C. et al. (2019). Study on the synthesis of poly(pyrrole methane)s with the hydroxyl in different substituent position and their selective adsorption for Pb2+. *Chemical Engineering Journal* 361: 528–537.

67 Wu, Y., Ma, J., and Bao, Y. (2018). Advances in interfacial interaction within polymer matrix nanocomposites. *Materials Review* 32 (2): 434–442.

68 Chen, H., Ling, M., Hencz, L. et al. (2018). Exploring chemical, mechanical, and electrical functionalities of binders for advanced energy-storage devices. *Chemical Reviews* 118 (18): 8936–8982.

69 Jung, J., Chang, M., and Yoon, H. (2018). Interface engineering strategies for fabricating nanocrystal-based organic-inorganic nanocomposites. *Applied Sciences-Basel* 8 (8): 1376.

70 Nie, K., Hong, Y., Qiu, J. et al. (2018). Interfaces between cathode and electrolyte in solid state lithium batteries: challenges and perspectives. *Frontiers in Chemistry* 6: 616.

71 Li, Y., Yi, X., Yu, T., and Xian, G. (2018). An overview of structural-functional-integrated composites based on the hierarchical microstructures of plant fibers. *Advanced Composites and Hybrid Materials* 1 (2): 231–246.

72 Agista, M.N., Guo, K., and Yu, Z. (2018). A state of the art review of manoparticles application in petroleum with a focus on enhanced oil recovery. *Applied Sciences* 8 (6): 871.

73 Raphael, N., Namratha, K., Chandrashekar, B.N. et al. (2018). Surface modification and grafting of carbon fibers: a route to better interface. *Progress in Crystal Growth and Characterization of Materials* 64 (3): 75–101.

74 Bauld, R., Choi, D.Y.W., Bazylewski, P. et al. (2018). Thermo-optical characterization and thermal properties of graphene-polymer composites: a review. *Journal of Materials Chemistry C* 6 (12): 2901–2914.

75 Li, X., Li, Y., and Hu, C. (2018). Progress in preparation and application of polypyrrole nanocomposites. *Chemical Industry and Engineering Progress* 37 (4): 1489–1500.

76 Ansari, M.T.A., Singh, K.K., and Azam, M.S. (2018). Fatigue damage analysis of fiber-reinforced polymer composites—a review. *Journal of Reinforced Plastics and Composites* 37 (9): 636–654.

77 Monsalve-Bravo, G.M. and Bhatia, S.K. (2018). Modeling permeation through mixed-matrix membranes: a review. *Processes* 6 (9): 172.

78 Balla, V.K., Kate, K.H., Satyavolu, J. et al. (2019). Additive manufacturing of natural fiber reinforced polymer composites: processing and prospects. *Composites Part B: Engineering* 174: 106956.

79 Deshmukh, G.S. (2022). Advancement in hemp fibre polymer composites: a comprehensive review. *Journal of Polymer Engineering* 42 (7): 575–598.

80 Eyckens, D.J., Stojcevski, F., Hendlmeier, A. et al. (2021). Carbon fibre surface chemistry and its role in fibre to matrix adhesion. *Journal of Materials Chemistry A* 9 (47): 26528–26572.

81 Kenned, J.J., Sankaranarayanasamy, K., and Kumar, C.S. (2021). Chemical, biological, and nanoclay treatments for natural plant fiber-reinforced polymer composites: a review. *Polymers & Polymer Composites* 29 (7): 1011–1038.

82 Rogovina, S.Z., Prut, E.V., and Berlin, A.A. (2019). Composite materials based on synthetic polymers reinforced with natural fibers. *Polymer Science, Series A* 61 (4): 417–438.

83 Alshammari, B.A., Alsuhybani, M.S., Almushaikeh, A.M. et al. (2021). Comprehensive review of the properties and modifications of carbon fiber-reinforced thermoplastic composites. *Polymers* 13 (15): 2474.

84 Shao, Q., Hu, Z., Xu, X.R. et al. (2018). Mussel-inspired immobilization of BN nanosheets onto poly(p-phenylene benzobisoxazole) fibers: multifunctional interface for photothermal self-healing. *Applied Surface Science* 440: 1159–1165.

85 Tang, L., Yang, Z.Y., Tang, Y.S. et al. (2021). Facile functionalization strategy of PBO fibres for synchronous improving the mechanical and wave-transparent properties of the PBO fibres/cyanate ester laminated composites. *Composites Part A: Applied Science and Manufacturing* 150: 106622.

86 Chen, J., Zhu, Y., Ni, Q. et al. (2014). Surface modification and characterization of aramid fibers with hybrid coating. *Applied Surface Science* 321: 103–108.

87 Zhang, X., Shi, B., and Wang, Y. (2012). Influence of diameter on surface dispersive free energy of polyethersulfone nano-fibers. *Journal of Adhesion Science and Technology* 26 (1-3): 353–360.

88 Okamura, Y., Tagawa, M., Gotoh, F. et al. (1996). Influence of drawing and beat treatment on surface free energies of polyethylene terephthalate fibers. *Colloid and Polymer Science* 274 (7): 628–633.

89 Ahmed, S., Wang, X., Li, K., and Ulven, C.A. (2018). Investigation on the surface free energy of a single flax fiber through adhesion measurement by atomic force microscopy. *Journal of Composite Materials* 52 (23): 3233–3239.

90 Rubio, J., Rubio, F., Fierro, J.L. et al. (2002). Surface characterization of carbon fibers by inverse gas chromatography at low pressures. *Journal of Materials Research* 17 (2): 413–422.

91 Stachewicz, U., Dijksman, J.F., Soudani, C. et al. (2017). Surface free energy analysis of electrospun fibers based on Rayleigh-Plateau/Weber instabilities. *European Polymer Journal* 91: 368–375.

92 Deng, R., Li, M., Zhang, Z., and Jia, J. (2007). Surface free energy of basalt and glass fibers by contact angle method. *Journal of Beijing University of Aeronautics and Astronautics* 33 (11): 1349–1352.

93 Emig, G., Popovska, N., Edie, D.D., and Rhee, B.S. (1995). Surface free-energy of carbon fibers with circular and noncircular cross-section coating with silicon-carbide. *Carbon* 33 (6): 779–782.

94 Gotoh, K., Yasukawa, A., Ohkita, M. et al. (1995). Wettability and surface free energy of electrolytically oxidized graphite fibers. *Colloid and Polymer Science* 273 (12): 1144–1150.

95 Tze, W.T. and Gardner, D.J. (2001). Contact angle and IGC measurements for probing surface-chemical changes in the recycling of wood pulp fibers. *Journal of Adhesion Science and Technology* 15 (2): 223–241.

96 Lindsay, B., Abel, M.-L., and Watts, J.F. (2007). A study of electrochemically treated PAN based carbon fibres by IGC and XPS. *Carbon* 45 (12): 2433–2444.

97 Huang, X., Li, B., Shi, B. et al. (2008). Surface characterization of glass fiber by inverse gas chromatography. *Journal of Wuhan University of Technology Materials Science Edition* 23 (5): 687–690.

98 Bledzki, A.K., Lieser, J., Wacker, G., and Frenzel, H. (1997). Characterization of the surfaces of treated glass fibres with different methods of investigation. *Composite Interfaces* 5 (1): 41–53.

99 Papirer, E., Brendle, E., Balard, H., and Vergelati, C. (2000). Inverse gas chromatography investigation of the surface properties of cellulose. *Journal of Adhesion Science and Technology* 14 (3): 321–337.

100 Price, G.J., Pastor, M.L., Towo, A.N. et al. (2007). Shear strength at Sisal fibre-polyester resin interfaces: use of inverse gas chromatography to study pretreatment effects. *Composite Interfaces* 14 (1): 21–31.

101 Li, G., Chen, B., and Liu, J. (2017). Research progress of surface modification technology of carbon fiber. *New Chemical Materials* 45 (7): 39–41.

102 Song, W., Gong, Y., Sheng, X., and Hao, Z. (2019). Research progress of hemp fiber reinforced polymer composites. *Polymer Bulletin* 4: 27–32.

103 Song, J., Xiao, C., An, S., and Jia, G. (2007). Surface treatment of UHMWPE fibers for adhesion promotion in epoxy polymers. *Journal of Donghua University* (3): 329–332.

104 Liu, Z., Fan, X.L., Han, M.Y. et al. (2022). Significantly improved interfacial properties and wave-transparent performance of PBO fibers/cyanate esters laminated composites via introducing a polydopamine/ZIF-8 hybrid membrane. *Composites Science and Technology* 223: 109426.

105 Sun, Z., Zhao, X., and Ma, J. (2014). Capillary effect in the impregnation of jute fiber mat reinforced polypropylene composites. *Journal of Composite Materials* 48 (4): 447–454.

106 Zhu, L., Zhang, Z., Li, M. et al. (2010). Contact angle and action of adhesion between epoxy resin and fibers at processing temperatures. *Acta Materiae Compositae Sinica* 27 (5): 41–46.

107 Mao, Y., Pan, Q., Wang, W., and Yang, K. (2017). Study on wetting behavior between phenolic resin system and glass fiber. *Fiber Reinforced Plastics/Composites* 4: 66–69.

108 Zhu, M.M. and Ma, J.X. (2019). Basalt fiber modified with lanthanum-ethylenediaminetetraacetic acid as potential reinforcement of cyanate matrix composites. *Applied Surface Science* 464: 636–643.

109 Tao, Y., Liu, Z., Ding, G., and Wang, Y. (2018). Preparation and property of modified basalt fiber/natural rubber composite. *New Chemical Materials* 46 (7): 60–63.

110 Peng, H. and Lu, Q. (2007). The research on character of modified melamine resin fiber. *Polymer Materials Science & Engineering* 23 (1): 165–168.

111 Pasto, A.E., Braski, D.N., Watkins, T.R. et al. (1999). Characterization techniques for composites and other advanced materials. *Composites Part B: Engineering* 30 (7): 631–646.

112 Hassan, S.F. and Gupta, M. (2002). Development of ductile magnesium composite materials using titanium as reinforcement. *Journal of Alloys and Compounds* 345 (1-2): 246–251.

113 Liu, L., Jia, C., He, J. et al. (2015). Interfacial characterization, control and modification of carbon fiber reinforced polymer composites. *Composites Science and Technology* 121: 56–72.

114 Molins, R., Bartout, J.D., and Bienvenu, Y. (1991). Microstructure and analytical characterization of Al_2O_3 (Al-Mg) composites interfaces. *Materials Science &*

Engineering, A: Structural Materials: Properties, Microstructure and Processing 135: 111–117.

115 Rokhlin, S.I. and Matikas, T.E. (1996). Ultrasonic characterization of surfaces and interphases. *MRS Bulletin* 21 (10): 22–29.

116 Acharya, S.K., Mishra, P.P., and Mehar, S.K. (2009). The influence of fiber treatment on the performance of bagasse fiber-reinforced polymer composite. *Journal of Reinforced Plastics and Composites* 28 (24): 3027–3036.

117 Hua, Y., Li, F., Liu, Y. et al. (2017). Positive synergistic effect of graphene oxide/carbon nanotube hybrid coating on glass fiber/epoxy interfacial normal bond strength. *Composites Science and Technology* 149: 294–304.

118 Jia, C.X., Chen, P., Li, B. et al. (2010). Effects of Twaron fiber surface treatment by air dielectric barrier discharge plasma on the interfacial adhesion in fiber reinforced composites. *Surface & Coatings Technology* 204 (21–22): 3668–3675.

119 Shao, Q., Lu, F., Yu, L. et al. (2021). Facile immobilization of graphene nanosheets onto PBO fibers via MOF-mediated coagulation strategy: multi-functional interface with self-healing and ultraviolet-resistance performance. *Journal of Colloid and Interface Science* 587: 661–671.

120 Tang, L., Zhang, J.L., Wu, C.L. et al. (2021). UV etched random copolymer membrane coated PBO fibers/cyanate ester wave-transparent laminated composites. *Composites Part B: Engineering* 212: 108680.

121 Shi, L., Song, G., Li, P. et al. (2021). Enhancing interfacial performance of epoxy resin composites via in-situ nucleophilic addition polymerization modification of carbon fibers with hyperbranched polyimidazole. *Composites Science and Technology* 201: 108522.

122 Zheng, Y., Chen, L., Wang, X., and Wu, G. (2020). Modification of renewable cardanol onto carbon fiber for the improved interfacial properties of advanced polymer composites. *Polymers* 12 (1): 45.

123 Zhang, X., Xu, Y., Zhang, X. et al. (2019). Progress on the layer-by-layer assembly of multilayered polymer composites: strategy, structural control and applications. *Progress in Polymer Science* 89: 76–107.

124 Chen, L., Hu, Z., Wu, G.S. et al. (2017). POSS-bound ZnO nanowires as interphase for enhancing interfacial strength and hydrothermal aging resistance of PBO fiber/epoxy resin composites. *Composites Part A: Applied Science and Manufacturing* 96: 1–8.

125 Hu, Z., Lu, F., Liu, Y. et al. (2018). Construction of anti-ultraviolet "shielding clothes" on poly(p-phenylene benzobisoxazole) fibers: metal organic framework-mediated absorption strategy. *ACS Applied Materials & Interfaces* 10 (49): 43262–43274.

126 Jing, J.H., Huang, Y.D., Liu, L. et al. (2011). Characterization of fiber-bundle/matrix interfacial strength in carbon/carbon composites. *Materials Science and Technology* 19 (3): 60–65.

127 Qiu, B., Xing, S., and Dong, Q. (2019). Characterization of interfacial bonding strength of particles reinforced metal matrix composites: theory model, finite element simulation and experimental test. *Materials Review* 33 (3A): 862–870.

128 Latour, R.A., Black, J., and Miller, B. (1992). Fiber matrix interfacial bond ultimate and fatigue-strength characterization in a 37°C day environment. *Journal of Composite Materials* 26 (2): 256–273.

129 Lee, S., Nguyen, T., Chin, J., and Chuang, T.J. (1998). Analysis of the single-fiber fragmentation test. *Journal of Materials Science* 33 (21): 5221–5228.

130 Liu, Y.N., Li, M., Gu, Y., and Zhang, Z. (2014). Characterization of torsion behavior and fracture morphology of single carbon fiber. *Journal of Composite Materials* 48 (16): 1993–1999.

131 Lu, D., Yu, W., and Pan, N. (2021). Determination of the strength and elongation distribution of single wool through fiber bundle testing based on acoustic emissions. *Textile Research Journal* 91 (11–12): 1263–1273.

132 Furukawa, Y., Hatta, H., and Kogo, Y. (2003). Interfacial shear strength of C/C composites. *Carbon* 41 (9): 1819–1826.

133 Gu, Y., Liu, H., Li, M. et al. (2014). Macro- and micro-interfacial properties of carbon fiber reinforced epoxy resin composite under hygrothermal treatments. *Journal of Reinforced Plastics and Composites* 33 (4): 369–379.

134 Goto, K., Kawahara, I., Hatta, H. et al. (2005). Measurement of fiber/matrix interface properties of C/C composites by single fiber and fiber bundle push-out methods. *Composite Interfaces* 12 (7): 603–616.

135 Shao, J., Li, W., Wang, R. et al. (2018). Modeling the temperature and test rate dependent fracture strength of zirconia and alumina single crystal fibers. *Composites Part B: Engineering* 133: 26–34.

136 Sui, X., Wu, K., Li, Y. et al. (2020). Study on interfacial properties of single fiber composites by acoustic emission. *Materials Science and Technology* 28 (1): 53–59.

137 Hampe, A., Kalinka, G., Meretz, S., and Schulz, E. (1995). An advanced equipment for single fiber pull-out test designed to monitor the fracture process. *Composites* 26 (1): 40–46.

138 Fu, S.Y., Yue, C.Y., Hu, X., and Mai, Y.W. (2000). Analyses of the micromechanics of stress transfer in single- and multi-fiber pull-out tests. *Composites Science and Technology* 60 (4): 569–579.

139 Bannister, D.J., Andrews, M.C., Cervenka, A.J., and Young, R.J. (1995). Analysis of the single fiber pull-out test by Raman spectroscopy: 2. Micromechanics of deformation for an aramid/epoxy system. *Composites Science and Technology* 53 (4): 411–421.

140 Yue, C., Looi, H.C., and Quek, M.Y. (1995). Assessment of fiber matrix adhesion and interfacial properties using the pull-out test. *International Journal of Adhesion and Adhesives* 15 (2): 73–80.

141 Chandran, M.S., Padmanabhan, K., Raj, D.K.D., and Chebiyyam, Y. (2020). A comparative investigation of interfacial adhesion behaviour of polyamide based self-reinforced polymer composites by single fibre and multiple fibre pull-out tests. *Journal of Adhesion Science and Technology* 34 (5): 511–530.

142 Fu, S.Y. and Lauke, B. (2000). Comparison of the stress transfer in single- and multi-fiber composite pull-out tests. *Journal of Adhesion Science and Technology* 14 (3): 437–452.

143 Marshall, P. and Leonard, J. (1992). The dependence of data from single fiber pull-out tests on the loading direction. *Journal of Materials Science Letters* 11 (21): 1401–1402.

144 Zhamu, A., Zhong, W.H., and Stone, J.J. (2006). Experimental study on adhesion property of UHMWPE fiber/nano-epoxy by fiber bundle pull-out tests. *Composites Science and Technology* 66 (15): 2736–2742.

145 Won, J., Lee, J., and Lee, S. (2015). Predicting pull-out behaviour based on the bond mechanism of arch-type steel fibre in cementitious composite. *Composite Structures* 134: 633–644.

146 Nieuwoudt, P.D., Babafemi, A.J., and Boshoff, W.P. (2017). The response of cracked steel fibre reinforced concrete under various sustained stress levels on both the macro and single fibre level. *Construction and Building Materials* 156: 828–843.

147 Wen, S., Zeng, J., Xiao, J. et al. (2007). Study on the interfacial shear strength of BF/epoxy resin composites by single fiber pull-out Test. *Journal of Hunan University Natural Sciences* 34 (5): 53–57.

148 Liu, J., Chen, M., Yang, J., and Wu, Z. (2022). Study on mechanical properties of basalt fibers superior to E-glass fibers. *Journal of Natural Fibers* 19 (3): 882–894.

149 Zhang, L., Kong, H., Qiao, M. et al. (2019). Growing nano-SiO_2 on the surface of aramid fibers assisted by supercritical CO_2 to enhance the thermal stability, interfacial shear strength, and UV resistance. *Polymers* 11 (9): 1397.

150 Tang, L., Zhang, J., and Gu, J. (2021). Random copolymer membrane coated PBO fibers with significantly improved interfacial adhesion for PBO fibers/cyanate ester composites. *Chinese Journal of Aeronautics* 34 (2): 659–668.

151 Genna, S., Trovalusci, F., and Tagliaferri, V. (2017). Indentation test to study the moisture absorption effect on CFRP composite. *Composites Part B: Engineering* 124: 1–8.

152 Guruprasad, T.S., Keryvin, V., Charleux, L. et al. (2021). On the determination of the elastic constants of carbon fibres by nanoindentation tests. *Carbon* 173: 572–586.

153 Hinoki, T., Kohyama, A., Wen, Z. et al. (1997). Effect of interfacial shear strength on mechanical property of SiC/SiC. *Science Reports of the Research Institutes Tohoku University Series A-Physics Chemistry and Metallurgy* 45 (1): 133–136.

154 Jiang, G., Dai, Y., and Ji, J. (2006). Singularity study of interface end in micro-indentation test. *Chinese Quarterly of Mechanics* 27 (4): 578–584.

155 Kharrat, M. and Chateauminois, A. (1997). On the interfacial behaviour of a glass/epoxy composite during a micro-indentation test: assessment of interfacial shear strength using reduced indentation curves. *Composites Part A: Applied Science and Manufacturing* 28 (1): 39–46.

156 Kuboki, T., Jar, P.Y.B., and Cheng, J.J.R. (2003). Energy absorption for indentation damage in glass fiber reinforced polymers (GFRP). *Journal of Materials Science Letters* 22 (7): 561–563.

157 Sakaue, K., Isawa, T., Ogawa, T., and Yoshimoto, T. (2012). Evaluation of viscoelastic characteristics of short-fiber reinforced composite by indentation method. *Experimental Mechanics* 52 (8): 1003–1008.

158 Shih, C., Katoh, Y., Leonard, K.J. et al. (2013). Determination of interfacial mechanical properties of ceramic composites by the compression of micro-pillar test specimens. *Journal of Materials Science* 48 (15): 5219–5224.

159 Yue, Z. and Lu, Z. (2003). A numerical determination of the fiber/matrix interlayer creep properties from the indentation creep testing in fiber reinforced composites. *Materials Science & Engineering, A: Structural Materials: Properties, Microstructure and Processing* 352 (2): 266–272.

160 Cheng, X., Zhang, J., Wang, H. et al. (2019). Improving the interlaminar shear strength and thermal conductivity of carbon fiber/epoxy laminates by utilizing the graphene-coated carbon fiber. *Journal of Applied Polymer Science* 136 (7): 47061.

161 Gayle, A.J. and Cook, R.F. (2016). Mapping viscoelastic and plastic properties of polymers and polymer-nanotube composites using instrumented indentation. *Journal of Materials Research* 31 (15): 2347–2360.

162 Ho, E. and Marcolongo, M. (2005). Effect of coupling agents on the local mechanical properties of bioactive dental composites by the nano-indentation technique. *Dental Materials* 21 (7): 656–664.

163 Luiz, B.K.M., Quintella, C.M., Friedrich, L.A. et al. (2007). Effect of drinks on the surface properties of dental resin composites. *Polymer Testing* 26 (7): 855–861.

164 Pochiraju, K.V., Tandon, G.P., and Schoeppner, G.A. (2008). Evolution of stress and deformations in high-temperature polymer matrix composites during thermo-oxidative aging. *Mechanics of Time Dependent Materials* 12 (1): 45–68.

165 Young, T.J., Crocker, L.E., Broughton, W.R. et al. (2013). Observations on interphase characterisation in polymer composites by nano-scale indentation using AFM and FEA. *Composites Part A: Applied Science and Manufacturing* 50: 39–43.

166 Young, T.J., Monclus, M., Broughton, W.R. et al. (2012). Observations on interphase characterisation in polymer composites by nanoscale indentation and use of AFM cantilever torsion to identify measurement artefacts. *Plastics, Rubber and Composites* 41 (6): 240–246.

167 Khanna, S.K., Winter, R.M., Ranganathan, P. et al. (2003). Sample preparation techniques for nano-mechanical characterization of glass fiber reinforced polyester matrix composites. *Composites Part A: Applied Science and Manufacturing* 34 (1): 53–65.

168 Song, B., Liu, Z.D., Chen, L. et al. (2022). Enhancing interfacial adhesion and mechanical performance of PBO fibers composites through mussel-inspired molecular regulation of interphase structure. *Polymer Composites* 43 (1): 454–466.

169 Ge, C., Wang, G., Zhao, G., and Li, B. (2021). Strong breathable membrane with excellent self-cleaning, wave-transparent, and heat dissipation performances. *Journal of Applied Polymer Science* 138 (45): e51338.

170 Mao, Y., Yao, Z., Zhou, J. et al. (2020). Fabrication and investigations on BMI/OMMT nanocomposites with superior high-temperature wave-transparent performance. *Journal of Materials Science: Materials in Electronics* 31 (18): 16073–16086.

171 Sun, Z., Huang, P., Gu, A. et al. (2012). Novel high-performance wave-transparent aluminum phosphate/cyanate ester composites. *Journal of Applied Polymer Science* 123 (3): 1576–1583.

172 Cao, S., Zhang, D., Guan, Y. et al. (2009). Surface modification technology of glass fiber. *Aerospace Materials & Technology* 39 (1): 5–7.

173 Karger, J., Mahmood, H., and Pegoretti, A. (2015). Recent advances in fiber/matrix interphase engineering for polymer composites. *Progress in Materials Science* 73: 1–43.

174 Luo, S. and van Ooij, W.J. (2002). Surface modification of textile fibers for improvement of adhesion to polymeric matrices: a review. *Journal of Adhesion Science and Technology* 16 (13): 1715–1735.

175 Zhang, X., He, X., and Cheng, J. (2018). Research progress on surface modification of aramid fibers and preparation of their reinforced resin matrix composites. *Engineering Plastics Application* 46 (8): 149–153.

176 Zhang, Y., Li, P., and Wen, B. (2014). Research progress on surface modification of plant fibers used for reinforcing polymer. *Engineering Plastics Application* 42 (7): 118–121.

177 Sepe, R., Bollino, F., Boccarusso, L., and Caputo, F. (2018). Influence of chemical treatments on mechanical properties of hemp fiber reinforced composites. *Composites Part B: Engineering* 133: 210–217.

178 Vinod, A., Gowda, T.G.Y., Vijay, R. et al. (2021). Novel Muntingia Calabura bark fiber reinforced green-epoxy composite: a sustainable and green material for cleaner production. *Journal of Cleaner Production* 294: 126337.

179 Jia H, Wang Y, Zhang Z, Cheng L, Tan J, Yang W. Research progress on plasma surface modification of CF for composite. *New Chemical Materials* 2022;50(5):62-5,70.

180 Sun, Y., Liang, Q., Chi, H. et al. (2014). The application of gas plasma technologies in surface modification of aramid fiber. *Fibers and Polymers* 15 (1): 1–7.

181 Canal, C., Gallinetti, S., and Ginebra, M.P. (2014). Low-pressure plasma treatment of polylactide fibers for enhanced mechanical performance of fiber-reinforced calcium phosphate cements. *Plasma Processes and Polymers* 11 (7): 694–703.

182 Zhang, Q., Jiang, Y., Yao, L. et al. (2015). Hydrophobic surface modification of ramie fibers by plasma-induced addition polymerization of propylene. *Journal of Adhesion Science and Technology* 29 (8): 691–704.

183 Xu, Z., Wu, X., Liu, L. et al. (2009). Surface grafting modification of carbon fibers by gamma-ray pre-irradiation and plasma technique. *Journal of Solid Rocket Technology* 32 (4): 456–460.

184 Cech, V., Prikryl, R., Balkova, R. et al. (2002). Plasma surface treatment and modification of glass fibers. *Composites Part A: Applied Science and Manufacturing* 33 (10): 1367–1372.

185 Zhang, L.X., Chang, Q., Sun, Z. et al. (2019). Wetting of AgCuTi alloys on quartz fiber reinforced composite modified by vertically aligned carbon nanotubes. *Carbon* 154: 375–383.

186 Kim, M.T., Kim, M.H., Rhee, K.Y., and Park, S.J. (2011). Study on an oxygen plasma treatment of a basalt fiber and its effect on the interlaminar fracture property of basalt/epoxy woven composites. *Composites Part B: Engineering* 42 (3): 499–504.

187 Fan, W., Tian, H.X., Wang, H.H. et al. (2018). Enhanced interfacial adhesion of aramid fiber III reinforced epoxy composites via low temperature plasma treatment. *Polymer Testing* 72: 147–156.

188 Brown, J.R., Chappell, P.J.C., and Mathys, Z. (1992). Plasma suface modification of advanced organic fibers: 3. Effects on the mechanical properties of aramid vinylester and extended-chain polyethylene vinylester composites. *Journal of Materials Science* 27 (23): 6475–6480.

189 Brown, J.R. and Mathys, Z. (1997). Plasma surface modification of advanced organic fibres. 5. Effects on the mechanical properties of aramid/phenolic composites. *Journal of Materials Science* 32 (10): 2599–2604.

190 Deng, T., Zhang, G., Dai, F., and Zhang, F. (2017). Mild surface modification of para-aramid fiber by dilute sulfuric acid under microwave irradiation. *Textile Research Journal* 87 (7): 799–806.

191 Jia, C., Chen, P., Wang, Q. et al. (2015). On-line modification of continuous fibers by atmospheric air plasma. *Chinese Journal of Materials Research* 29 (1): 10–16.

192 Mathur, A. and Netravali, A.N. (1996). Modification of mechanical properties of Kevlar fibre by polymer infiltration. *Journal of Materials Science* 31 (5): 1265–1274.

193 Liu, Z., Chen, P., Han, D.B. et al. (2013). Atmospheric air plasma treated PBO fibers: wettability, adhesion and aging behaviors. *Vacuum* 92: 13–19.

194 Liu, H.J., Pei, Y.N., Xie, D. et al. (2010). Surface modification of ultra-high molecular weight polyethylene (UHMWPE) by argon plasma. *Applied Surface Science* 256 (12): 3941–3945.

195 Matta, A., Sedlacek, T., Kadleckova, M., and Lengalova, A. (2022). The effect of surface substrate treatments on the bonding strength of aluminium inserts with glass-reinforced poly(phenylene) sulphide. *Materials* 15 (5).

196 Wang, S., Zhang, S., Yang, Y. et al. (2022). Direct electrochemical grafting of crystalline PAEK macromolecule on carbon fiber to enhance the interfacial properties of PEEK/CF composites. *Composites Science and Technology* 220: 109262.

197 Cho, K., Wang, G., Raju, R. et al. (2019). Influence of surface treatment on the interfacial and mechanical properties of short S-glass fiber reinforced dental composites. *ACS Applied Materials & Interfaces* 11 (35): 32328–32338.

198 Li, J., Shen, S., Li, W., and Yuan, H. (2014). Effects of acid modification on coupling agent amount of basalt fiber surface and mechanical property of BF/epoxy composites. *Acta Materiae Compositae Sinica* 31 (4): 888–894.

199 Li, J., Shen, S., Yuan, H., and Zhang, L. (2013). Influence of surface treatment on the interface bonding strength of basalt fiber reinforced phenolic resin composite. *Materials Review* 27 (10B): 71–73.

200 Machavaram, V.R., Badcock, R.A., and Fernando, G.F. (2007). Fabrication of intrinsic fibre Fabry-Perot sensors in silica fibres using hydrofluoric acid etching. *Sensors and Actuators A: Physical* 138 (1): 248–260.

201 Parsons, A.J., Felfel, R.M., Wadge, M.D., and Grant, D.M. (2020). Improved phosphate-based glass fiber performance achieved through acid etch/polydopamine treatment. *International Journal of Applied Glass Science* 11 (1): 35–45.

202 Akin, G.E., Akin, H., Sipahi, C. et al. (2014). Evaluation of surface roughness and bond strength of quartz fiber posts after various pre-treatments. *Acta Odontologica Scandinavica* 72 (8): 1010–1016.

203 Haili, W., Shubin, W., and Yue, Z. (2008). Preparation of BN coating on quartz fibers by a low-temperature reaction. *Journal of Synthetic Crystals* 37 (2): 475–479. 83.

204 Chen, H., Wang, X., Xue, T. et al. (2021). Surface treatment of fluorozirconate glass fiber preform using nonaqueous solution. *Chinese Journal of Lasers* 48 (7): 0258-7025(2021)48:72.0.TX;2-5.

205 Vasil'ev, M.G., Vasil'ev, A.M., Golovanov, V.V. et al. (2016). Method for stepped etching of optical glass fibers. *Russian Journal of Inorganic Chemistry* 61 (9): 1160–1162.

206 Jin, T., Shen, S., Li, J., and Li, W. (2014). Feasibility study of a new method of basalt fiber surface treatment acid etching treatment. *Materials Review* 28 (6B): 116–118.

207 Tomao, V., Siouffi, A.M., and Denoyel, R. (1998). Influence of time and temperature of hydrothermal treatment on glass fibers surface. *Journal of Chromatography A* 829 (1-2): 367–376.

208 Cho, K., Wang, G., Raju, F.J. et al. (2019). Selective atomic level etching on short S-glass fibres to control interfacial properties for restorative dental composites. *Scientific Reports* 9 (1): 3851.

209 Karunagaran, N. and Rajadurai, A. (2016). Effect of surface treatment on mechanical properties of glass fiber/stainless steel wire mesh reinforced epoxy hybrid composites. *Journal of Mechanical Science and Technology* 30 (6): 2475–2482.

210 Li, C., Wang, H., Zhao, X. et al. (2021). Investigation of mechanical properties for basalt fiber/epoxy resin composites modified with La. *Coatings* 11 (6): 666.

211 Song, W., Sun, Y., Lu, X. et al. (2018). Study on the surface treatment of aramid fiber and its reinforced natural rubber composite. *New Chemical Materials* 46 (11): 104–107.

212 Li, G., Zhang, C., Wang, Y. et al. (2008). Interface correlation and toughness matching of phosphoric acid functionalized Kevlar fiber and epoxy matrix for filament winding composites. *Composites Science and Technology* 68 (15–16): 3208–3214.

213 Zhao, J. (2013). Effect of surface treatment on the structure and properties of para-aramid fibers by phosphoric acid. *Fibers and Polymers* 14 (1): 59–64.

214 Wu, G.M. and Shyng, Y.T. (2004). Surface modification and interfacial adhesion of rigid rod PBO fibre by methanesulfonic acid treatment. *Composites Part A: Applied Science and Manufacturing* 35 (11): 1291–1300.

215 Liu, L., Li, W., Gu, S. et al. (2020). A comparison of two different surface modifications on the properties of poly (p-phenylene benzobisoxazole) fibers. *IOP Conference Series: Earth and Environmental Science* 446 (2): 022029.

216 Gu, J., Bai, T., Dang, J. et al. (2013). Improvement of surface property of HMPBO fibers by polyphosphoric acid/absolute alcohol under ultrasonic vibration. *Fibers and Polymers* 14 (5): 781–785.

217 Belgacemi, R., Derradji, M., Mouloud, A. et al. (2020). On the mechanical and morphological properties of highly performant composite laminates based on epoxy resin and oxidized ultrahigh-molecular-weight polyethylene fibers. *High Performance Polymers* 32 (9): 992–1000.

218 de Moura, A.S., Pereira, R.D., Abi Rached-Junior, F.J. et al. (2017). Influence of root dentin treatment on the push-out bond strength of fibre-reinforced posts. *Brazilian Oral Research* 31: e29.

219 Li, Z., Yang, D., Liu, S. et al. (2017). Broadband gradient impedance matching using an acoustic metamaterial for ultrasonic transducers. *Scientific Reports* 7: 42863.

220 Deng, et al. (2015). Surface modification of para-aramid fiber by phosphoric acid. *Journal of Textile Research* 36 (11): 12–19. (WOS: CSCD:5566535).

221 Luo, L., Hong, D., Zhang, L. et al. (2018). Surface modification of PBO fibers by direct fluorination and corresponding chemical reaction mechanism. *Composites Science and Technology* 165: 106–114.

222 Jin, X., Wang, W., Bian, L. et al. (2011). The effect of polypyrrole coatings on the adhesion and structure properties of UHMWPE fiber. *Synthetic Metals* 161 (11): 984–989.

223 Wang, X., Zhu, X., Jiang, C. et al. (2019). Influence of boric acid on radial structure of oxidized polyacrylonitrile fibers. *E-Polymers* 19 (1): 79–86.

224 Choi, J., Son, G., and Yu, K. (2019). A study on the radius of curvature of concave optical fiber tips fabricated by laser-induced photothermal effect. *The Journal of The Korea Institute of Electronic Communication Sciences* 14 (5): 871–876.

225 Li, J., Huang, Y., Fu, S. et al. (2010). Study on the surface performance of carbon fibres irradiated by gamma-ray under different irradiation dose. *Applied Surface Science* 256 (7): 2000–2004.

226 Nambiar, S. and Yeow, J.T.W. (2012). Polymer composite materials for radiation protection. *ACS Applied Materials & Interfaces* 4 (11): 5717–5726.

227 Yuan, B., Tan, B., Hu, Y. et al. (2019). Improving impact resistance and residual compressive strength of carbon fibre composites using un-bonded non-woven short aramid fibre veil. *Composites Part A: Applied Science and Manufacturing* 121: 439–448.

228 Zhang, Y., Huang, Y., He, J. et al. (2008). Influence of γ-ray radiation grafting on interfacial properties of aramid fibers and epoxy resin composites. *Composite Interfaces* 15 (6): 611–628.

229 Zhang, C., Zhang, C., Zhang, Q. et al. (2018). Synergy modification of the microstructure and the property of PBO fiber by γ-ray radiation. *Polymer Engineering & Science* 58 (3): 272–279.

230 Qiu, J., Wang, Z., Sun, Q., and Gao, X. (2015). Effect of surface modification and hybridization of UHMWPE fibers on performance of their composites with epoxy resin. *Chinese Journal of Materials Research* 29 (11): 807–813.

231 Suzuki, Y., Cousins, D.S., Dorgan, J.R. et al. (2019). Dual-energy X-ray computed tomography for void detection in fiber-reinforced composites. *Journal of Composite Materials* 53 (17): 2349–2359.

232 Fiore, V., Scalici, T., Di Bella, G., and Valenza, A. (2015). A review on basalt fibre and its composites. *Composites Part B: Engineering* 74: 74–94.

233 Kabir, M.M., Wang, H., Lau, K.T., and Cardona, F. (2012). Chemical treatments on plant-based natural fibre reinforced polymer composites: an overview. *Composites Part B: Engineering* 43 (7): 2883–2892.

234 Collier, B.J. and Collier, J.R. (1992). Use of coupling agents for internally reinforced rayon fibers. *Journal of Adhesion Science and Technology* 6 (1): 99–107.

235 Hu, P., Jiang, M., Huang, C., and Yu, Q. (2004). Research on interface property of silicon coupling agent. *Surface Technology* 33 (5): 19–21.

236 Huang, C., Chen, G., Wang, Q. et al. (2022). Modifying carbon nanotubes-carbon fiber mutiscale reinforcement surface by silane coupling agent. *Engineering Plastics Application* 50 (6): 136–143.

237 Sun, P., Jin, T., Tian, S. et al. (2014). Selection of the coupling agent types for glass fiber reinforced PA6. *China Plastics Industry* 42 (11): 80–84.

238 Li, W., Shen, S., and Zhang, L. (2013). Effect on coupling agent adsorption of basalt fiber with acid treatment. *Aerospace Materials & Technology* 43 (2): 83–87.

239 Rajulu, A.V., Devi, L.G., and Devi, R.R. (2008). Effect of orientation of fibers and coupling agent on tensile properties of phenolic/silica composites. *Journal of Reinforced Plastics and Composites* 27 (6): 599–603.

240 Sun, P., Jin, T., Tian, S. et al. (2016). Study on reinforcement of PA6 by glass fiber treated with silane coupling agent. *New Chemical Materials* 44 (1): 205–207.

241 Wang, B., Jin, Z., Qiu, Z., and Liu, X. (2002). Effect of coupling agent on interfacial adhesion of poly (p-phenylene benzobisoxazole) fibre/epoxy matrix composites. *Journal of Xi'an Jiaotong University-Natural Science Edition* 36 (9): 975–978.

242 Dieu, T.V., Phai, L.T., Ngoc, P.M. et al. (2004). Study on preparation of polymer composites based on polypropylene reinforced by jute fibers. *JSME International Journal Series A-Solid Mechanics and Material Engineering.* 47 (4): 547–550.

243 Fabris, F.W., Cardozo, N.S.M., Mauler, R.S., and Nachtigall, S.M.B. (2009). Improving the properties of LDPE/glass fiber composites with silanized LDPE. *Polymer Composites* 30 (7): 872–879.

244 Park, S.J. and Jin, J.S. (2003). Effect of silane coupling agent on mechanical interfacial properties of glass fiber-reinforced unsaturated polyester composites. *Journal of Polymer Science Part B: Polymer Physics* 41 (1): 55–62.

245 St John, N.A. and Brown, J.R. (1998). Flexural and interlaminar shear properties of glass-reinforced phenolic composites. *Composites Part A: Applied Science and Manufacturing* 29 (8): 939–946.

246 Li, P., Zhang, X., Chen, Y. et al. (2014). A sequential solid phase microextraction system coupled with inductively coupled plasma mass spectrometry for speciation of inorganic arsenic. *Analytical Methods* 6 (12): 4205–4211.

247 Wang, L., Luo, F., Zhou, W., and Zhu, D. (2014). Influence of the coupling agent on the mechanical properties of SiCf/poly(phenylene benzobisoxazole) composites. *Journal of Applied Polymer Science* 131 (2): 39805.

248 Yu, L., He, R., Zhang, Y., and Gao, J. (2018). Effect of surface treatment on flexural and tribological properties of poly(p-phenylene benzobisoxazole)/polyimide composites under normal and elevated temperature. *Materials* 11 (11): 2131.

249 Zhang, X., Wang, Y., and Cheng, S. (2013). Properties of UHMWPE fiber-reinforced composites. *Polymer Bulletin* 70 (3): 821–835.

250 Jian, Y. and Ming, L. (2018). Improving the interfacial strength of PMMA resin composites by chemically grafting graphene oxide on UHMWPE fiber. *Composite Interfaces* 25 (11): 949–958.

251 Tian, Y. and Guo, L. (2021). Adhesion performance of UHMWPE fiber treated with polyethylene wax grafted methyl methacrylate alone or in conjunction with silane coupling agent. *Journal of Adhesion Science and Technology* 35 (11): 1219–1235.

252 Shen, T., Zhou, X., and Lin, Q. (2011). Influence of grafting density of diblock copolymers on interfacial assembly behavior and interfacial shear strength of glass fiber/polystyrene composites. *Composite Interfaces* 18 (8): 701–714.

253 Gu, Y.B., Hu, Y.H., Du, L. et al. (2017). Reinforcement and tougheness of quartz fiber/silicon-containing arylacetylene with heat-resistant polyetherimide macromolecular coupling agent. *Chemical Journal of Chinese Universities* 38 (9): 1670–1677.

254 Liu, L., Liang, S., Wang, D. et al. (2022). Effect of secondary modification of silane coupling agent and graphene oxide on interfacial properties of aramid fiber. *Journal of Materials Engineering* 50 (1): 145–153.

255 Rallis, G., Tarantili, P.A., and Andreopoulos, A.G. (2000). Epoxy resin composites with surface modified aramid fibres. *Advanced Composites Letters* 9 (2): 127–133.

256 Holloway, J.L., Lowman, A.M., VanLandingham, M.R., and Palmese, G.R. (2013). Chemical grafting for improved interfacial shear strength in UHMWPE/PVA-hydrogel fiber-based composites used as soft fibrous tissue replacements. *Composites Science and Technology* 85: 118–125.

257 Kikuchi, M., Nakaso, K., Murata, A. et al. (2022). Influence of covalently-bound graphene oxide on the mechanical properties of carbon fiber composite materials. *Diamond and Related Materials* 121: 108730.

258 Li, Y.W., Zhao, F., Song, Y.J. et al. (2013). Interfacial microstructure and properties of poly (phenylene benzobisoxazole) fiber grafted with graphene oxide via solvothermal method. *Applied Surface Science* 266: 306–312.

259 Lv, Y. and Peng, C. (2018). Chemically grafting carbon nanotubes onto carbon fibers for enhancing interfacial strength in carbon fiber/HDPE composites. *Surface and Interface Analysis* 50 (5): 552–557.

260 Singha, A.S. and Rana, A.K. (2012). Effect of graft copolymerization on mechanical, thermal, and chemical properties of grewia optiva/unsaturated polyester biocomposites. *Polymer Composites* 33 (8): 1403–1414.

261 Xie, F., Xing, L., Liu, L. et al. (2018). A facile method to prepare nanoscale polyacrylonitrile particles grafted aramid fibers for superior interfacial and mechanical properties of epoxy composites. *Polymer Composites* 39: E2436–E2444.

262 Zhang, B., Liang, T., Shao, X. et al. (2021). Nondestructive grafting of ZnO on the surface of aramid fibers followed by silane grafting to improve its interfacial adhesion property with rubber. *ACS Applied Polymer Materials* 3 (9): 4587–4594.

263 Xing, L., Zhang, Y., Liu, L., and Huang, Y. (2016). Co-irradiation grafting of aramid fiber F3A surface. *Chemistry and Bonding* 38 (3): 153–156. 64.

264 Benrashid, R. and Tesoro, G.C. (1990). Effect of surface-limited reactions on the properties of Kevlar® fibers. *Textile Research Journal* 60 (6): 334–344.

265 Chen, M., Lin, H., Yang, C., and Li, L. (2022). Research progress on modification of ultra high molecular weight polyethylene(UHMWPE) fiber. *Applied Chemical Industry* 51 (2): 520–524.

266 Shang, Q., Zhao, H., Zhao, N. et al. (2020). Research progress on surface modification of ultra-high molecular weight polyethylene fiber. *Polymer Bulletin* 4: 22–29.

267 Zhao, Y., Han, G., Hou, J. et al. (2013). Surface modification of UHMWPE fiber by low temperature plasma treatment and grafting using acrylic acid. *Polymer Materials Science & Engineering* 29 (6): 62–65.

268 Mohammadalipour, M., Masoomi, M., Ahmadi, M., and Safi, S. (2016). Interfacial shear strength characterization of GMA-grafted UHMWPE fiber/epoxy/nano clay hybrid nanocomposite materials. *RSC Advances* 6 (48): 41793–41799.

269 Yuan, R., Chu, Z., Wu, X. et al. (2022). Surface modification of UHMWPE fibers with chlorination and polydopamine for enhancing the mechanical properties of composites. *Journal of Applied Polymer Science* 139 (41): e52988.

270 Liu, X. and Wang, K. (2020). Interfacial microstructure and properties between epoxy resin and novel organic hybrid graphene oxide modification ultra high molecular weight polyethylene fiber. *Polymer* 197: 122472.

271 Wang, L., Gao, S., Wang, J. et al. (2018). Surface modification of UHMWPE fibers by ozone treatment and UV grafting for adhesion improvement. *Journal of Adhesion* 94 (1): 30–45.

272 Xie, W., Cheng, H., Chu, Z. et al. (2007). Effect of surface modification on microwave absorbing properties of magnetic metal fibers. *Journal of Wuhan University of Technology Materials Science Edition* 22 (2): 218–220.

273 Yang, C., Yu, J., Wang, X., and Huang, P. (2012). Friction and wear performance of carbon fiber reinforced thermoplastic polyimide composites prepared by different surface modification methods. *Materials for Mechanical Engineering* 36 (1): 90–93.

274 Yuan, X., Zhu, B., Cai, X. et al. (2018). Influence of different surface treatments on the interfacial adhesion of graphene oxide/carbon fiber/epoxy composites. *Applied Surface Science* 458: 996–1005.

275 Kuzmin, K.L., Timoshkin, I.A., Gutnikov, S.I. et al. (2016). Effect of silane/nano-silica on the mechanical properties of basalt fiber reinforced epoxy composites. *Composite Interfaces* 24 (1): 13–34.

276 Li, Z., Li, H., Liu, L. et al. (2018). Surface modification of poly-p-phenylenebenzobisthiazole fibers and preparation of hydrophobic coatings. *Polymer Materials Science & Engineering* 34 (5): 161–165.

277 Cech, V., Marek, A., Knob, A. et al. (2019). Continuous surface modification of glass fibers in a roll-to-roll plasma-enhanced CVD reactor for glass fiber/polyester composites. *Composites Part A: Applied Science and Manufacturing* 121: 244–253.

278 Bulatov, M.I. and Shatsov, A.A. (2021). Strength and fracture resistance of quartz fibers with polyimide coatings. *Russian Journal of Non-Ferrous Metals* 62 (6): 756–762.

279 Nasser, J., Lin, J., Steinke, K., and Sodano, H.A. (2019). Enhanced interfacial strength of aramid fiber reinforced composites through adsorbed aramid nanofiber coatings. *Composites Science and Technology* 174: 125–133.

280 Cai, J., Li, Q., and Wu, C. (2022). Surface modification of recycled coir fibers with hybrid coating and its effect on the properties of ABS composites. *Materials Research Express* 9 (2).

281 Cao, H., Zhang, C., Zhang, Z., and Gao, Y. (2007). Study on sizing modification of basalt fibers. *Journal of Aeronautical Materials* 27 (5): 77–82.

282 Yi, M., Sun, H., Zhang, H. et al. (2016). Flexible fiber-reinforced composites with improved interfacial adhesion by mussel-inspired polydopamine and poly(methyl methacrylate) coating. *Materials Science & Engineering, C: Materials for Biological Applications* 58: 742–749.

283 Zhu, X., Yuan, L., Liang, G., and Gu, A. (2014). Unique UV-resistant and surface active aramid fibers with simultaneously enhanced mechanical and thermal properties by chemically coating $Ce_{0.8}Ca_{0.2}O_{1.8}$ having low photocatalytic activity. *Journal of Materials Chemistry A* 2 (29): 11286–11298.

284 Yu, L., Lu, F., Huang, X. et al. (2019). Facile interface design strategy for improving the uvioresistant and self-healing properties of poly(p-phenylene benzobisoxazole) fibers. *ACS Applied Materials & Interfaces* 11 (42): 39292–39303.

285 Wu, W., Zhang, X., Qin, L. et al. (2020). Enhanced MPBR with polyvinylpyrrolidone-graphene oxide/PVDF hollow fiber membrane for efficient ammonia nitrogen wastewater treatment and high-density Chlorella cultivation. *Chemical Engineering Journal* 379: 122368.

286 Zhang, C., Zhang, X., Ling, Y. et al. (2022). A novel eco-friendly strategy on the interfacial modification of a carbon-fiber-reinforced polymer composite via chitosan encapsulation. *Materials Chemistry Frontiers* 6 (6): 765–774.

287 Kang, G. and Cao, Y. (2014). Application and modification of poly(vinylidene fluoride) (PVDF) membranes: a review. *Journal of Membrane Science* 463: 145–165.

288 Xing, R., Wang, W., Jiao, T. et al. (2017). Bioinspired polydopamine sheathed nanofibers containing carboxylate graphene oxide nanosheet for high-efficient dyes scavenger. *ACS Sustainable Chemistry & Engineering* 5 (6): 4948–4956.

289 Yan, J., Huang, Y., Miao, Y. et al. (2015). Polydopamine-coated electrospun poly(vinyl alcohol)/poly(acrylic acid) membranes as efficient dye adsorbent with good recyclability. *Journal of Hazardous Materials* 283: 730–739.

290 Sa, R., Yan, Y., Wei, Z. et al. (2014). Surface modification of aramid fibers by bio-inspired poly(dopamine) and epoxy functionalized silane grafting. *ACS Applied Materials & Interfaces* 6 (23): 21730–21738.

291 Zeng, L., Liu, X., Chen, X., and Soutis, C. (2018). Surface modification of aramid fibres with graphene oxide for interface improvement in composites. *Applied Composite Materials* 25 (4): 843–852.

292 Tang, L., Dang, J., He, M. et al. (2019). Preparation and properties of cyanate-based wave-transparent laminated composites reinforced by dopamine/POSS functionalized Kevlar cloth. *Composites Science and Technology* 169: 120–126.

293 Ai, X., Xie, A., Cheng, J. et al. (2022). Fabrication of robust and high resilient polythiophene conductive polyamides fibers based on tannic acid modification. *Industrial and Engineering Chemistry Research* 61 (30): 11025–11033.

294 Gao, J., Ma, J., Wang, S. et al. (2021). Tannic acid bionic modification and polyamine controllable grafting surface modification of ultra-high molecular weight polyethylene fiber. *Polymer Materials Science & Engineering* 37 (8): 131–137.

295 Ai, X., Cheng, J., Hou, X. et al. (2022). Fabrication of robust silver plated conductive polyamide fibres based on tannic acid modification. *RSC Advances* 12 (29): 18585–18593.

296 Wang, L., Zhang, B., Li, X. et al. (2021). Enhanced adhesion property of aramid fibers by polyphenol-metal iron complexation and silane grafting. *Journal of Adhesion* 97 (4): 346–360.

297 Wang, Z., Han, M., Zhang, J. et al. (2020). Investigating and significantly improving the stability of tannic acid (TA)-aminopropyltriethoxysilane (APTES) coating for enhanced oil-water separation. *Journal of Membrane Science* 593: 117383.

298 Liu, Z., Fan, X.L., Cheng, L. et al. (2022). Hybrid polymer membrane functionalized PBO fibers/cyanate esters wave-transparent laminated composites. *Advanced Fiber Materials* 4 (3): 520–531.

299 Damodaran, S., Anand, K., and Razumovsky, L. (1998). Competitive adsorption of egg white proteins at the air-water interface: direct evidence for electrostatic complex formation between lysozyme and other egg proteins at the interface. *Journal of Agricultural and Food Chemistry* 46 (3): 872–876.

300 de Eulate, E.A., Silvester, D.S., and Arrigan, D.W.M. (2012). Behavior of lysozyme at the electrified water/room temperature ionic liquid interface. *Chemistry An Asian Journal* 7 (11): 2559–2561.

301 Skrabkova, H.S., Chernysheva, M.G., Arutyunyan, A.M., and Badun, G.A. (2020). Complex of lysozyme and Myramistin: formation and adsorption at the water-xylene interface. *Mendeleev Communications* 30 (5): 645–646.

302 Zhong, X., Song, Y.J., Yang, P. et al. (2016). Titanium surface priming with phase-transited lysozyme to establish a silver nanoparticle-loaded chitosan/hyaluronic acid antibacterial multilayer via layer-by-layer self-assembly. *PLoS One* 11 (1): e0146957.

303 Gu, J., Li, Y., Liang, C. et al. (2018). Synchronously improved dielectric and mechanical properties of wave-transparent laminated composites combined with outstanding thermal stability by incorporating iysozyme/POSS functionalized PBO fibers. *Journal of Materials Chemistry C* 6 (28): 7652–7660.

304 Liu, Y., Liang, G., Gu, A., and Yuan, L. (2012). Enzyme-mediated grafting of AGE to kevlar fiber using horseradish peroxidase. *Advanced Materials Research* 430–432: 233–241.

305 Wang, J., Liang, G., Zhao, W., and Zhang, Z. (2007). Enzymatic surface modification of PBO fibres. *Surface and Coatings Technology* 201 (8): 4800–4804.

306 Yan, H., Feng, S., Zhang, Z., and Hu, S. (2014). Improvement of interfacial adhesion between PBO fibers and cyanate ester matrix. *Journal of Applied Polymer Science* 131 (9): 40204.

307 Pu, Y., Ma, Z., Liu, L. et al. (2022). Improvement on strength and toughness for CFRPs by construction of novel "soft-rigid" interface layer. *Composites Part B: Engineering* 236: 109846.

308 Shao, Y., Zhang, X., and Zhou, Y. (2011). Preparation of POSS/PI hybridized film on quartz fiber surface and study of acid corrosion resistance. *China Materials Seminar* 21: 301.

309 Li, C., Su, R., Li, X., and Wang, L. (2022). Interfacial properties of aramid fiber composites reinforced with SiO_2-coated ZnO nanoparticles. *Fibers and Polymers* 23: 1050–1058.

310 Li, X., Li, C., and Su, R. (2020). Effect of dopamine doped carbon nanotubes on the interfacial properties of aramid fibers. *Materials Guides* 34 (S2): 1562–1566.

311 Zhu, P., Liu, B., and Bao, L. (2018). Evaporation-induced surface coating of poly(p-phenylene benzobisoxazole) fibers with polyetherimide encapsulated nano-TiO_2. *Progress in Organic Coatings* 116: 43–50.

312 Lv, J., Wang, B., Ma, Q. et al. (2018). Interfacially enhancement of PBO/epoxy composites by grafting MWCNTs onto PBO surface through melamine as molecular bridge. *Materials Research Express* 5 (6): 065006.

313 Chen, L., Du, Y., Huang, Y. et al. (2016). Hierarchical poly(p-phenylene benzobisoxazole)/graphene oxide reinforcement with multifunctional and biomimic middle layer. *Composites Part A: Applied Science and Manufacturing* 88: 123–130.

314 Tang, L., Fan, X., Tang, Y. et al. (2021). Calcia-doped ceria hybrid coating functionalized PBO fibers with excellent UV resistance and improved interfacial compatibility with cyanate ester resins. *Applied Surface Science* 569: 151124.

315 Sharma, M., Gao, S., Maeder, E. et al. (2014). Carbon fiber surfaces and composite interphases. *Composites Science and Technology* 102: 35–50.

316 Guo, Y., Qiu, H., Ruan, K. et al. (2022). Flexible and insulating silicone rubber composites with sandwich structure for thermal management and electromagnetic interference shielding. *Composites Science and Technology* 219: 109253.

317 Yang, P., Yuan, J., and He, L. (2017). Carbon fibers surface modification and effects on the interfaces between fibers and resin matrices: a review. *Materials Review* 31 (4): 129–136.

318 Sarr, M.M., Inoue, H., and Kosaka, T. (2021). Study on the improvement of interfacial strength between glass fiber and matrix resin by grafting cellulose nanofibers. *Composites Science and Technology* 211: 108853.

319 Huang, Y., Zhang, M., Meng, J., and Xu, L. (2011). Glass fiber acrylic grafting modification study. *Journal of Soochow University* 27 (2): 90–94.

320 Tang, L., He, M., Na, X. et al. (2019). Functionalized glass fibers cloth/spherical BN fillers/epoxy laminated composites with excellent thermal conductivities and electrical insulation properties. *Composites Communications* 16: 5–10.

321 Zhang, J. and Teng, C. (2020). Nondestructive growing nano-ZnO on aramid fibers to improve UV resistance and enhance interfacial strength in composites. *Materials & Design* 192: 108774.

322 Lu, C., Qiu, S., Lu, X. et al. (2019). Enhancing the interfacial strength of carbon fiber/poly(ether ether ketone) hybrid composites by plasma treatments. *Polymers* 11 (5): 753.

323 Zhang, H. and Li, W. (2015). Plasma-grafting polymerization on carbon fibers and its effect on their composite properties. *Applied Surface Science* 356: 492–498.

324 Jin, X., Wang, W., Xiao, C. et al. (2016). Improvement of coating durability, interfacial adhesion and compressive strength of UHMWPE fiber/epoxy composites through plasma pre-treatment and polypyrrole coating. *Composites Science and Technology* 128: 169–175.

325 Li, J., Wen, Y., Xiao, Z. et al. (2022). Holey reduced graphene oxide scaffolded heterocyclic aramid fibers with enhanced mechanical performance. *Advanced Functional Materials* 32: 2200937.

326 Li, H., Tang, Y., Gu, J. et al. (2013). Structures and properties of HMPBO fibers treated by oxygen plasma/polyhedral oligomeric silsesquioxane. *Polymer Composites* 34 (12): 2026–2030.

327 Zhao, M., Meng, L., Ma, L. et al. (2018). Layer-by-layer grafting CNTs onto carbon fibers surface for enhancing the interfacial properties of epoxy resin composites. *Composites Science and Technology* 154: 28–36.

328 Fang, C., Wu, J., Wang, J., and Zhang, T. (2015). Modification of carbon fiber surfaces via grafting with Meldrum's acid. *Applied Surface Science* 356: 9–17.

329 Peng, Q., Li, Y., He, X. et al. (2013). Interfacial enhancement of carbon fiber composites by poly(amido amine) functionalization. *Composites Science and Technology* 74: 37–42.

330 Maity, J., Jacob, C., Alam, S., and Singh, R.P. (2009). Direct fluorination of twaron fiber and preparation of PP/twaron fiber composites using MA-g-PP as a compatibilizer. *Journal of Composite Materials* 43 (7): 709–723.

331 Cui, J., Mei, C., Jia, C. et al. (2011). Acrylamide-formaldehyde-urea copolymer as a novel compatibilizer for high density polyethylene/plant fiber composite. *Polymer Bulletin* 67 (3): 375–382.

332 Karsli, N.G., Aytac, A., Akbulut, M. et al. (2013). Effects of irradiated polypropylene compatibilizer on the properties of short carbon fiber reinforced polypropylene composites. *Radiation Physics and Chemistry* 84: 74–78.

333 Kim, J.S. and Kim, D.H. (2015). Compatibilizing effects of maleic anhydride-grafted-polypropylene (PP)on long carbon fiber-reinforced PP composites. *Journal of Thermoplastic Composite Materials* 28 (11): 1599–1611.

334 Liu, H.Y., Chen, F.Q., Guo, R.B. et al. (2015). Effect of compatibilizer on the properties of PBS/lignin composites prepared via a vane extruder. *Journal of Polymer Engineering* 35 (9): 829–837.

335 Seo, Y. (1997). Interfacial adhesion and deformation of thermotropic liquid crystal polymers in engineering thermoplastics: blends of a poly(ester amide) with nylon 6 and a polyester with PBT. *Journal of Applied Polymer Science* 64 (2): 359–366.

336 Wang, B., Sun, C., Wang, Y. et al. (2021). Improved mechanical properties of in situ microfibrillar polypropylene/polyamide6 composites through constructing strong interfacial adhesion. *Polymers for Advanced Technologies* 32 (8): 3343–3357.

337 Ye, M., Zhu, N., Li, X. et al. (2017). Monofunctional compatibilizer with long alkyl end for fabrication of superior tensile wood flour-polyolefin composites. *Journal of Applied Polymer Science* 134 (5): 44429.

338 Zhang, C., Li, K.C., and Simonsen, J. (2004). Improvement of interfacial adhesion between wood and polypropylene in wood-polypropylene composites. *Journal of Adhesion Science and Technology* 18 (14): 1603–1612.

339 Basalp, D., Tihminlioglu, F., Sofuoglu, S.C. et al. (2020). Utilization of municipal plastic and wood waste in industrial manufacturing of wood plastic composites. *Waste and Biomass Valorization* 11 (10): 5419–5430.

340 Ding, Y., Ni, G., Xu, P. et al. (2017). Investigation on the synergistic effect of c-aminopropyltriethoxysilane and polyethylene-grafted glycidylmethacrylate on the properties of high-density polyethylene/poplar wood flour composites and their synergistic mechanism. *Journal of Composite Materials* 51 (7): 955–964.

341 Abdul Razak, N.C., Inuwa, I.M., Hassan, A., and Samsudin, S.A. (2013). Effects of compatibilizers on mechanical properties of PET/PP blend. *Composite Interfaces* 20 (7): 507–515.

342 Akbari, M., Zadhoush, A., and Haghighat, M. (2007). PET/PP blending by using PP-g-MA synthesized by solid phase. *Journal of Applied Polymer Science* 104 (6): 3986–3993.

343 Catto, A. and Santana, R. (2013). Influence of coupling agent in polyolefinic composites from post-consumer waste with eucalyptus grandis flour. *Journal of Materials Science and Engineering* 3: 641–652.

344 Yi, X., Xu, L., Wang, Y. et al. (2010). Morphology and properties of isotactic polypropylene/poly(ethylene terephthalate) in situ microfibrillar reinforced blends: influence of viscosity ratio. *European Polymer Journal* 46 (4): 719–730.

345 Zhang, L., Deng, H., Jing, Y. et al. (2018). An effective compatibilizer for tin fluorophosphate glass/polymer composites obtained from "one pot" KF-RAFT polymerization. *Composites Science and Technology* 168: 336–345.

346 Zhang, K., Zhang, G., Liu, B. et al. (2014). Effect of aminated polyphenylene sulfide on the mechanical properties of short carbon fiber reinforced polyphenylene sulfide composites. *Composites Science and Technology* 98: 57–63.

347 Jia, H., Liu, C., Qiao, Y. et al. (2021). Enhanced interfacial and mechanical properties of basalt fiber reinforced poly(aryl ether nitrile ketone) composites by amino-silane coupling agents. *Polymer* 230: 124028.

348 Fu, X., He, B., and Chen, X. (2010). Effects of compatibilizers on mechanical properties of long glass fiber-reinforced polypropylene. *Journal of Reinforced Plastics and Composites* 29 (6): 936–949.

349 Liu, R., Han, Z., Li, W. et al. (2016). Improvement of the interfacial shear strength of poly(p-phenylene benzobisoxazole) fiber/epoxy resin composite via a novel surface coating agent. *Polymer Composites* 37 (4): 1198–1205.

350 Wu, S., Li, C., Yu, Z. et al. (2020). Nondestructive strategy to effectively enhance the interfacial adhesion of PBO/epoxy composites. *ACS Applied Materials & Interfaces* 12 (40): 45383–45393.

351 Dammak, M., Fourati, Y., Tarres, Q. et al. (2020). Blends of PBAT with plasticized starch for packaging applications: mechanical properties, rheological behaviour and biodegradability. *Industrial Crops and Products* 144: 112061.

352 Kun, D. and Pukanszky, B. (2017). Polymer/lignin blends: interactions, properties, applications. *European Polymer Journal* 93: 618–641.

353 Gu, J.W., Dong, W.C., Xu, S. et al. (2017). Development of wave-transparent, light-weight composites combined with superior dielectric performance and desirable thermal stabilities. *Composites Science and Technology* 144: 185–192.

354 Liu, Z., Zhang, J., Tang, L. et al. (2019). Improved wave-transparent performances and enhanced mechanical properties for fluoride-containing PBO precursor modified cyanate ester resins and their PBO fibers/cyanate ester composites. *Composites Part B: Engineering* 178: 107466.

355 Zhu, B. (2011). Preparation and characterization of interpenetrating polymer networks of BADCy/PSt. *Applied Chemical Industry* 40 (8): 1366–1370.

356 Wang, S. and Zhu, B. (2012). Preparation of interpenetrating polymer networks of BADCy/PAN and the structures and properties. *Fiber Reinforced Plastics/Composites* 4: 13–18.

357 Zhu, B. (2012). Preparation and characterization of BADCy/MMA semi-IPN. *Thermosetting Resin* 27 (1): 9–13. 20.

358 Liu, J., Fan, W., Lu, G. et al. (2019). Semi-Interpenetrating polymer networks based on cyanate ester and highly soluble thermoplastic polyimide. *Polymers* 11 (5): 862.

359 Liu, Z., Zhang, J., Tang, Y. et al. (2021). Optimization of PBO fibers/cyanate ester wave-transparent laminated composites via incorporation of a fluoride-containing linear interfacial compatibilizer. *Composites Science and Technology* 210: 108838.

360 Chen, S., Zhang, J., Zhou, J. et al. (2018). Dramatic toughness enhancement of benzoxazine/epoxy thermosets with a novel hyperbranched polymeric ionic liquid. *Chemical Engineering Journal* 334: 1371–1382.

361 Liu, Z., Yuan, L., Liang, G., and Gu, A. (2015). Tough epoxy/cyanate ester resins with improved thermal stability, lower dielectric constant and loss based on unique hyperbranched polysiloxane liquid crystalline. *Polymers for Advanced Technologies* 26 (12): 1608–1618.

362 Liu, Z., Fan, X., Zhang, J. et al. (2021). Improving the comprehensive properties of PBO fibres/cyanate ester composites using a hyperbranched fluorine and epoxy containing PBO precursor. *Composites Part A: Applied Science and Manufacturing* 150: 106596.

6

The Molding Technologies of Polymer Matrix Wave-Transparent Composites

6.1 Structural Design of Polymer Matrix Wave-Transparent Composites

The structural design of polymer matrix wave-transparent composites is an indispensable part of industrial production. It is necessary to consider various factors such as material properties, preparation technologies, structure qualities, production cost, and design experience [1, 2]. The integrated process of the structural design for polymer matrix wave-transparent composites is shown in Figure 6.1.

The structural design contains the following three steps:

1) Defining the design requirements, such as performance requirements, load conditions, environmental conditions, and shape limitation.
2) The materials design, including the selecting of raw materials, determination of layer performance, and prepreg design.
3) The shape design mainly focuses on the design of typical structural parts of polymer matrix wave-transparent composites, such as conical radomes, oval radomes, and spherical radomes, which needed to be designed in advance for shape structures [3, 4].

In addition, strain, stress, and deformation analysis, as well as failure analysis, are involved in the above structural design to ensure the strength and the stiffness of the integral structures [5, 6]. The materials and structures of polymer matrix wave-transparent composites are usually formed simultaneously, and the raw materials are also designable. Therefore, the structural design of polymer matrix wave-transparent composites differs from the conventional integral structural design, being a specific structural design method including materials design and structural design [7–9].

6.1.1 Structural Design Condition of the Polymer Matrix Wave-Transparent Composites

Structural design conditions, including load requirement, environmental condition, geometry, and size of polymer matrix wave-transparent composites should be

Polymer Matrix Wave-Transparent Composites: Materials, Properties, and Applications, First Edition. Junwei Gu, Yusheng Tang, Jie Kong, and Jing Dang.

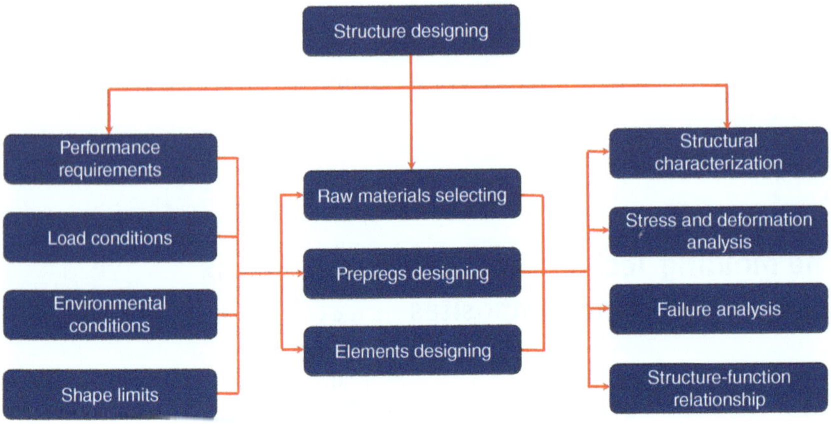

Figure 6.1 The integrated process of the structural design for polymer matrix wave-transparent composites.

first considered. For example, the shape of the structures can be assumed in advance according to the performance requirements to determine whether they can withstand the load under certain environmental conditions. Such a process must be repeated several times to determine the reasonable structural shape [10, 11].

6.1.1.1 Requirements of Structural Properties

Generally, the main contents of structural properties are as follows:

1) The designed structures of products can bear various loads to ensure safety during the service life;
2) The shapes and sizes of the structures need to be reasonable in order to install various accessories, instruments, and other accessories;
3) It is necessary to isolate the external environment to avoid interference with the structural properties [12, 13].

The structural properties of products are closely related to their qualities. Especially for polymer matrix wave-transparent composites applied in airborne radomes, they should be reduced in overall mass as much as possible to meet the requirements for lightweight and high strength under the premise of meeting the basic structural properties [14, 15].

6.1.1.2 Load Condition

Appropriate polymer matrix and reinforced fibers can be selected according to actual load conditions. The load conditions of polymer matrix wave-transparent composites contain the static and dynamic load. The static load slowly increases from zero to a certain value and then remains unchanged. The self-gravity and corresponding inertial force of the polymer matrix wave-transparent composites can be ignored. The dynamic load can cause the component to generate large acceleration, and the inertial force cannot be ignored. Under the action of dynamic load, the stress generated in the products is called dynamic stress, and dynamic load can be divided into instantaneous, impact, and alternating loads [16, 17].

The instantaneous load increases from zero to the maximum value in a fraction of a second, such as the generated load when cars start suddenly. The impact load is the moment when the load is applied. The object subjected to the load presents a certain kinetic energy, such as the radomes impacted by birds. The alternating load is the load that changes continuously and periodically, such as the load on various shafts and connecting rods when the train is running [18, 19].

When the load condition is only static load, the corresponding structure of polymer matrix wave-transparent composites is not required to possess high resistance to damage and deformation. For example, the ground and 5G communication radomes can choose reinforced fibers (glass fibers, and so on.) with relatively low mechanical properties. When the load condition includes the instantaneous and impact loads, the structure of polymer matrix wave-transparent composites is required to possess sufficient resistance to instantaneous and impact loads. For example, aircraft and airborne radomes need to choose reinforced fibers (aramid fibers and PBO fibers) with excellent mechanical properties. When the load condition contains alternating load, the structure of polymer matrix wave-transparent composites is required to present excellent fatigue resistance, and the structure should be designed according to fatigue strength and life. Meanwhile, the reinforced fibers should present excellent interfacial compatibility with the polymer matrix [20, 21].

6.1.1.3 Environmental Condition

Structural design needs to specify the environment, that is, storage, packaging, and transportation, as well as the time and cycle times of the above-mentioned process. In general, various possible environmental condition needs to be fully considered, including the following environmental condition:

1) Mechanical condition: acceleration, impact, vibration, thermodynamics, and so on.
2) Physical condition: pressure, temperature, humidity, and so on.
3) Meteorological conditions: wind, rain, hail, ice, snow, ultraviolet rays, acid rain, salt fog, and so on [22, 23].

Herein, both the mechanical and physical conditions affect the strength and stiffness of the polymer matrix wave-transparent composites. The meteorological condition mainly affects the corrosion, wear, aging, and other physicochemical properties of the polymer matrix wave-transparent composites [24–26].

Polymer matrix wave-transparent composites are subjected to two or more environments simultaneously. The effects of various environmental are not simple additions but complex interactions. Therefore, it is necessary to carry out various environmental conditions simultaneously in the environmental testing of the polymer matrix wave-transparent composites to ensure the optimization of the structural design [27, 28].

6.1.1.4 Reliability and Economy of the Products

With the rapid development of science and technology, structural design in the field of aerospace puts forward the requirements of structural reliabilities. Structural reliabilities refer to the ability of the structure to fully realize the expected properties and achieve normal operation within the specified service life under the specified load and environmental condition. Since the structural damages are generally static load and fatigue fracture damage, the structural reliabilities are also divided into static strength and fatigue life reliabilities [29, 30].

In addition, the rationality of the structural design is mainly manifested in two aspects:reliability and economy. The relationship between initial cost, maintenance cost, total cost, and reliability of the structural design is shown in Figure 6.2. Generally, in order to improve reliability, it has to increase the initial cost, while the maintenance cost decreases as reliability and initial cost increase. As reliability increases, there is a minimum value for the total cost where the structural design is most reasonable [31].

6.1.2 Materials Design

Polymer matrix wave-transparent composites with different properties can be fabricated by adopting different polymer matrix and reinforced fibers, as well as different lay-up directions. For example, the single layer of wave-transparent laminated composites is the basic unit, mainly composed of polymer matrix and reinforced fibers, and the structural unit is the laminate composed of a single layer. Studying the properties of the single layer and laminates is particularly important for preparing wave-transparent laminated composites. The following takes wave-transparent laminated composites as an example to introduce the content of the materials design, including the selection of raw materials, the properties of the single layers, and the design of the laminates [32]. Other wave-transparent composites also follow the corresponding principles.

6.1.2.1 Selection of Raw Materials

The raw materials determine the upper limit of the properties for wave-transparent composites. Therefore, selecting suitable raw materials can meet the actual applications [33, 34].

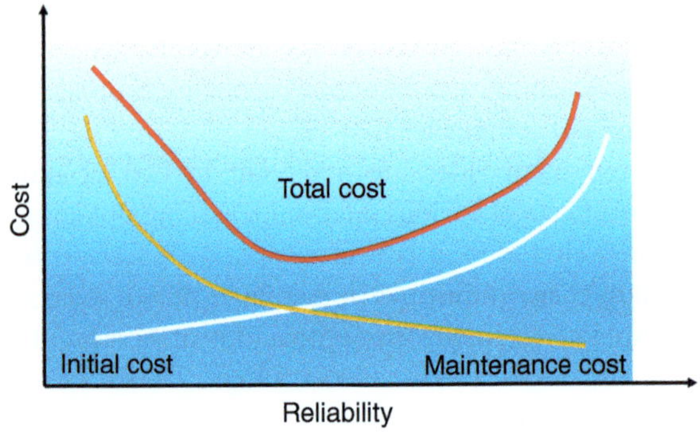

Figure 6.2 Relationship between initial cost, maintenance cost, total cost, and reliability of the structural design.

6.1.2.1.1 Selective Principle of Raw Materials

1) The principle of high specific strength and stiffness. For polymer matrix wave-transparent composites applied in aerospace, the structures should be as light as possible under the premise of meeting the requirements for strength, stiffness, durability, and damage tolerance.
2) The principle of raw materials being compatible with the applied environment. Generally, the decline rate of the main properties for raw materials should be less than 10% under long-term environmental conditions. The main environmental conditions of temperature and humidity would cause performance degradation. Especially in high temperature and humidity environments, the corresponding mechanical properties of the polymer matrix would be seriously affected. Therefore, it is important to select a suitable polymer matrix for different environments to meet the requirements of the actual applied environment.
3) The principle of satisfying the special requirement of the structures. Polymer matrix wave-transparent composites not only need to satisfy structural stiffness and strength but also excellent wave-transparent performance. During the material selection and manufacturing, if any of the variables, such as the type and amount of polymer matrix and reinforced fibers, the properties of a single layer, the orientation of prepreg laying, and the curing process, are changed, the corresponding ε and $\tan\delta$ values of the whole system would be changed. Therefore, it is necessary to clarify the requirements of the applied environment for dielectric properties to select suitable polymer matrix, reinforced fibers, and molding processes for the preparation of polymer matrix wave-transparent composites.
4) The principle of meeting technological requirements. The technologies of polymer matrix wave-transparent composites include three aspects :prepreg, curing, and repair technology. The prepreg technology includes volatiles content, viscosity, resin efflux, prepreg storage period, and treatment period. The curing technology includes the pressing time, curing temperature and pressure, and the shrinkage of cured polymer matrix wave-transparent composites, and so on. The repair technology mainly refers to the ability of the cured wave-transparent composites to adhere to other matrix or adhesives.
5) The principle of low cost and high efficiency. The cost includes initial and maintenance costs. The initial cost includes the cost of raw materials and manufacturing. The efficiency refers to the economic efficiency in terms of material and energy savings obtained through the weight reduction [35, 36].

6.1.2.2 Determination of the Performances for Single Layer

The single layer of the polymer matrix wave-transparent composites comprises the polymer matrix and reinforced fibers. In the initial stage of the structural design for the laminates, it is necessary to provide the performances of the single layer, especially the parameters of modulus and strength. Initially, the performances of the single layer are simply determined by the prediction equation derived from the micromechanical analysis method. In the final design stage, it is necessary to carry out an actual inspection to verify the real reliability of the properties for a single layer [37–39]. The prediction equation for the modulus and strength of the single layer can be described below.

6.1.2.2.1 Prediction Equations of the Stiffness for Single Layer

The prediction equations for the longitudinal (E_L) and transverse (E_T) elastic modulus of the single layer are shown in Eqs. (6.1) and (6.2):

$$E_L = E_f V_f + E_m(1 - V_f) \tag{6.1}$$

$$E_T = \frac{E_f E_m}{E_m V_f + E_f(1 - V_f)} \tag{6.2}$$

where E_f is the elastic modulus of the reinforced fibers, V_f is the volume fraction of the reinforced fibers, and E_m is the elastic modulus of the polymer matrix.

6.1.2.2.2 Prediction Equations of the Strength for Single Layer

The prediction equations for the longitudinal tensile strength (X_T) and compressive strength (X_C) of the single layer are shown in Eqs. 6.3–6.5:

$$X_T = \sigma_{fmax} V_f + \sigma_{mfmax}(1 - V_f) \rightarrow (V_f \geq V_{fmin}) \tag{6.3}$$

$$X_T = \sigma_{mmax}(1 - V_f) \rightarrow (V_f \leq V_{fmin}) \tag{6.4}$$

where, σ_{fmax} is the maximum tensile stress of the reinforced fibers; σ_{mfmax} is the stress when the strain of the polymer matrix is equal to the maximum tensile strain of the reinforced fibers; σ_{mmax} is the maximum tensile stress of the polymer matrix; V_f is the volume fraction of the reinforced fibers; V_{fmin} is the minimum volume fraction where the strength is controlled by the reinforced fibers, which needs to be further verified.

$$X_C = \min\left\{\left[2V_f\sqrt{\frac{V_f E_f E_m}{3(1 - V_f)}}\right] or \frac{G_m}{1 - V_f}\right\} \tag{6.5}$$

where, E_f is the elastic modulus of the reinforced fibers; G_m is the shear elastic modulus of the polymer matrix; E_m is the elastic modulus of the polymer matrix; V_f is the volume fraction of the reinforced fibers [40–42].

6.1.2.3 Laminates Design

The laminates design of polymer matrix wave-transparent composites, also known as lay-up design, determines the orientation of each layer in the laminates, the order of laying, and the percentage and thickness of each oriented layer according to the properties of the single layer.

6.1.2.3.1 General Principles of the Laminates Design

1) The principle of the directional lay-up. Excessive lay-up orientation of the laminates would complicate the design of the wave-transparent composites. Currently, four lay-up directions of 0°, 45°, 90° and ⊥45° are widely applied. In the isotropic design of the laminates, in addition to the [0/45/90/−45/0] laminates, the [60/0/60] laminates can also be used to reduce the number of layers.

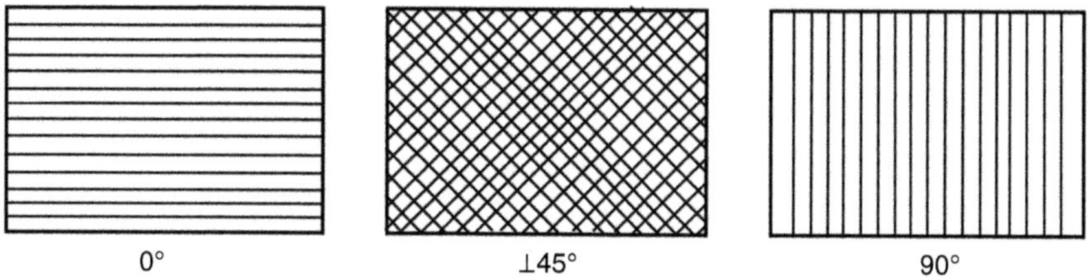

Figure 6.3 Schematic diagram of layers in different directions.

2) The principle of balanced and symmetrical laying. Except for special needs, it is generally designed as a balanced and symmetrical layer to avoid deformation caused by tension-shearing and tension-bending.
3) The determined orientation of the lay-up according to the loading direction. If the tensile or compressive load is applied, the orientation of the lay-up would be laid according to the direction (0°) of the load; if the shear load is applied, the orientation of the lay-up would be laid in pairs in the 45° direction; if the load is biaxial, the orientation of the lay-up would be laid orthogonality according to the load direction of 0°, 45° and 90°; if the multiple loads are applied, the layers would be laid in multiple directions of 0°, ⊥45°, 90°. Figure 6.3 shows the common layering directions.
4) The principle of minimum proportion for layers. To reduce hygrothermal stresses in the inner polymer matrix, the Poisson's ratio of each component should be harmonized. Especially for the direction of 0°, 45°, and 90°, the minimum proportion of the layer in any direction should be 6–10% [43].

6.1.2.3.2 Design Methods of the Laminates

The number of layers in each direction for wave-transparent laminated composites is determined according to the design requirement. At present, the replacement design is generally used to determine the order and the number of layers. The replacement design generally refers to replacing the laminates with other materials of the same shape under the same condition. During the replacement design, if the loads include tensile and compressive loads, the lay-up sequence of [0°/90°/90°/0° or 90°/0°/0°/90°] would be generally applied; if the loads include tensile and shearing loads, the lay-up sequence of [45°/−45°/−45°/45°] or [−45°/45°/45°/−45°] would be generally applied; if the loads include tensile, compressive and shearing loads, the lay-up sequence of [0°/45°/90°/−45°/−45°/90°/45°/0°] would be generally applied [44–46].

6.1.3 Structural Design

In addition to materials design, the structural design of the polymer matrix wave-transparent composites is also an indispensable part. It mainly includes three aspects: structural design principles, technological requirements, and external factors.

6.1.3.1 Structural Design Principles

The structural design principles of the polymer matrix wave-transparent composites need to satisfy the strength and stiffness, mainly including the following points:

1) The structure can withstand a load that is generally not less than the minimum service load;
2) The failure criterion of the polymer matrix wave-transparent composites only applies to the single layer. When the failure criterion is not specified, the Tsai-Wu failure criterion is generally adopted;
3) For areas with less stringent requirements on stiffness, the average value of test data can be used for the structural elastic constant, but for important areas with strict requirements on stiffness, the B reference value needs to be selected for the calibration of the structural elastic constant [47–49].

6.1.3.2 Technological Requirements

The main technological requirements that should be considered in the structural design of the polymer matrix wave-transparent composites are listed as follows:

1) The corners of the structures should have a large fillet radius to avoid defects such as fiber breakages, resin-rich areas, and crevices (incomplete bonding between layers).
2) Complex structural design should consider the difficulty of the manufacturing process. The surface with large curvature should be paved with fabric; The surface at the shape mutation should be smoothed; The stepped thickness should be adopted to avoid abrupt changes in wall thickness.
3) The surface should be coated with a protective film to meet the high-quality requirements of the components.
4) In principle, the wall thickness of the structures should be controlled below 7.5 mm. For areas with a wall thickness greater than 7.5 mm, the corresponding mechanical properties should be appropriately reduced during design.
5) The structure connected with metal materials should be covered with a layer of fabric as far as possible.
6) To reduce the assembly workload, the structure of the products should be designed as a whole as much as possible [50, 51].

6.1.3.3 External Factor

It is also necessary to consider external factors, such as the generated thermal stress during curing and the impact of the external environment [52].

6.1.3.3.1 Thermal Stress

There are large differences in coefficients of thermal expansion between the polymer matrix wave-transparent composites and other parts. In some extreme environments with large temperature differences, it is easy to generate thermal stress inside the composites, leading to warping and deformation or even cracking. In order to reduce the thermal stress, the coefficients of the thermal expansion for the polymer matrix should be similar to that of the reinforced fibers. Besides, symmetrical structures should be considered during the structural design [53, 54].

6.1.3.3.2 Impact Resistance

Especially for aircraft and airborne radomes in high-speed motion, the impact damage is the main form of damage. The most typical one is the bird impact damage. The impact damage can be divided into three categories according to the impact energy and structural defects: 1) The high energy impact causes the penetrating damage on the structure with a small amount of fracture delamination; 2) The medium-energy impact causes the outer sag and fibers fracture of the surfaces, and internal delamination in the impact area; 3) The low-energy impact causes internal delamination, but only minor damage on the surfaces.

The impact damage can reduce the mechanical properties of the polymer matrix wave-transparent composites, especially the compressive strength. Therefore, the effect of impact damage on the mechanical strength can be reduced by limiting the allowable strain or stress during the structural design. From the aspect of the raw materials, aramid fibers with excellent impact resistance can be selected as reinforced fibers. In addition, the impact resistances of the laminates composed of fiber fabrics are better than that of the unidirectional laminates [55–57].

6.2 Molding Process of the Polymer Matrix Wave-Transparent Composites

After the structural design of the polymer matrix wave-transparent composites is determined, the quality of the products largely depends on the molding process. Shaping, impregnating, and curing are the three main elements of the molding process for polymer matrix wave-transparent composites.

6.2.1 Shaping

The process of shaping is the uniform distribution of the reinforced fibers' inner polymer matrix or the orientation of the reinforced fibers in a set direction, also called preforming [58].

6.2.2 Impregnating

Impregnating refers to replacing the air between the reinforced fibers with the polymer matrix. The main influencing factors of impregnating are the viscosity of the polymer matrix, the ratio of the polymer matrix to reinforced fibers, and the types of reinforced fibers [59].

6.2.3 Curing

Curing refers to the process in which the polymer matrix is crosslinked to change from the linear structure into the network structure. What's more, initiators, accelerators, and so on, are usually used to promote the curing of the polymer matrix. The easiest way to determine the degree of the curing is to test the surface hardness [60, 61].

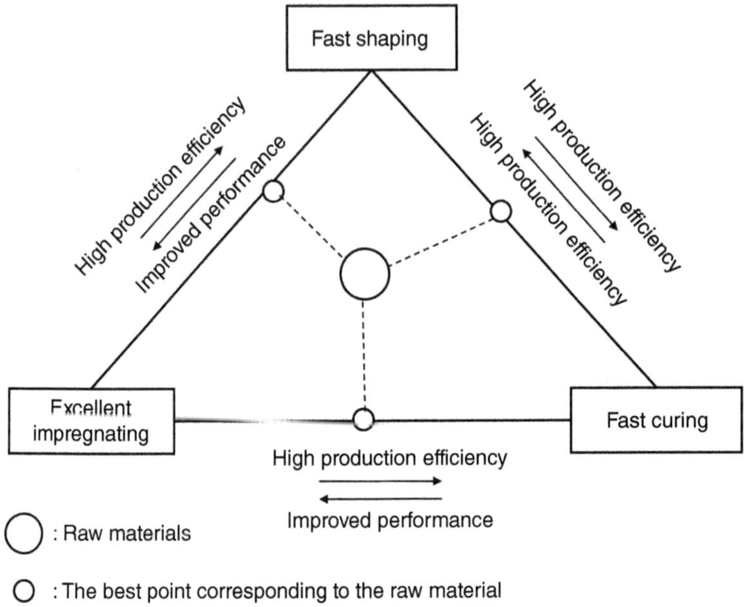

Figure 6.4 The relationship between raw materials and the three elements.

For different molding processes, it is necessary to select an appropriate process according to the three elements of molding and comprehensively control the quality of shaping, the speed of impregnating, and the curing time. In addition, shaping, impregnating, and curing affect each other. Figure 6.4 shows the relationship between the three elements of molding and the raw materials. In order to reduce the cost, shortening the molding cycle would sacrifice the part properties of the products. On the contrary, if the properties of the products would be improved, the corresponding cost would be increased. According to different raw materials, the correlation diagram of three elements for each molding method must be made to optimize the most reasonable solution.

The main molding methods of the polymer matrix wave-transparent composites include hand paste molding, bag molding, laminated molding, resin transfer molding (RTM), and filament winding. During the molding process, three points must be followed: 1) Reinforced fibers must be evenly distributed. 2) Polymer matrix is guaranteed to be fully impregnated with the reinforced fibers and cured according to the curing procedure. 3) During the molding process, it is necessary to reduce the generation of air bubbles and porosities to improve the compactness of the products. Several molding methods of the polymer matrix wave-transparent composites are introduced in detail below [62, 63].

6.2.4 Hand Paste Molding

Hand paste molding is also known as contact molding or lay-up molding. In 1940, the U.S. military prepared glass fibers reinforced polyester wave-transparent composites applied to military aircraft radomes by hand paste molding [64].

6.2.4.1 Types of Hand Paste Molding

Hand paste molding is divided into wet and dry molding [65].

6.2.4.1.1 Wet Molding

Wet molding is the simultaneous completion of impregnation and preforming. During the wet molding, the mold is first coated with a layer of release agents, followed by painting and spraying a layer of the polymer glue with curing agents as a gelcoat layer [66, 67]. Then, the cut reinforced fiber cloths are laid on the gelcoat layer. Subsequently, the polymer glue is evenly immersed in the reinforced fiber cloths, and the next layer of reinforced fiber cloths is laid after the air bubbles are removed. Repeat the above steps several times until the required layers has been laid. Finally, the wave-transparent composites are cured according to the corresponding curing process [68, 69]. Figure 6.5 shows the schematic diagram of the wet molding.

Demolding by adopting air stripping, top stripping and water stripping, and so on, is the key step and is related to the final quality of the products. El-baky et al. [70] prepared glass fibers/halloysite nanotubes/epoxy resins laminated composites by molding wet hand paste. Results showed that the method of wet hand paste molding could mass-produce laminated composites. When the mass fraction of the halloysite nanotubes was 1 wt%, the obtained glass fibers/halloysite nanotubes/epoxy resins laminated composites presented the optimal mechanical properties.

6.2.4.1.2 Dry Molding

For dry molding, the reinforced fiber cloths are firstly dipped into the polymer glues to obtain the prepregs, which are cut into a certain shape and softened by heating to remove the air bubbles. Finally, the wave-transparent laminated composites are cured according to the corresponding curing process, followed by demolding and trimming [71].

The prepregs by solution impregnation are intermediate products for the preparation of the polymer matrix wave-transparent composites by dry molding. Firstly, the reinforced fibers are heated to remove the moisture and paraffin sizing agent. Next, the polymer matrix is dissolved in an appropriate solvent, followed by poured into the dipping tank. The pretreated reinforced fibers are immersed in the dipping tank

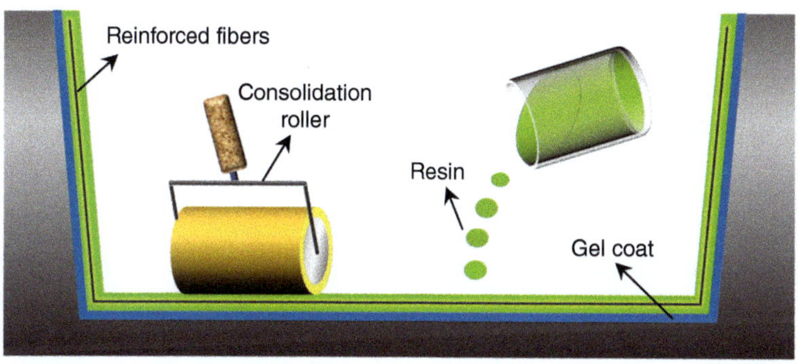

Figure 6.5 Schematic diagram of the wet molding.

and fully impregnated to obtain the prepregs [72]. The quality of prepregs should be strictly controlled in the manufacturing process, including the content of the volatile and polymer matrix, the gel time, and the viscosity of the prepregs [73, 74].

1) The volatiles content of the prepregs

The volatiles content refers to the proportion of the volatiles mass to the total mass of the prepregs. According to the *Standard Test Method for Volatiles Content of Composite Material Prepreg* (ASTM D3530-20), the volatiles content is generally measured by the balance weighing method. The details are described as follows: the prepregs are cut into small pieces, weighed and heated to the curing temperature and kept for a period of time. Afterwards, the heated prepregs are reweighed. The volatile content of the prepregs is calculated by Eq. (6.6):

$$V_c = \frac{M_i - M_f}{M_i} \times 100\% \tag{6.6}$$

where, V_c is the volatile content, M_i is the initial mass of the prepregs, and M_f is the mass of the heated prepregs.

The prepregs with appropriate volatiles can endow the polymers with excellent fluidity during the molding. However, excessive volatiles are difficult to remove, leading to the formation of voids in the wave-transparent composites and reducing the mechanical properties [75, 76].

2) The polymer content of the prepregs

The polymer content of the prepregs, also known as the glue content, refers to the mass fraction of various additives (polymers, curing agents, toughening agents, and so on) in the prepregs. Currently, the most common method for measuring the glue content is solvent extraction. According to *Standard Test Methods for Constituent Content of Composite Prepreg* (ASTM D3529-16), the sample of 80 mm × 80 mm is placed in a good solvent (acetone or butanone) for full extraction. The samples before and after extraction are weighed to calculate the corresponding glue content according to Eqs. (6.7) and (6.8).

$$\mathrm{MC} = \frac{M_i - M_f}{M_i} \times 100\% \tag{6.7}$$

$$\mathrm{MS} = \mathrm{MC} - V_c \tag{6.8}$$

where, MC is the polymer content containing volatile matter, M_i is the initial mass of the prepregs before being solvent extracted, and M_f is the mass of the prepregs after being solvent extracted. MS is the glue content, and V_c is the volatile content.

3) The gel time of the polymers

The gel time of the polymers determines the curing process of the polymer matrix wave-transparent composites. According to *Composites-Prepregs-Determination of gel time* (ISO 15040:1999), the prepregs are clamped between coverslips and placed on an electric plate at the specified temperature. Then, appropriate pressure is applied to the coverslips, allowing the polymers to flow out from between the two coverslips. Finally, stop heating when the outflow of polymers is in a wire drawing state. The time consumed by the whole process is the gel time of the prepregs [77].

4) The viscosity of the prepregs

The viscosity of the prepregs is the property of adhering and fitting to each other at a certain temperature, which is related to the softening point of the polymer matrix. If the prepregs are stored beyond the storage period, it would lead to degradation of performance and viscosity and further decrease the mechanical properties of the products. Therefore, it is necessary to test the viscosity of the prepregs before fabricating the polymer matrix wave-transparent composites. According to the *Standard Test Method for Characterizing Tack of Prepregs Using a Continuous Application-and-Peel Procedure* (ASTM D8336-21), the prepregs with 80 mm × 216 mm are laid on a clean metal substrate, and the peel strength is tested to characterize the viscosity [78].

6.2.4.2 Raw Materials of Hand Paste Molding

6.2.4.2.1 Polymer Matrix

The polymer matrix of hand paste molding should present a suitable viscosity to completely impregnate the reinforced fibers, and the viscosity should be in the range of 0.2–0.8 Pa s; The polymer matrix can be gelled and cured at room temperature and no low molecular substances are released during the curing process; In addition, the polymer matrix should possess low toxicity and cost. The common polymer matrix includes unsaturated polyester resins and epoxy resins [79].

6.2.4.2.2 Reinforced Fibers

The reinforced fibers of hand paste molding should be easily impregnated by a polymer matrix, presenting a certain flexibility. There are several types of reinforced fibers for hand paste molding: 1) The twist-free roving fabric, as the main type for hand paste molding, presents the following advantages of excellent deformability, easy to impregnated by polymer matrix, and high thickening efficiency; 2) Twisted fabrics include the types of plain weave, twill weave, and satin weave, and the surface is generally covered with the wax. The obtained wave-transparent composites with twisted fabric present a smooth surface and excellent air tightness, but the twisted fabrics are difficult to be impregnated by the polymer matrix, resulting in low production efficiency; 3) The chopped strand mats present the best wettability to the polymer matrix and possess outstanding deformability. The products by chopped strand mat present a high glue content, which is widely applied in the anti-leakage layer of waterproof and corrosion-resistant [80–82].

6.2.4.3 Release Agents of the Hand Paste Molding

The release agents of the hand paste molding include three types of polyvinyl alcohol (PVA), wax and ointment, and liquid. 1) PVA release agents with 5 vol% PVA possess low cost and toxicity, and the using temperature is below 150 °C; 2) Wax and ointment release agents are easy to operate, nontoxic, and low corrosive to molds; 3) Liquid release agents are the polymer solution without wax, such as the polystyrene release agent and perchloroethylene release agent. Liquid release agents present short film-molding time but are highly toxic. In addition, the silicone and methyl silicone oil can be used as release agents for metal molds, and the cellulose acetate can be used as release agents for wooden molds [83–85].

6.2.4.4 Features of the Hand Paste Molding

Advantages: 1) Hand paste molding can prepare polymer matrix wave-transparent composites with multi-shape, multi-size, and multi-variety to meet the needs of complex shape products; 2) Hand paste molding does not require complicated equipment, is easy to operate, and suitable for small batch production.

Disadvantages: 1) The quality of the products by hand paste molding is difficult to control, which is affected by the technical level of the operators; 2) Hand paste molding presents low production efficiency and long cycle; 3) Solvents used to dissolve the polymers tend to evaporate toxic gases, endangering the health of people [86].

6.2.5 Bag Molding

Bag molding is the method of adopting the negative pressure formed by vacuuming between the elastic forming bag and the mold or applying fluid pressure outside the bag to make the blank close to the mold and then curing and forming. The molding process only adopts a positive or negative mold to obtain products with complex shapes, excellent quality, and large dimensions. According to the different ways of pressurization, bag molding can be divided into three methods: vacuum bag molding, pressure bag molding, and vacuum bag-autoclave molding [87, 88].

6.2.5.1 Vacuum Bag Molding

Vacuum bag molding is a pressure molding method that generates atmospheric negative pressure by vacuuming. First, the preformed blanks are sealed between the vacuum bag and the mold, and then the gas inside the bag is pumped to form a negative pressure. Finally, the preformed blanks are pressurized and cured [87, 89]. The device of the vacuum bag molding is shown in Figure 6.6.

The vacuum bag is made of nylon film with excellent ductility and high strength, which is bonded to the mold with a sticky sealing strip. The vacuum bag is connected to the air duct with a breathable felt. American Air Tech Company produced a series of nylon vacuum bags. The BW900 vacuum bag could be used at 176 °C for 10 hours, and Wrightcast 8500Y and Wrightcast 7500R vacuum bags can be used at 230 °C for 16 hours, 285 °C for 15 hours, respectively. The nylon-polyester composite

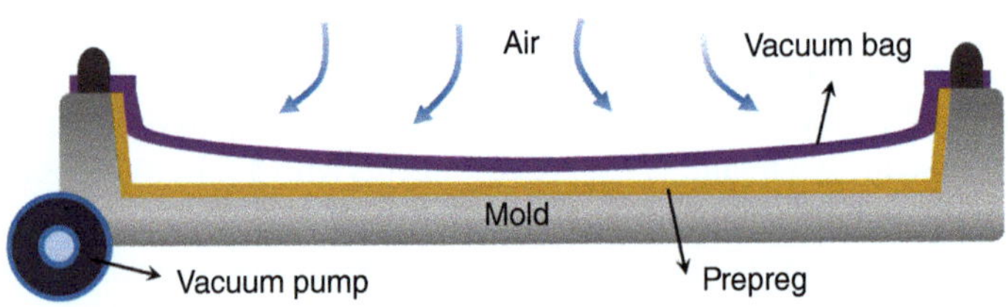

Figure 6.6 Schematic diagram of the vacuum bag molding.

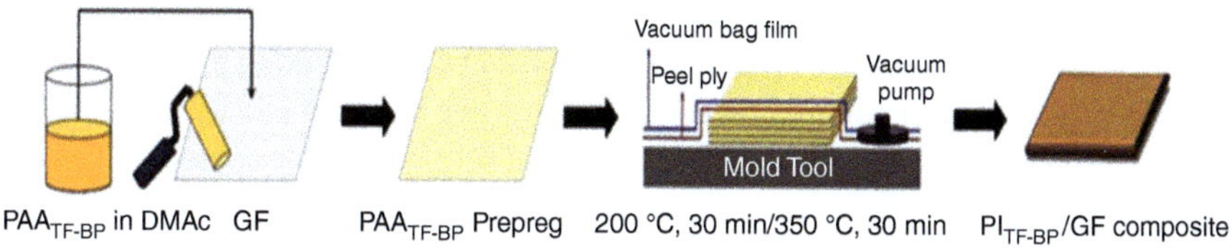

Figure 6.7 Schematic diagram of the fabrication process of the PI$_{TF-BP}$/GF composites. Source: [90]/Springer Nature.

vacuum bag of CVF120 from China Colas Company can be vacuum bagged in a large area below 120 °C, and the maximum length can reach 16 m. Park et al. [90] adopted 2,2′-bis(trifluoromethyl)-4,4′-diaminobiphenyl (TFDB) and biphenyl tetracarboxylic diandhydride (BPDA) to synthesize poly (amic acid) (PAA$_{TF-BP}$), which was mixed with silicon (Si) inorganic particles, followed by impregnating with glass fibers (GF) to obtain PI$_{TF-BP}$/GF/Si blanks. Next, the PI$_{TF-BP}$/GF/Si blanks were formed by vacuum bag molding to prepare PI$_{TF-BP}$/GF/Si composites (Figure 6.7). Results showed that the PIT$_{F-BP}$/GF/Si composites presented excellent impact strength (102.23 kJ/m^2), low ε (1.91), and $\tan\delta$ (0.024, 13 MHz, 25 °C), which could be used as aircraft radomes.

Vacuum bag molding presents a simple process and equipment, low cost, which is commonly used to form large-size products at room curing temperature. Since the generated pressure by the vacuum bag can only reach 0.1 MPa at most, it is only suitable for polymer matrix wave-transparent composites with a thickness of less than 1.5 mm and low forming pressure [91].

To solve the issues of low forming pressure for vacuum bags, the National Aeronautics and Space Administration (NASA) and Naval Air Warfare Center developed a double vacuum bag molding, which sealed the lay-up billet connected to the vacuum system on top of a stacked blank [92]. The corresponding schematic diagram of the double vacuum bag is shown in Figure 6.8. During the molding process, the

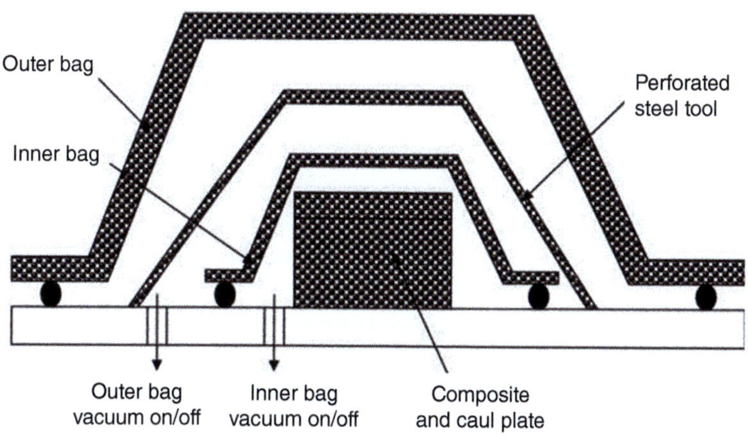

Figure 6.8 Schematic diagram of the double vacuum bag. Source: [92]/John Wiley & Sons.

inner vacuum bag is evacuated without applying pressure so that the low molecular volatiles and air are removed. The outer vacuum bag is evacuated so that the product is pressurized and formed through the perforated steel diaphragm. This process can provide enough pressure to obtain products with little or no void content.

6.2.5.2 Pressure Bag Molding

Pressure bag molding is a molding method developed based on vacuum bag molding. Pressure bag molding presents the advantages of simple equipment and easy operation, mainly used to produce thin skins and honeycomb sandwich structures. The pressure bag is composed of a rubber bag, which the pressure can reach 0.25–0.5 MPa. Figure 6.9 shows the schematic diagram for pressure bag molding. The pressure bag molding requires high strength, stiffness, and thermal efficiency due to the high pressure [93].

6.2.5.3 Vacuum Bag-Autoclave Molding

Vacuum bag-autoclave molding is a method of using the static gas pressure inside the autoclave to complete the curing and molding of the composite prepregs at a certain temperature and pressure. The autoclave comprises a tank body, a vacuum pump, a compressor, a gas storage tank, an electric heating device, and a control cabinet. The vacuum pump can pre-pressure the prepregs and suck the glue to create low-pressure environment; the compressor and the gas storage tank are inflatable systems, and the inside is generally inert gas; the electric heating device provides constant temperature in the autoclave [94].

The process of vacuum bag-autoclave molding is described as follows: First, the prepregs are laid-up and encapsulated; then, the air and volatiles inside the prepregs are removed through a vacuum system. Finally, the whole system is applied to a certain pressure after being heated, followed by curing according to the corresponding curing process. The encapsulation of the prepregs is one of the most important links in the vacuum bag-autoclave molding. This link determines polymer effluent, entrained gas, and applied pressure of the prepregs to ensure glue penetration, absorption, and air permeability. Excessive polymer glue is usually removed by the absorbent paper, and absorbent felt. Figure 6.10 is a schematic diagram of

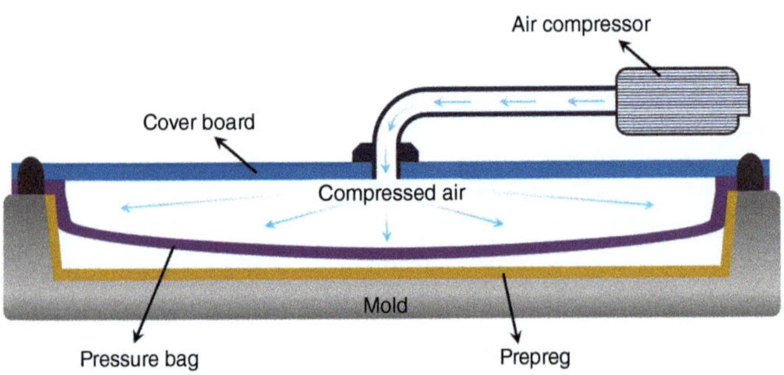

Figure 6.9 Schematic diagram of the pressure bag molding.

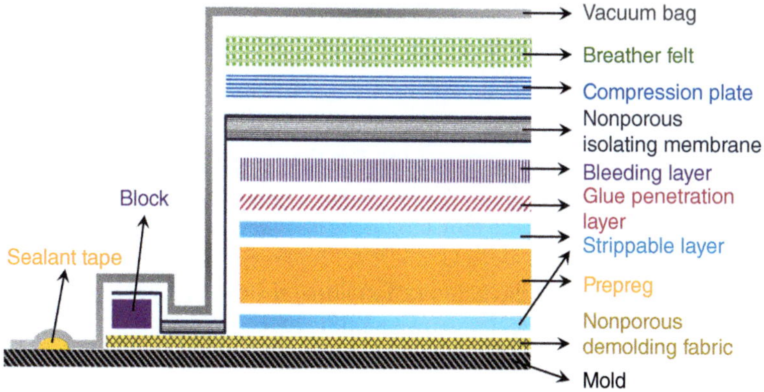

Figure 6.10 Schematic diagram of the stacking sequence for the vacuum bag-autoclave molding.

the stacking sequence for the vacuum bag-autoclave molding, and the stacking sequence is beneficial to extracting volatiles and gases from the prepregs [95].

The control of pressure in vacuum bag-autoclave molding is particularly important. During the curing process of the polymer matrix wave-transparent composites, the polymer matrix can undergo a phase transition (a change from a fluid state to a high-elastic state), when pressure is applied to obtain dense and uniform wave-transparent composites. Too early pressurization would cause a large amount of polymer matrix to be lost. Too late pressurization would cause the polymer matrix to enter the high-elastic state with many voids and air bubbles [96].

Although vacuum bag-autoclave molding presents low energy utilization and expensive equipment investment, it can provide uniform temperature and pressure (1.5–2.5 MPa) for curing and molding the polymer matrix wave-transparent composites. Vacuum bag-autoclave molding is suitable for the preparation of large-area and complex polymer matrix wave-transparent composites [95], which present uniform polymer content, dense internal structure, and excellent mechanical properties.

6.2.6 Laminated Molding

Laminated molding is to cut the prepregs into certain shapes, which are laminated together, followed by applying certain pressure and temperature by the flat vulcanizing press to obtain the products after curing [97, 98].

6.2.6.1 Laminated Molding Process

The drying process of the laminated molding is to remove the volatile solvents in the prepregs, and the speed of drying air is usually controlled at 3–4 m/s. The glue content and fluidity of the prepregs on the upper and lower surfaces should be slightly higher than that of the inner layer to improve the moisture resistance and smoothness of the products. In addition, the 10–20 layers of prepregs adjacent to the upper layer should be smooth and without damage. In order to make the thickness of the products accurate, the mass method is generally used to determine the thickness of the products, as shown in Eq. (6.9).

$$m = \frac{Fhd(1+a)}{1000} \tag{6.9}$$

where, m is the total mass of the prepregs (g), F is the area of the products (cm^2), h is the thickness of the products (cm), d is the density of the products (g/cm^3), and a is the correction factor: when $h < 5$ mm, a is taken as 0.02–0.03; when $h \geq 5$ mm, a is taken as 0.03–0.08.

The hot-pressing process of the laminated molding needs to pay attention to the three parameters of temperature, pressure, and time [99].

6.2.6.1.1 Temperature

The heating process of the laminated molding can be divided into five stages:

The first stage is the preheating stage from room temperature to the significant reaction temperature of the prepregs. At this time, the polymer matrix starts to flow, and part of the volatiles are discharged. The pre-added pressure is generally 1/3–1/2 of the total pressure.

The second stage is the holding stage. At this time, the polymer matrix penetrates with each other, but the curing reaction rate is low. The holding time depends on the gel time of the prepregs and the thickness of the products. The pressure and temperature should be increased as the outflowing polymer matrix is close to the gelling state.

The third stage is the heating stage, which increases the reaction rate and the degree of curing for the polymer matrix.

The fourth stage is the heat preservation stage, aiming to fully cure the polymer matrix. The curing temperature depends on the curing characteristics of the polymer matrix, and the curing time also depends on the thickness of the products.

The fifth stage is the cooling stage. When the end of the fourth stage is reached, the heating can be stopped, followed by slow cooling. However, the required maximum pressure is still maintained. The cooling speed should not be too fast, and the temperature should be dropped below 50 °C before demolding [100].

6.2.6.1.2 Pressure

The pressure is applied to overcome the vapor pressure of the volatiles and allow close contact between the prepreg layers. The amount of the molding pressure is determined according to the curing characteristics of the polymer matrix. If a large amount of volatiles escapes during curing, the pressure should be high. The higher the polymer curing temperature is, the higher the molding pressure is. In addition, the amount of pressure also needs to consider the thickness of the products, the glue content of the prepregs, and the heating rate [101].

6.2.6.1.3 Time

The time from the start of preheating to the removal of the products is the total time of the laminated molding, mainly including pre-pressing time, hot-pressing time, and cooling time. The pre-pressing time and hot-pressing time are related to the curing speed of the polymer matrix, the thickness of the products, and the pressing temperature. The pre-pressing time depends on the properties of the prepregs. If the insoluble resin content of the prepregs is high and the volatile content is low, the pre-pressing time would be short, and vice versa would be relatively longer. The hot-pressing time should ensure that the polymer matrix in the prepregs is fully cured. The cooling time is the last link to ensure the quality of the products. If the cooling time is too short, it would easily cause the products to warp and crack, and if the cooling time is too long, the production efficiency would be reduced.

In addition, the heat source can be turned off after curing, followed by naturally cooled or cooled by cold air and cold water, and so on. The pressure should not be lowered too early; otherwise, it would foam or warp the product surface. The demolding temperature is generally lower than 50 °C. Some products need to be post-cured to eliminate part of the internal stress. For example, the post-cured process of the epoxy resins products is 120–150 minutes at 120–130 °C, and 120–150 minutes at 220–230 °C for cyanate resins products [102, 103].

The laminated molding can efficiently fabricate the wave-transparent laminated composites with glass fiber cloths, Kevlar fiber cloths, and quartz fiber cloths as reinforced fibers. For example, the laminated molding was performed to prepare the wave-transparent composites reinforced by quartz fiber cloths for radomes of F15 and F22 fighter jets [104, 105]. Yang et al. [106] adopted 2,2′-bis (trifluoromethyl)benzidine and 3,3′,4,4′-oxydiphthalic anhydride as monomers, cis-5-norbornene-endo-2,3-dicarboxylic anhydride as the endcap to synthesize the A-stage polyimide. Then, the A-stage polyimide solution was impregnated with quartz fiber cloths to obtain quartz fiber cloths/A-stage polyimide composites by laminated molding. Results showed that the quartz fiber cloths/A-stage polyimide composites presented low ε (<3.2, 1–12 GHz) and $\tan\delta$ ($<5\times10^{-3}$, 1–12 GHz), which could be used as packaging materials for integrated circuits. In our previous work, Gu et al. [98] adopted bisphenol A cyanate (BADCy) resins and epoxy resins (E-51) to prepare the prepolymers, followed by impregnated with Kevlar fiber cloths to obtain the prepregs. Kevlar fiber cloths/BADCy wave-transparent composites were fabricated, curing at 10 MPa through laminated molding (Figure 6.11). The wave-transparent performance of the Kevlar fiber cloths/BADCy wave-transparent composites could reach 92.0% (1 MHz), and the flexural strength and interlaminar shear strength (ILSS) were 236.3 MPa and 28.5 MPa, respectively.

6.2.6.2 Common Defects and Solutions of the Laminated Molding

The laminated molding is relatively simple but difficult to control the high quality of the products. The defects, such as uneven surface, intermediate cracking, blackening of the laminate core, and thickness deviation, often occur during the laminated molding.

6.2.6.2.1 Surface Inhomogeneity

Surface inhomogeneity often occurs inner thin wave-transparent laminated composites, mainly due to the poor flow of the polymer matrix. During the preparation of the

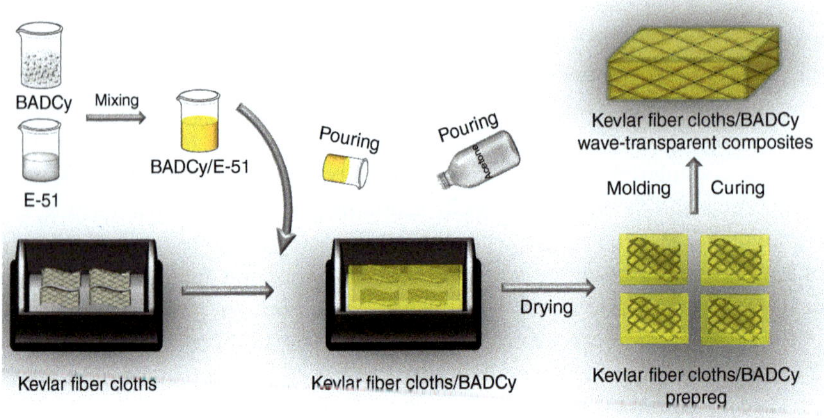

Figure 6.11 Schematic diagram of the preparation for Kevlar fiber cloths/BADCy wave-transparent composites [98]. Source: [10]/with permission of Elsevier.

prepregs, the content of insoluble components in the inner polymer matrix should be reduced, and the fluidity of the polymers should be improved. The preheating time should not be too long, and the final pressure should be applied in time to ensure the excellent mechanical properties of the wave-transparent laminated composites.

6.2.6.2.2 Intermediate Cracking

The main reason for intermediate cracking is that the blanks contain partially aged prepregs. If the glue contents of the prepregs and the pressure during molding are too low, the polymer matrix of the wave-transparent laminated composites would be uneven, which makes it easy to generate large internal stress and cracks. During the production, the quality of the prepregs, the time of pressing, and the holding time should be strictly controlled.

6.2.6.2.3 Blackening of Laminate Core

The main reason for the blackening of the laminate core is that the proportion of soluble polymers and volatile components in the prepregs is too large, leading the polymer matrix to flow around during pressing. In addition, the volatile components evaporate around the board, while there are too many residual volatile components in the laminate core. The mass fraction of insoluble polymers should be appropriately increased, and the mass fraction of volatile components should be reduced to prevent the prepregs from moisture during storage.

6.2.6.2.4 Thickness Deviation

The main reason for the thickness deviation is the deviation of the upper and lower sides of the mold during lamination, the uneven glue of the prepregs, and the high fluidity of the polymer glues. During the molding process, it should check whether the molds are parallel, control the quality of the prepregs, and maintain uniform temperature distribution during hot pressing [107–109].

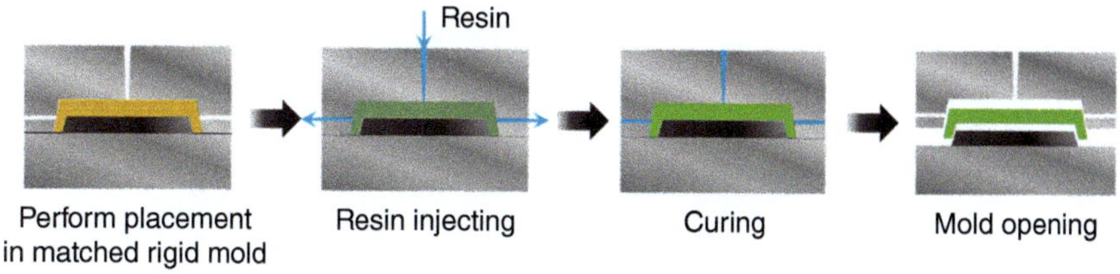

Figure 6.12 Schematic diagram of main process flow for RTM.

6.2.7 RTM

RTM, also known as injection-compression molding, is the molding process that evolved from wet lay-up and injection molding. RTM is the method of injecting polymers into a closed mold cavity by pressure, infiltrating the reinforced fiber fabrics, and then molding according to a certain curing process [110, 111]. Figure 6.12 is the schematic diagram of the main process flow for RTM.

6.2.7.1 Process Flow of RTM

RTM is the advanced molding process of injecting polymers into a closed mold to infiltrate reinforced fibers and further curing, which is suitable for the molding of advanced composites with multi-species, medium batch, and high quality [112]. RTM presents the advantages of low volatile components in the molding process, low environmental pollution, low cost, and high production efficiency, and so on, which have been widely applied in the automotive industry, aerospace, mechanical equipment, and electronic products [113, 114]. The process flow of RTM is described in detail below: the manufacturing of preformed blank, injection and filling, and curing [115, 116].

6.2.7.1.1 Manufacturing of Preformed Blank

Manufacturing of preformed blank is the design of reinforced fibers into a specific shape with uniform density, and the size of the preformed blank should be slightly smaller than the mold sealing area for the mold to be closed and sealed [117].

The preformed blank of the polymer matrix wave-transparent composites is generally prepared by laying up two-dimensional (2D) fabrics or 3D woven preforms. The interlaminar properties and impact resistances of the 3D woven preforms are better than those of the 2D fabrics [118–120]. Figure 6.13 shows the structures of the 3D woven preforms, mainly including 3D four-directional, 3D five- directional, 3D six-directional, 3D seven-directional, and other structures.

Figure 6.14a shows a 3D braiding machine produced by 3Tex (USA), which can weave 50 cell layers with a maximum fabric thickness of 25.4 mm. Figure 6.14b–b′ shows the preformed blanks of the radomes prepared by the 3D braiding machine. The radomes using the preformed blanks prepared by RTM present the characteristics of structural integrity and high mechanical properties [121].

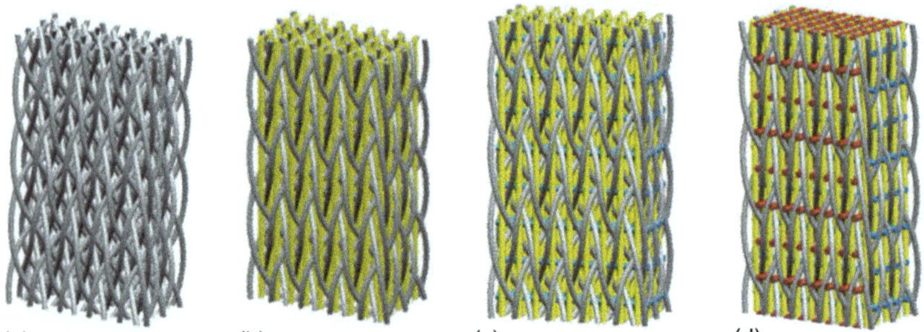

(a) (b) (c) (d)

Figure 6.13 3D four-direction (a), 3D five-direction (b), 3D six-direction (c), 3D seven-direction (d) structures of the 3D woven preforms [121].

(a) (b) (b′)

Figure 6.14 3D braiding machine produced by 3Tex, USA (a); the preformed blanks of radomes prepared by 3D braiding machine (b–b′) [121].

6.2.7.1.2 Injection and Filling

The process of injection and filling is to inject the polymer matrix into the mold to impregnate the reinforced fibers and expel the air. The injection pressure, injection rate, and injection temperature are the main parameters of RTM.

1) Injection pressure

The injection pressure is related to the mold materials and the structural design. The higher the injection pressure is, the higher the strength and stiffness of the mold. In order to reduce cost, RTM generally adopts low injection pressure, which can be reduced by using a polymer matrix with low viscosity, appropriate mold injection and exhaust port design, and specific arrangement design of the reinforced fibers.

2) Injection rate

The injection rate depends on the wettability of the polymer matrix to the reinforced fibers as well as the surface tension and viscosity of the polymer matrix. The injection rate is also related to the activity period of the polymer matrix, mold stiffness, product size, and the content of the reinforced fibers. A high injection rate is beneficial to reducing the bubbles in the composites and improving the production efficiency. A low injection rate provides sufficient time to improve the wettability of the polymer matrix to reinforced fibers, thereby enhancing the mechanical properties of the products.

3) Injection temperature

The injection temperature depends on the active period and the minimum viscosity-temperature of the polymer matrix. High injection temperature can reduce the surface tension of the polymers, which is conducive to the ejection of the bubbles. However, low temperature can increase the viscosity of the polymer matrix and injection pressure and reduce the ability of the polymer to infiltrate the reinforced fibers. In order to make the polymer matrix fully infiltrate the reinforced fibers under the minimum pressure, the injection temperature should be close to the minimum viscosity-temperature of the polymer matrix [122–124].

6.2.7.1.3 Curing

The curing process should start when the polymer matrix is just filled with the mold, and premature curing would make the polymer matrix cannot completely infiltrate the reinforced fibers, resulting in the existence of voids and the degradation of the mechanical properties of the products [124, 125].

6.2.7.2 Structures of RTM Machine

RTM machine generally comprises a hydraulic pump, injection gun, mixer, cleaning device, vacuum system, pressure system and control system. According to the different injection systems of the RTM machine, there are two categories: single-pump injection and double-pump injection [126].

6.2.7.2.1 Single Pump Injection of RTM

The single pump injection is the RTM machine with only one hydraulic pump, which is suitable for a one-component polymer system with low-reaction. During the single pump injection, the polymer needs to be cured at a high temperature, and the pressure system needs to be cleaned after each injection in order to prevent the residual polymer matrix from clogging the orifice.

6.2.7.2.2 Double Pump Injection of RTM

There are three types of RTM molding machines with double pump pressurization systems: two-component pressurized type, two-component pump type, and double pump with added catalyst type. A highly reactive polymer matrix with two components starts to cure quickly once mixed, so the double pump press system should be chosen to control the curing rate. Two components of the polymer matrix are placed in different tanks of the double pump injection system, respectively. Each component is pressed into the pipeline by a pressure pump or a piston pump then mixed thoroughly by the mixing head. Finally, the polymer mixtures are entered into the mold through the injection nozzle. When the mold is heated, the injection nozzle must be cooled. To ensure the polymer mixtures enter the mold smoothly, the injection nozzle should be cleaned and dried in time after each injection to prevent the

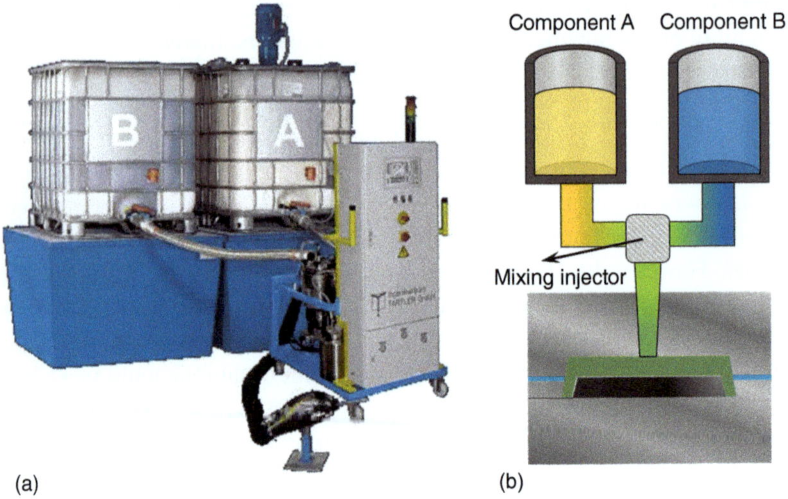

Figure 6.15 Molding machine of the two-component pump RTM (Nodopur B 1000) from TARTLER, Germany (a) and the process diagram (b). Source: (a) TARTLER GmbH.

fast-curing polymer matrix from blocking the line [127]. Figure 6.15 shows the molding machine of the two-component pump RTM (Nodopur B 1000) from TARTLER, Germany (a), and the process diagram of the molding machine (b).

6.2.7.3 Characteristics of RTM

The products by RTM present the advantages of high dimensional accuracy, easy operation, and no post-treatment processing. However, RTM has a long injection cycle, and the quality of injection is hard to control well. RTM process is suitable for the preparation of the polymer matrix wave-transparent composites with low viscosity (0.1–1 Pa s) at room temperature, no solvent or volatile, and no release of small molecules [128, 129].

6.2.8 Filament Winding

Filament winding is suitable for reinforced fibers with smooth surfaces and is difficult to weave into fabrics. During filament winding, the reinforced fibers are completely impregnated with the polymer matrix, then followed by wound onto the core mold under a certain tension, and finally cured by heating or at room temperature [130, 131].

6.2.8.1 Classification of Filament Winding

According to the different physical and chemical states of the polymer matrix during winding, the filament winding can be divided into dry winding, wet winding, and semi-dry winding [132].

6.2.8.1.1 Dry Winding

Dry winding is to impregnate the polymer matrix with continuous reinforced fibers, followed by removing the solvent to obtain spindles. Subsequently, the spindles are wound directly on the core mold to make them into shape. Dry winding is easy to control, and the quality of the products is stable, which is suitable for aviation/aerospace and military equipment and other fields. However, the dry winding presents disadvantages of complicated equipment and high cost, and the curing agent of the dry winding does not sublimate and volatilize sublimation and volatilization [133].

6.2.8.1.2 Wet Winding

Wet winding is the process of winding the impregnated reinforced fibers directly onto the core mold under a certain tension (Figure 6.16). The reinforced fibers used in the wet winding are generally continuous reinforced fibers such as aramid and PBO fibers, and the polymer matrix mainly includes unsaturated polyester, epoxy resins, and cyanate resins with lower viscosity and longer service life. The wet winding presents the advantages of simple molding equipment, low cost, excellent parallelism of reinforced fibers, and high production efficiency. However, the wet winding makes it difficult to control the amount of polymer matrix and the quality of the finished products [134, 135].

6.2.8.1.3 Semi-Dry Winding

Semi-dry winding is adding a set of drying equipment based on wet molding to remove the solvent of the reinforced fibers by heating. The quality of the final products by the semi-dry winding is also difficult to control. Compared with dry winding, the semi-dry winding saves the process of the prepregs [136].

6.2.8.2 Winding Rules

The winding rules refer to the rule of relative movement between the winding head and the core mold. The winding rules include hoop winding, longitudinal winding, and polar winding.

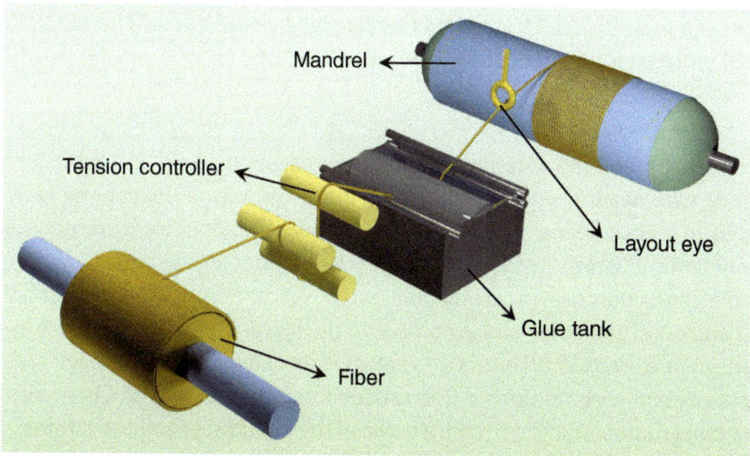

Figure 6.16 Schematic diagram of the wet winding.

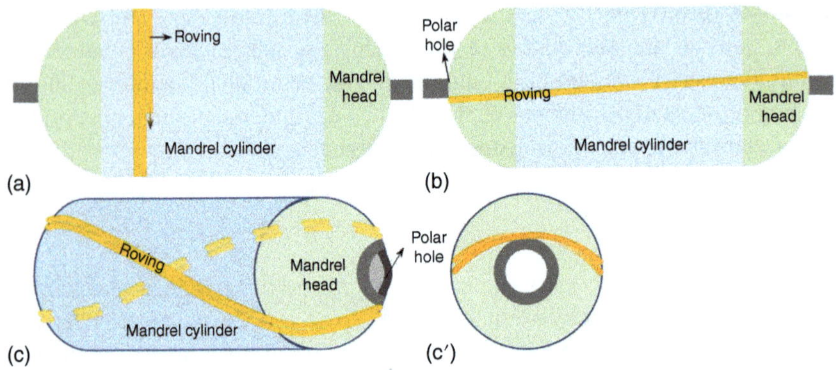

Figure 6.17 Schematic diagram of hoop winding (a), longitudinal winding (b), and helical winding (c).

6.2.8.2.1 Hoop Winding

Hoop winding is winding along the circumference of the core mold. During hoop winding, the core mold rotates at a constant speed along the axis, and the guide wire head moves slowly in the cylindrical section parallel to the axis of the core mold. When the core mold rotates once, the guide wire head moves a fiber bundle width in the direction of the core axis [137]; the schematic diagram of hoop winding is shown in Figure 6.17a.

6.2.8.2.2 Longitudinal Winding

During longitudinal winding, the guide wire head makes a uniform circular motion in a fixed plane, and the core mold rotates slowly and intermittently around its longitudinal axis. The guide wire head rotates once, the core mold rotates at a small angle, and the angle of the longitudinal winding reinforced fibers and the longitudinal axis of the core mold is 0–25°. The schematic diagram of longitudinal winding is shown in Figure 6.17b. The reinforced fibers in each revolution can be kept on the same plane, and their trajectory is a single-circle plane closed curve. Therefore, the longitudinal winding is also called plane winding.

6.2.8.2.3 Helical Winding

During helical winding, the core mold rotates at a constant speed around its axis, and the guide wire head moves back and forth along the axis of the core mold at a certain speed to realize helical winding on the cylinder section. The corresponding winding schematic diagram is shown in Figure 6.17c–c′. During the helical winding, the reinforced fibers start from a certain point on the circumference of the polar hole at one end and enter the other end according to the spiral trajectory, followed by a return to the cylinder section. The above process is repeated until the surface of the core mold is covered with reinforced fibers. Each bundle of reinforced fibers corresponds to a tangent point on the circumference of the polar hole, and the reinforced fibers in different directions intersect to be evenly wound on the surface of the core mold [138].

6.2.8.3 Winding Process

The winding process generally includes preparing raw materials and polymer glue, equipment inspection, program inputting, core mold installation, winding, curing and demolding, and so on.

6.2.8.3.1 Preparation of Raw Materials

Before winding, it is necessary to review the name, specification, and manufacturer of the polymer matrix, reinforced fibers, and other auxiliary materials according to the specific requirements of the relevant molding process guidance documents. For example, epoxy resins usually need to be reviewed for the epoxy value, hydroxyl value, chlorine content, viscosity, and so on. The indicators that unsaturated polyester resins usually need to be retested for some properties, including the viscosity, acid value, solid content, hydroxyl value, reactivity, and gel time. Reinforced fibers need to be dried in an oven at 80 °C for 48 hours in advance. It is worth noting that aramid fibers should be sealed and heated to isolate them from moisture before winding due to their high water absorption rate [139].

6.2.8.3.2 Preparation of Polymer Glue

According to the formula of the polymer glue, appropriate solvent, curing agent, accelerator, and other auxiliary materials are added into the polymer matrix, in turn,to obtain polymer glue after full stirring. Particularly, before preparing the glue of unsaturated polyester and epoxy resins, it is necessary to adjust the amount of curing agent according to the ambient temperature and test the gel time, endowing them possessing the most suitable service life [140].

6.2.8.3.3 Equipment Inspection and Program Input

Before winding, it is necessary to inspect, debug, and input the program of the winding machine.

1) Equipment inspection

Start the winding machine and check the operation of the mechanical system (winding frame, motor, transmission system, and so on), control system, auxiliary system (yarn frame, glue tank, heater, and so on), and tension control system (sensor, controller, measurement, and control system).

2) Designing and debugging of winding line type

When the winding line type is debugged, the core mold is installed on the winding machine, and the predetermined line pattern is wound to ensure that there is no yarn piece leaving the seam or slipping line.

3) The installation and debugging of the auxiliary equipment

The auxiliary equipment, such as creel, glue tank, wire winding nozzle, and heater. are checked to ensure normal operation and smooth yarn passing path [141].

6.2.8.3.4 Installation of the Core Mold

1) Installation of the metal core mold

The oil on the surface of the metal core mold is removed by using acetone or ethyl acetate before winding. If there is rust, sandpaper is used to polish the surface of the metal core mold, followed by cleaning it with ethanol. Subsequently, the surface of the cleaned metal core mold is applied a layer of release agent.

2) Installation of the gypsum core mold

The surface of the gypsum core mold is coated with a layer of polymer glue to close the small pores inside, followed by coating with a layer of release agent. However, the gypsum core mold is not suitable for products with curing temperatures above 150 °C.

3) Preparation of the water-soluble core mold

The commonly used binders for making water-soluble core molds are PVA and sodium silicate. When the curing temperature of the products is lower than 150 °C, the PVA binder is commonly used. When the curing temperature is higher than 150 °C, the sodium silicate binder is commonly used. The water-soluble core mold is treated similarly to the gypsum core mold before use [142].

6.2.8.3.5 Impregnating and Winding

The glue content of the reinforced fibers during the winding process is determined by the impregnation method. The impregnation methods include the dipping method and the rubber roller contact method. The dipping method controls the glue content of reinforced fibers by adjusting the pressure of the squeezing roller, and Figure 6.18a shows the diagram of the dipping method. The rubber roller contact method is used to control the glue content by adjusting the distance between the scraper and the rubber roller and changing the thickness of the glue layer on the surface of the rubber roller. And Figure 6.18b shows the diagram of the rubber roller contact method.

The tension on the reinforced fibers during winding effects the mechanical properties, denseness, and glue content of the polymer matrix wave-transparent composite

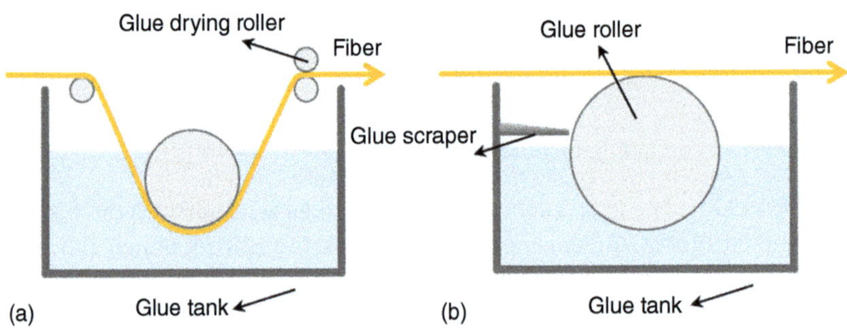

Figure 6.18 Diagram of the dipping method (a) and the rubber roller contact method (b).

products. The tension during winding should be decreasing layer by layer regularly so that the initial stress of the inner and outer reinforced fibers can be kept the same.

The main processes of the filament winding are described as follows.

1) The tension of the reinforced fibers is measured with a tensioner before winding and adjusted continuously to achieve the target tension.
2) The polymer glue is then poured into the glue tank, and the reinforced fibers pass through the glue tank and the extrusion roller, followed by introducing into the wire winding nozzle.
3) The reinforced fibers are winded according to the design requirement, and the dipping device is adjusted at any time to control the glue content of the reinforced fibers. And then, the excessive polymer glue on the surface of the products is constantly scratched off. If the reinforced fibers appear to slip, overlap, or gap situation, the winding should be suspended in time.
4) When the winding is coming to an end, the thickness of the sample is measured until it reaches the design requirement and can be stopped.
5) After winding, the formed fibers are placed in the curing oven or left at room temperature for curing [143, 144].

6.2.8.3.6 *Curing*

The curing of the fiber products should be carried out in strict accordance with the curing process. The products are placed and cured in ovens, curing ovens, vacuum tanks, or at room temperature. During the curing process, the curing temperature needs to be checked and debugged at any time. If the curing temperature is too high, too low, or heating too fast, the curing process should be stopped immediately and timely repair the equipment. It is strictly forbidden for the products to be taken out from the furnace at high temperature, which would make the products seriously shrink and cracks.

6.2.8.3.7 *Demolding*

The core mold should be removed after the product is cured. The metal core mold is demolded mechanically. The combined mold should be disassembled first, and then removed carefully, taking care not to damage the products. The sand core mold needs to be partially flushed out with water first, and then the metal shaft is removed.

6.2.8.4 Characteristics of the Filament Winding

The filament winding presents the following advantages.

1) Compared with those of other molding processes, the obtained wave-transparent composites by filament winding present higher neatness and accuracy of fiber arrangement, and the final products can give full play to the strength of fibers.
2) Filament winding presents the advantages of low manufacturing cost, high product quality, and reproducibility. Furthermore, filament winding is easy to realize mechanization and automation and is convenient for mass production [145].

However, there are some limitations of the filament winding.

1) During the winding process, especially in the wet winding process, it is easy to form air bubbles, thereby reducing the ILSS, compressive strength, and anti-instability ability of the final products.
2) The opening of the hole would cause stress concentration around the hole of the wave-transparent composites by filament winding, and the corresponding ILSS would also be reduced. Therefore, it is necessary to design the structures of the products reasonably and try to avoid destructive operations such as cutting and drilling the products.
3) The filament winding presents certain limitations in the shape of the products, which is not suitable for the manufacture of parts with concave curves. The obtained products by filament winding are mostly cylinders, spheres, and some positive curvature revolving bodies, such as some oval and spherical radar radomes [146]. PBO fibers present very low surface energy and are difficult to prepare reinforced fabrics, and the corresponding polymer matrix wave-transparent composites are usually prepared by filament winding. In our previous work, Gu et al. [147] adopted "Reversible Addition-Fragmentation Chain Transfer" polymerization to prepare P(S-*co*-BCB-*co*-GMA), followed by forming homogeneous membrane to coat the surface of PBO fibers (PBO@P-GMA fibers) through thermal cross-linking. Then, 2, 2-bis(3-amino-4-hydroxyphenyl)hexafluoropropane (AHHFP) was grafted on the PBO@P-GMA fibers to obtain 6F-PBO@P-GMA fibers. Finally, the 6F-PBO@P-GMA fibers/modified cyanate laminated composites were prepared by winding and laminated molding process. Results showed that the flexural strength and ILSS of the 6F-PBO@P-GMA fiber/modified cyanate laminated composites were 701.5 and 46.4 MPa, respectively. In the meantime, the obtained composites presented excellent wave-transparent performance at the X-band (wave-transparent coefficient higher than 93.5%).

6.3 Summary

This chapter mainly summarizes the design methods of the polymer matrix wave-transparent composites and five typical molding processes, including hand paste molding, pressure bag molding, laminated molding, RTM, and winding molding. According to different application environments, there are different requirements for the molding of the composites. Appropriate polymer matrix, reinforced fibers, and molding processes can be adopted to meet practical applications through forming, dipping, curing, and other links. In general, improving the preparation process and reducing defects of the polymer matrix wave-transparent composites are beneficial to improving the wave-transparent performances, mechanical properties, and heat resistances. On the premise of ensuring the quality of polymer matrix wave-transparent composites, reducing the preparation cost, and producing high-quality products in a green and efficient molding process is the development trend of preparing high-performance polymer matrix wave-transparent composites.

References

1 Kumar, A., Sharma, S., and Singh, G. (2007). Measurement of dielectric constant and loss factor of the dielectric material at microwave frequencies. *Progress in Electromagnetic Research-Pier* 69: 47–54.

2 Campbell, F.C. (2004). Chapter 1 - Introduction to composite materials and processes: unique materials that require unique processes. In: *Manufacturing Processes for Advanced Composites* (ed. F.C. Campbell), 1–37. Amsterdam: Elsevier Science.

3 Galos, J. (2020). Thin-ply composite laminates: a review. *Composite Structures.* 236: 111920.

4 Öziş, E., Osipov, A.V., and Eibert, T.F. (2017). Metamaterials for microwave radomes and the concept of a metaradome: review of the literature. *International Journal of Antennas and Propagation.* 2017: 1356108.

5 Hasan, Z. and Muliana, A. (2012). Failure and deformation analyses of smart laminated composites. *Mechanics of Composite Materials* 48 (4): 391–404.

6 Meltem, H.G. (2018). A review on machinability of carbon fiber reinforced polymer (CFRP) and glass fiber reinforced polymer (GFRP) composite materials. *Defence Technology* 14 (4): 318–326.

7 Koronis, G., Silva, A., and Fontul, M. (2013). Green composites: a review of adequate materials for automotive applications. *Composites Part B: Engineering* 44 (1): 120–127.

8 Llorca, J., González, C., Molina-Aldareguía, J.M. et al. (2011). Multiscale modeling of composite materials: a roadmap towards virtual testing. *Advanced Materials* 23 (44): 5130–5147.

9 Panwar, R. and Lee, J.R. (2018). Performance and non-destructive evaluation methods of airborne radome and stealth structures. *Measurement Science and Technology* 29 (6): 062001.

10 Feng, C., Bei, Y., Changrui, Z. et al. (2011). Progress of broadband wave-transparent radome. *Ordnance Material Science and Engineering* 34 (4): 85–89.

11 MacDonald, M.E. (2019). An overview of radomes for large ground-based antennas. *IEEE Aerospace and Electronic Systems Magazine* 34 (10): 36–43.

12 Nag, A., Rao, R.R., and Panda, P.K. (2021). High temperature ceramic radomes (HTCR): a review. *Ceramics International* 47 (15): 20793–20806.

13 Davis, J., Bilsborrow, R., and Gray, C. (2015). Delayed fertility transition among indigenous women in the ecuadorian amazon. *International Perspectives on Sexual and Reproductive Health* 41 (1): 1–10.

14 Van Tooren, M., Kasapoglou, C., and Bersee, H. (2011). Composite materials, composite structures, composite systems. *The Aeronautical Journal* 115 (1174): 779–787.

15 Hueber, C., Fischer, G., Schwingshandl, N., and Schledjewski, R. (2019). Production planning optimisation for composite aerospace manufacturing. *International Journal of Production Research* 57 (18): 5857–5873.

16 Ray, C. and Majumder, S. (2014). Failure analysis of composite plates under static and dynamic loading. *Structural Engineering and Mechanics* 52 (1): 137–147.

17 Choi, Y., Kang, W.M., Moon, J., and Kim, C.W. (2022). Optimum design of sandwich composite structures under various dynamic loads using the equivalent static load method. *Composite Structures* 298: 115991.

18 Andrew, J.J., Srinivasan, S.M., Arockiarajan, A., and Dhakal, H.N. (2019). Parameters influencing the impact response of fiber-reinforced polymer matrix composite materials: a critical review. *Composite Structures* 224: 111007.

19 Bhowmik, C., Gupta, K., and Chakraborti, P. (2020). Stability analysis of glass fibre polyamide composite transmission tower. *Journal of Reinforce Plastics and Composites* 39 (23–24): 918–931.

20 Yu, J., Chen, L., Lu, Y. et al. (2013). Progress of broadband wave-transparent materials for radome. *Aerospace Materials & Technology* 43 (2): 1–7.

21 Bhudolia, S.K., Kam, K.K.C., and Joshi, S.C. (2018). Mechanical and vibration response of insulated hybrid composites. *Journal of Industrial Textiles* 47 (8): 1887–1907.

22 De Simone, M.E., Ciampa, F., Boccardi, S., and Meo, M. (2017). Impact source localisation in aerospace composite structures. *Smart Materials and Structures* 26 (12): 125026.

23 Lei, Y., Wang, Q., and Huo, J. (2014). Fabrication of durable superhydrophobic coatings with hierarchical structure on inorganic radome materials. *Ceramics International* 40 (7): 10907–10914.

24 Cheng, Z., Hong, D.W., Dai, Y. et al. (2018). Highly improved UV resistance and composite interfacial properties of aramid fiber via iron (III) coordination. *Applied Surface Science* 434: 473–480.

25 Qamar, Z., Aboserwal, N., and Salazar-Cerreno, J.L. (2020). An accurate method for designing, characterizing, and testing a multi-layer radome for mm-wave applications. *IEEE Access* 8: 23041–23053.

26 Zhou, L., Liu, Z., Tang, L., and Pei, Y. (2017). Design and characterization for a high-temperature dual-band radome wall structure for airborne applications. *Materials & Design* 114: 264–270.

27 Al Ahsan, M., Rahman, T., Ludwick, A., and Aglan, H. (2020). Effect of salt fog spray and ultraviolet exposure on the fracture surface morphology of fiber reinforced polymer composites. *Microscopy and Microanalysis* 26 (S2): 2666–2668.

28 Zu, Q., Song, W., Zeng, H. et al. (2020). Glass resistance to radiation–Part I: preliminary investigation of three commercial glass fibers. *International Journal of Applied Glass Science* 11 (3): 522–536.

29 Antonio, C.C. and Hoffbauer, L.N. (2009). An approach for reliability-based robust design optimisation of angle-ply composites. *Composite Structures* 90 (1): 53–59.

30 Cabrera-Rios, M. and Castro, J.M. (2007). The balance between durability, reliability, and affordability in structural composites manufacturing. *Polymer Composites* 28 (2): 233–240.

31 Sohouli, A., Yildiz, M., and Suleman, A. (2018). Cost analysis of variable stiffness composite structures with application to a wind turbine blade. *Composite Structures* 203: 681–695.

32 Xu, Y., Zhu, J., Wu, Z. et al. (2018). A review on the design of laminated composite structures: constant and variable stiffness design and topology optimization. *Advanced Composites and Hybrid Materials* 1 (3): 460–477.

33 Li, W., Huang, W., Kang, Y. et al. (2019). Fabrication and investigations of G-POSS/cyanate ester resin composites reinforced by silane-treated silica fibers. *Composites Science and Technology* 173: 7–14.

34 Oliveira, J.R., Kotzebue, L.R.V., Freitas, D.B. et al. (2020). Towards novel high-performance bio-composites: polybenzoxazine-based matrix reinforced with Manicaria saccifera fabrics. *Composites Part B: Engineering* 194: 108060.

35 Boyce, C., Landi, Z., and Mufit, A. (2016). Low cost composites for vacuum insulation core material. *Vacuum* 131: 120–126.

36 Guo, F., Wang, N., Cheng, Q. et al. (2016). Low-cost coir fiber composite with integrated strength and toughness. *ACS Sustainable Chemistry & Engineering* 4 (10): 5450–5455.

37 Múgica, J.I., Lopes, C.S., Naya, F. et al. (2019). Multiscale modelling of thermoplastic woven fabric composites: from micromechanics to mesomechanics. *Composite Structures* 228: 111340.

38 Jain, A. (2019). Micro and mesomechanics of fibre reinforced composites using mean field homogenization formulations: a review. *Materials Today Communications* 21: 100552.

39 Swolfs, Y., Gorbatikh, L., and Verpoest, I. (2014). Fibre hybridisation in polymer composites: a review. *Composites Part A: Applied Science and Manufacturing* 67: 181–200.

40 González, A., Graciani, E., and París, F. (2008). Prediction of in-plane stiffness properties of non-crimp fabric laminates by means of 3D finite element analysis. *Composites Science and Technology* 68 (1): 121–131.

41 Song, N., Jackson, M., Wu, S., and Souza, F. (2021). Micromechanical analysis of mechanical response for unidirectional fiber-reinforced plies. *Integrating Materials and Manufacturing Innovation* 10 (4): 542–550.

42 Xu, H., Gu, X., and Li, C. (2016). Tensile stiffness of hybrid fiber composites. *Acta Materiae Compositae Sinica* 33 (2): 394.

43 Lakshmi Kumari, N.B.V., Mehar, A., Abdulrahman, M. et al. (2018). Performance analysis of ply orientation in composite laminates. *Materials Today: Proceedings* 5 (2): 5984–5992.

44 Wang, G.D., Wang, J., Hossain, S.K.M., and Chen, H. (2015). Research on design rules for composite laminate. *Science and Engineering of Composite Materials* 22 (3): 315–323.

45 Rydarowski, H. and Koziol, M. (2014). Repeatability of glass fiber reinforced polymer laminate panels manufactured by hand lay-up and vacuum-assisted resin infusion. *Journal of Composite Materials* 49 (5): 573–586.

46 Yamaguchi, K., Phenisee, S.E., Chen, Z. et al. (2020). Ply-drop design of non-conventional laminated composites using bayesian optimization. *Composites Part A: Applied Science and Manufacturing* 139: 106136.

47 Zixun, W., Weicheng, J., and Xiaodong, H. (2021). Research progress in calculation method of composite material's B basis. *Acta Materiae Compositae Sinica* 38 (6): 1669.

48 Van Paepegem, W. and Degrieck, J. (2003). Calculation of damage-dependent directional failure indices from the Tsai–Wu static failure criterion. *Composites Science and Technology* 63 (2): 305–310.

49 Daniel, I.M., Werner, B.T., and Fenner, J.S. (2011). Strain-rate-dependent failure criteria for composites. *Composites Science and Technology* 71 (3): 357–364.

50 Türk, D.-A., Rüegg, F., Biedermann, M., and Meboldt, M. (2019). Design and manufacture of hybrid metal composite structures using functional tooling made by additive manufacturing. *Design Science* 5. e16.

51 Kim, P.C., Lee, D.G., Seo, I.S., and Kim, G.H. (2008). Low-observable radomes composed of composite sandwich constructions and frequency selective surfaces. *Composites Science and Technology* 68 (9): 2163–2170.

52 Wang, H., Cao, M., Siddique, A. et al. (2017). Numerical analysis of thermal expansion behaviors and interfacial thermal stress of 3D braided composite materials. *Computational Materials Science* 138: 77–91.

53 Hancox, N.L. (1998). Thermal effects on polymer matrix composites: part 1. *Thermal cycling. Materials & Design* 19 (3): 85–91.

54 Li, Y., Xiao, Y., Yu, L. et al. (2022). A review on the tooling technologies for composites manufacturing of aerospace structures: materials, structures and processes. *Composites Part A: Applied Science and Manufacturing* 154: 106762.

55 Cherukattu Gopinathapanicker, J., Inamdar, A., Anand, A. et al. (2020). Radar transparent, impact-resistant, and high-temperature capable radomecomposites using polyetherimide-toughened cyanate ester resins for high-speed aircrafts through resin film infusion. *Industrial & Engineering Chemistry Research* 59 (16): 7502–7511.

56 Safri, S.N.A., Sultan, M.T.H., Jawaid, M., and Jayakrishna, K. (2018). Impact behaviour of hybrid composites for structural applications: a review. *Composites Part B: Engineering* 133: 112–121.

57 Wang, F., Zhang, J., Liu, Y. et al. (2013). Comprehensive optimization design of a composite radome for its anti-bird strike and electromagnetic performance. *Journal of Vibration and Shock* 32 (13): 93–98.

58 Fauster, E., Schillfahrt, C., Hueber, C., and Schledjewski, R. (2017). Automated profile preforming for structural components. *Science and Engineering of Composite Materials* 24 (5): 631–650.

59 Yeager, M., Simacek, P., and Advani, S.G. (2022). Prediction of micro impregnation phenomena in additively manufactured composite materials. *Composites Part A: Applied Science and Manufacturing* 156: 106861.

60 Abdel-Raheem, N.A., Halim, S.F., and Al-Khoribi, A.H. (2017). The effect of different curing conditions on hardness, thickness, and residual stress of carbon fiber reinforced epoxy composites. *Journal of Composite Materials* 52 (14): 1959–1970.

61 Zarybnicka, L., Machotova, J., Kopecka, R. et al. (2021). Effect of cyclotriphosphazene-based curing agents on the flame resistance of epoxy resins. *Polymers* 13 (1): 8.

62 Varna, J., Joffe, R., Berglund, L.A., and Lundström, T.S. (1995). Effect of voids on failure mechanisms in RTM laminates. *Composites Science and Technology* 53 (2): 241–249.

63 Alves, F.C., Monticeli, F.M., Neves, R.M. et al. (2021). Influence of void content and morphology on the creep behavior on glass/epoxy composites. *Composites Communications* 25: 100712.

64 Mahavir, C., Ankush, S., Maheshwar, D., and Amar, P. (2019). A comparative study of the physical, mechanical and thermo-mechanical behavior of GFRP composite based on fabrication techniques. *Fibers and Polymers* 20: 823–831.

65 Yousef, D., Mostefa, B., Abderrezak, B. et al. (2016). Multi-axial mechanical characterization of jute fiber/polyester composite materials. *Composites Part B: Engineering* 90: 450–456.

66 Romi, S., Hitoshi, T., and Antonio, N.N. (2016). Strength evaluation of cross-ply green composite laminates reinforced by bamboo fiber. *Composites Part B: Engineering* 84: 9–16.

67 Lin, H.-J., Lai, W.-M., Huang, H.-D., and Kuo, Y.-M. (2010). Discussion on the cause of print-through phenomenon of FRP and several improvement methods. *Journal of Composite Materials* 44 (17): 2111–2126.

68 Singh, J., Kumar, M., Kumar, S., and Mohapatra, S.K. (2017). Properties of glass-fiber hybrid composites: a review. *Polymer-Plastics Technology and Engineering* 56 (5): 455–469.

69 Bigg, D.M., Hiscock, D.F., Preston, J.R., and Bradbury, E.J. (1993). The properties of wet-formed thermoplastic sheet composites. *Polymer Composites* 14: 26–34.

70 El-baky, M.A.A. and Attia, M.A. (2020). The mechanical performance of the laminated aluminum-epoxy/glass fiber composites containing halloysite nanotubes: an experimental investigation. *Journal of Industrial Textiles* 51: 8690S–8737S.

71 Chen, C., Poursartip, A., and Fernlund, G. (2020). Cure-dependent microstructures and their effect on elastic properties of interlayer toughened thermoset composites. *Composites Science and Technology* 197: 108241.

72 Yao, J., Zhan, H., and Zou, Z. (2017). Preparation and curing behaviour of epoxy based film for moderate temperature prepreg. *Polymers and Polymer Composites* 25 (8): 621–626.

73 Szpoganicz, E., Demleitner, M., Hübner, F. et al. (2022). Phenolic prepregs for automated composites manufacturing—correlation of rheological properties and environmental factors with prepreg tack. *Composites Science and Technology* 218: 109188.

74 Kim, D., Centea, T., and Nutt, S.R. (2014). In-situ cure monitoring of an out-of-autoclave prepreg: effects of out-time on viscosity, gelation and vitrification. *Composites Science and Technology* 102: 132–138.

75 Dong, C. (2016). Effects of process-induced voids on the properties of fibre reinforced composites. *Journal of Materials Science & Technology* 32 (7): 597–604.

76 Jiang, B., Huang, Y.D., Liu, L., and Hu, C.P. (2012). Use of near-infrared spectroscopy for online quality analysis of armos fiber/prepreg tape. *Journal of Applied Polymer Science* 125 (4): 2485–2488.

77 Hayaty, M., Beheshty, M.H., and Esfandeh, M. (2011). A new approach for determination of gel time of a glass/epoxy prepreg. *Journal of Applied Polymer Science* 120 (3): 1483–1489.

78 Mbotto Tonye, D. and Buet, G.K. (2021). Influence of the reinforcement on prepreg tack. *Polymer Composites* 42 (9): 4795–4803.

79 Roopa, T.S., Murthy, H.N., Sudarshan, K. et al. (2015). Mechanical properties of vinylester/glass and polyester/glass composites fabricated by resin transfer molding and hand lay-up. *Journal of Vinyl and Additive Technology* 21 (3): 166–173.

80 Seretis, G.V., Kouzilos, G., Manolakos, D.E., and Provatidis, C.G. (2017). On the graphene nanoplatelets reinforcement of hand lay-up glass fabric/epoxy laminated composites. *Composites Part B: Engineering* 118: 26–32.

81 Harish Kumar, M., Mathivanan, N.R., and Kumar, S. (2018). Experiment investigations of effect of laminate thickness on flexural properties of GLARE and GFRP laminates. *Materials Research Express* 6 (2): 025313.

82 Chen, Q., Linghu, T., Gao, Y. et al. (2017). Mechanical properties in glass fiber PVC-foam sandwich structures from different chopped fiber interfacial reinforcement through vacuum-assisted resin transfer molding (VARTM) processing. *Composites Science and Technology* 144: 202–207.

83 Wu, F., Gao, S., Zeng, Y. et al. (2021). Study on synthesis and demolding performance of polyethylene glycol fatty acid mold release agents. *Polymers for Advanced Technologies* 32 (10): 4061–4069.

84 Shields, A.J., Hepburn, D.M., Kemp, I.J., and Cooper, J.M. (2000). The absorption of mould release agent by epoxy resin. *Polymer Degradation and Stability* 70 (2): 253–258.

85 Mishra, R., Behera, B.K., Rajpurohit, A.N. et al. (2021). Effect of internal mold release agent on flexural and inter laminar shear properties of carbon and glass fabric reinforced thermoset composites. *Polymers for Advanced Technologies* 32 (1): 282–293.

86 Denkena, B., Dittrich, M.A., and Rahner, B.H. (2017). Smart and energy-efficient dust suction concept for milling of fibre-reinforced plastics. *Production Engineering* 11: 723–729.

87 Popineau, V., Célino, A., Le Gall, M. et al. (2021). Vacuum-bag-only (VBO) molding of flax fiber-reinforced thermoplastic composites for naval shipyards. *Applied Composite Materials* 28: 791–808.

88 Gibson, A.G., Ijaz, M., Dodds, N. et al. (2003). Vacuum bag moulding of large thermoplastic parts in commingled glass/PET. *Plastics, Rubber and Composites* 32 (4): 160–166.

89 Tamakuwala, V.R. (2021). Manufacturing of fiber reinforced polymer by using VARTM process: a review. *Materials Today: Proceedings* 44: 987–993.

90 Yim, D., Park, J., Jang, S. et al. (2021). Fabrication and properties of glass fiber-reinforced composites using polyimide prepregs with inorganic nanofillers. *Fibers and Polymers* 22 (3): 804–810.

91 Levy, A. and Hubert, P. (2019). Vacuum-bagged composite laminate forming processes: Predicting thickness deviation in complex shapes. *Composites Part A: Applied Science and Manufacturing* 126: 105568.

92 Hou, T.H. and Jensen, B.J. (2008). Double-vacuum-bag technology for volatile management in composite fabrication. *Polymer Composites* 29 (8): 906–914.

93 Forcellese, A., Marconi, M., Simoncini, M., and Vita, A. (2020). Life cycle impact assessment of different manufacturing technologies for automotive CFRP components. *Journal of Cleaner Production* 271: 122677.

94 Park, S.-b., Lee, D.-w., and Song, J.-i. (2018). Fabrication and evaluation of mechanical properties of carbon/epoxy square tube using pressure bag molding and compared with autoclave method. *International Journal of Precision Engineering and Manufacturing* 19 (3): 441–446.

95 Weber, T.A., Arent, J.-C., Steffens, L. et al. (2017). Thermal optimization of composite autoclave molds using the shift factor approach for boundary condition estimation. *Journal of Composite Materials* 51 (12): 1753–1767.

96 Mehdikhani, M., Gorbatikh, L., Verpoest, I., and Lomov, S.V. (2018). Voids in fiber-reinforced polymer composites: a review on their formation, characteristics, and effects on mechanical performance. *Journal of Composite Materials* 53 (12): 1579–1669.

97 Yaw, A.C., Ket, T.C., Wong, K.H., and Yang, J. (2022). Enhanced damping and stiffness trade-off of composite laminates interleaved with recycled carbon fiber and short virgin aramid fiber non-woven mats. *Composite Structures* 297: 115981.

98 Tang, L., Dang, J., He, M. et al. (2019). Preparation and properties of cyanate-based wave-transparent laminated composites reinforced by dopamine/POSS functionalized Kevlar cloth. *Composites Science and Technology* 196: 120–126.

99 Yan, S., Xue, Y., and Wang, S. (2019). Preparation and properties of aluminum phosphate-polyethersulfone laminated composites modified by aluminosilicate fiber fabric. *Polymer Composites* 40: 4102–4112.

100 White, S.R. and Kim, Y.K. (1996). Staged curing of composite materials. *Composites Part A: Applied Science and Manufacturing* 27 (3): 219–227.

101 Li, S.J., Zhan, L.H., Chen, R. et al. (2014). The influence of cure pressure on microstructure, temperature field and mechanical properties of advanced polymer-matrix composite laminates. *Fibers and Polymers* 15 (11): 2404–2409.

102 Megahed, M., Fathy, A., Morsy, D., and Shehata, F. (2019). Mechanical performance of glass/epoxy composites enhanced by micro- and nanosized aluminum particles. *Journal of Industrial Textiles* 51 (1): 68–92.

103 Ren, D., Li, B., Chen, S. et al. (2021). Investigation on curing reaction of phthalonitrile resin with nanosilica and the properties of their glass fiber-reinforced composites. *Journal of Applied Polymer Science* 138 (5): 49777.

104 Su, M., Gu, A., Liang, G., and Yuan, L. (2011). The effect of oxygen-plasma treatment on Kevlar fibers and the properties of Kevlar fibers/bismaleimide composites. *Applied Surface Science* 257 (8): 3158–3167.

105 Wu, Y., Xiao, Y., Zou, C. et al. (2022). High-temperature resistance and wave-transparent quartz-fibre/polyimide composite. *Plastics, Rubber and Composites* 51 (9): 489–496.

106 Liu, X., Yang, H., Liu, J. et al. (2015). Preparation and characterization of quartz-fiber-cloth-reinforced, polymerization-of-monomer-reactant-type polyimide substrates with a high impact strength. *Journal of Applied Polymer Science* 132 (36): 42358.

107 Hassan, M.H. (2021). A mini review on manufacturing defects and performance assessments of complex shape prepreg-based composites. *The International Journal of Advanced Manufacturing Technology* 115 (11): 3393–3408.

108 Lightfoot, J.S., Wisnom, M.R., and Potter, K. (2013). Defects in woven preforms: formation mechanisms and the effects of laminate design and lay-up protocol. *Composites Part A: Applied Science and Manufacturing* 51: 99–107.

109 Stadler, H., Kiss, P., Stadlbauer, W. et al. (2022). Influence of consolidating process on the properties of composites from thermosetting carbon fiber reinforced tapes. *Polymer Composites* 43 (7): 4268–4279.

110 Zhu, Y., Wang, Y., Chen, M. et al. (2020). Research progress and application of RTM for advanced resin matrix composite. *China Plastics Industry* 48 (5): 18–22.

111 Sun, Z., Xiao, J., Tao, L. et al. (2019). Preparation of high-performance carbon fiber-reinforced epoxy composites by compression resin transfer molding. *Materials* 12: 13.

112 Ballout, W., Coulon, B., Janssens, A. et al. (2016). Quantitative characterization of interdiffusion at the resin–resin and resin–prepreg interphases of epoxy systems processed by model SQ-RTM. *Polymer Engineering and Science* 2016 (56): 1061–1069.

113 Van Oosterom, S., Allen, T., Battley, M., and Bickerton, S. (2019). An objective comparison of common vacuum assisted resin infusion processes. *Composites Part A: Applied Science and Manufacturing* 125: 105528.

114 Murray, J.J., Robert, C., Gleich, K. et al. (2020). Manufacturing of unidirectional stitched glass fabric reinforced polyamide 6 by thermoplastic resin transfer moulding. *Materials & Design* 189: 108512.

115 Shojaei, A., Ghaffarian, S.R., and Karimian, S.M.H. (2003). Modeling and simulation approaches in the resin transfer molding process: a review. *Polymer Composites* 24 (4): 525–544.

116 Naik, N.K., Sirisha, M., and Inani, A. (2014). Permeability characterization of polymer matrix composites by RTM/VARTM. *Progress in Aerospace Sciences* 65: 22–40.

117 Vo, D.M.P., Kern, M., Hoffmann, G., and Cherif, C. (2017). 2D net shape weaving for cost effective manufacture of textile reinforced composites. *IOP Conference Series: Materials Science and Engineering* 254 (9): 092005.

118 Yang, Z., Jiao, Y., Xie, J. et al. (2022). Effect of weaving parameters on fiber structure of 3D woven preforms: a micro-CT investigation. *Journal of Composite Materials* 56 (16): 2609–2620.

119 Weimer, C., Preller, T., Mitschang, P., and Drechsler, K. (2000). Approach to net-shape preforming using textile technologies. Part I: edges. *Composites Part A: Applied Science and Manufacturing* 31 (11): 1261–1268.

120 Xiao, Y., Xu, J., Wang, M. et al. (2021). Multiscale model of the RTM process: from mesoscale anisotropic permeability of woven structures to macroscale resin impregnation. *Industrial & Engineering Chemistry Research* 60 (22): 8269–8279.

121 Chen, L., Zhao, S., and Wang, X. (2018). Development and application of 3D textile reinforcements in the aerospace field. *China Textile Leader* S1: 80–87.

122 Young, W.-B. and Tseng, C.-W. (1994). Study on the pre-heated temperatures and injection pressures of the RTM process. *Journal of Reinforced Plastics and Composites* 13 (5): 467–482.

123 Kaynak, C. and Kas, Y.O. (2006). Effects of injection pressure in resin transfer moulding (RTM) of woven carbon fibre/epoxy composites. *Polymers and Polymer Composites* 14 (1): 55–64.

124 Li, Y., Li, Q., and Ma, H. (2015). The voids formation mechanisms and their effects on the mechanical properties of flax fiber reinforced epoxy composites. *Composites Part A: Applied Science and Manufacturing* 72: 40–48.

125 Pupin, C., Ross, A., Dubois, C. et al. (2017). Predicting porosity formation in phenolic resins for RTM manufacturing: the porosity map. *Composites Part A: Applied Science and Manufacturing* 100: 294–304.

126 Dong, G. (2002). Development of a new low pressure premixing RTM machine. *Fiber Reinforced Plastics Composites* 0 (1): 23–25.

127 Miranda Campos, B., Bourbigot, S., Fontaine, G., and Bonnet, F. (2022). Thermoplastic matrix-based composites produced by resin transfer molding: a review. *Polymer Composites* 43 (5): 2485–2506.

128 He, S., Liang, G., Yan, H. et al. (2009). High performance toughened cyanate ester resin with low injection temperature for RTM process. *Polymers for Advanced Technologies* 20 (2): 143–146.

129 Gong, M., Zhang, D., Liu, Y. et al. (2020). Progresses of in-situ polymerization manufacturing technology of fiber reinforced thermoplastic composites: a review. *Materials Review* 34 (11A): 21180–21187.

130 Al-Furjan, M.S.H., Shan, L., Shen, X. et al. (2022). A review on fabrication techniques and tensile properties of glass, carbon, and Kevlar fiber reinforced polymer composites. *Journal of Materials Research and Technology* 19: 2930–2959.

131 Xu, J., Miao, Y., Qiao, M., and Li, S. (2014). Tension design of heated-core mold winding process based on analytical algorithm. *Journal of Reinforced Plastics and Composites* 33 (16): 1529–1541.

132 Wang, Y., Gai, D., and Song, Y. (2011). Present situation and development tendency of filament winding technology. *Materials Review* 25 (3): 110–113.

133 Sofi, T., Neunkirchen, S., and Schledjewski, R. (2018). Path calculation, technology and opportunities in dry fiber winding: a review. *Advanced Manufacturing: Polymer & Composites Science* 4 (3): 57–72.

134 Long, B., Yang, N., and Cao, X. (2022). Low-velocity impact damages of filament-wound composite overwrapped pressure vessel (COPV). *Journal of Engineered Fibers and Fabrics* 17: 1–16.

135 Wang, B., Qiu, Z., Liu, A., and Jin, Z. (2004). Study on manufacture processing parameter of high-performance PBO fiber composite. *Acta Aeronautica et Astronautica Sinica* 25 (2): 192–196.

136 Yu, C., Li, F., Wang, X., and Zhou, D. (2010). Research progress of impregnating technology for filament winding process. *Fiber Reinforced Plastics/Composites* 5: 84–88.

137 Lisbôa, T.V., Almeida, J.H.S. Jr., Dalibor, I.H. et al. (2020). The role of winding pattern on filament wound composite cylinders under radial compression. *Polymer Composites* 41 (6): 2446–2454.

138 Shaw-Stewart, D. (1985). Filament winding—materials & engineering. *Materials & Design* 6 (3): 140–144.

139 Li, G., Zhang, C., Wang, Y. et al. (2008). Interface correlation and toughness matching of phosphoric acid functionalized Kevlar fiber and epoxy matrix for filament winding composites. *Composites Science and Technology* 68 (15): 3208–3214.

140 Di, C., Yu, J., Wang, B. et al. (2019). Study of hybrid nanoparticles modified epoxy resin used in filament winding composite. *Materials* 12 (23): 3853.

141 Chen, J., Jang, S.-S., Wong, D.S.H. et al. (1999). Optimal design of filament winding using neural network experimental design scheme. *Journal of Composite Materials* 33 (24): 2281–2300.

142 Shen, F.C. (1995). A filament-wound structure technology overview. *Materials Chemistry and Physics* 42 (2): 96–100.

143 Azeem, M., Ya, H.H., Alam, M.A. et al. (2022). Application of filament winding technology in composite pressure vessels and challenges: a review. *Journal of Energy Storage* 49: 103468.

144 Quanjin, M., Rejab, M.R.M., Kaige, J. et al. (2018). Filament winding technique, experiment and simulation analysis on tubular structure. *Materials Science and Engineering.* 342 (1): 012029.

145 Soykok, I.F., Tas, H., Ozdemir, O., and Kandas, H. (2020). Effect of drop weight impact on the torsional-loading behavior of filament wound and prepreg-wrapped composite tubes. *Polymers and Polymer Composites* 29 (6): 617–628.

146 Asyraf, M.R.M., Ishak, M.R., Syamsir, A. et al. (2023). Filament-wound glass-fibre reinforced polymer composites: potential applications for cross arm structure in transmission towers. *Polymer Bulletin* 80: 1059–1084.

147 Tang, L., Yang, Z., Tang, Y. et al. (2021). Facile functionalization strategy of PBO fibres for synchronous improving the mechanical and wave-transparent properties of the PBO fibres/cyanate ester laminated composites. *Composites Part A: Applied Science and Manufacturing* 150: 106622.

7

Application of the Polymer Matrix Wave-Transparent Composites

7.1 Aircraft Radomes

Aircraft radomes mainly refer to the radomes for high Mach aircraft such as missiles, rockets, and satellites, which are important parts of aviation and aerospace weaponry [1–3]. The aircraft radomes face many challenges of high accelerating load, fast aerodynamic heating of the cover wall surface, and complex environment, which require polymer matrix wave-transparent composites for aircraft radomes with ε and $\tan\delta$, excellent mechanical properties, heat and environmental resistances [4–6].

At present, the polymer matrix of aircraft radomes with subsonic and low supersonic includes unsaturated polyester (UP) resins, epoxy resins, polyteflon (PTFE) resins, bismaleimide (BMI) resins, and polyimide (PI) resins [7, 8]. Reinforced fibers mainly adopt glass fibers and quartz fibers [9, 10]. Boeing Company (USA) adopted glass fibers/epoxy resins wave-transparent composites to fabricate the "Pomac" radomes for active homing guided missiles with a Mach number of 3. It adopted glass fibers/PTFE resins wave-transparent composites to fabricate AIM-71 radomes for Sparrow missile with a Mach number of 3. Raw materials of radomes for the submarine-launched "Tomahawk" missile (USA, a Mach number of 0.7) and Hamm missile (USA, a Mach number of 3.2) were, respectively, epoxy resins and PI resins as polymer matrix. In addition, the mud-cast fused quartz fibers reinforced composites prepared by the George Asia Institute of Technology (USA) were successfully applied in the United States "Patriot" air defense missile and Italy's Aspie air-to-air missile (Figure 7.1). Russia focused on developing radomes with quartz fibers as reinforcement, which presented stable properties below 1500–1800 °C. It was widely used in cruise missiles, anti-missile missiles, tactical missiles, and space shuttles. China Aerospace Materials and Technology Research Institute also developed radomes with quartz fiber fabrics as reinforcement, which were applied to some missile radomes.

In the future information warfare [11], multimode guidance technology (infrared/microwave, infrared/millimeter-wave) and stealth technology [12–14] put forward higher requirements for the comprehensive properties of the aircraft radomes.

Polymer Matrix Wave-Transparent Composites: Materials, Properties, and Applications, First Edition. Junwei Gu, Yusheng Tang, Jie Kong, and Jing Dang.

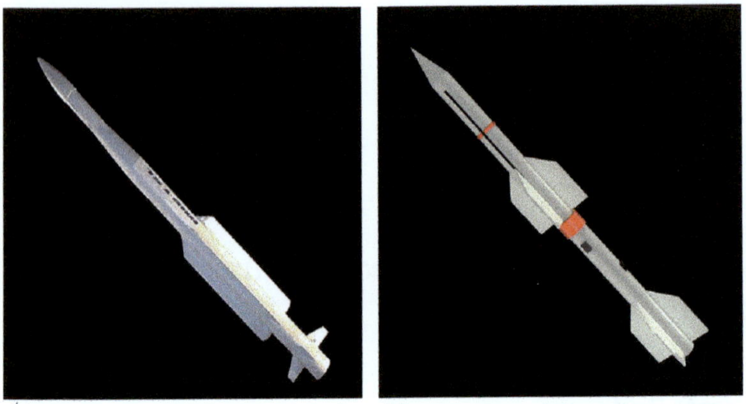

(a) (b)

Figure 7.1 Patriot air defense missile (PAC-3) (a) and Aspie air-to-air missile (b).

In order to improve the sudden defense capability of the aircraft, the multifunctional integration of wave-transparent/carrying/heat protection/stealth is the development trend of future aircraft radomes.

7.2 Radomes of Airborne, Shipborne, Ground, and Vehicle

7.2.1 Airborne Radomes

Airborne radomes (Figure 7.2), as functional composites with integrated aerodynamic/structural/wave-transparent properties [15–17], should not only ensure the airborne radar antenna system from heavy rain, strong wind, ice, snow, sand, solar radiation, and salt spray [18, 19] but also be deemed as the electromagnetic windows for receiving and transmitting antenna electromagnetic waves. At the same time, the airborne radomes, as an integral part of the aircraft shape, also play an important role in maintaining the aerodynamic shape of the whole aircraft. Therefore, the airborne radomes should present high strength and modulus, excellent heat and weather resistances, low ε and $\tan\delta$ to achieve the minimum reflection and the

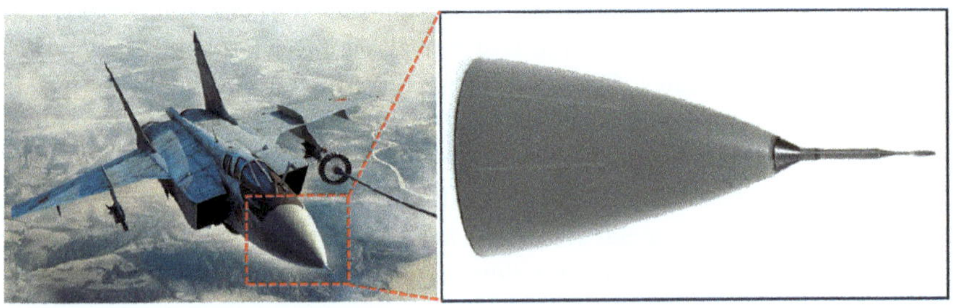

Figure 7.2 Typical airborne radomes. Credit: Courtesy of hobbyboss.

Table 7.1 The dielectric properties of the polymer matrix wave-transparent composites applied for airborne radomes.

Reinforced fibers	Polymer matrix	ε (composites)	Tan δ (composites)
E-glass fibers	Epoxy resin	4.40	0.016
	UP resin	4.15	0.015
	PI resin	4.70	0.014
D-glass fibers	CE resin	3.45	0.009
Quartz fibers	Epoxy resin	3.12	0.011
	UP resin	3.60	0.012
	PI resin	3.34	0.005
	CE resin	3.23	0.006
Aramid fibers	UP resin	3.50	0.050
High-modulus polyethylene fibers	Epoxy resin	2.80	0.004
	UP resin	2.52	0.007
	CE resin	2.65	0.003

maximum transmission of electromagnetic waves [20, 21]. The dielectric properties of the polymer matrix wave-transparent composites applied for airborne radomes are shown in Table 7.1.

The common polymer matrix for airborne radomes is PF resins, epoxy resins, UP resins, BMI resins, cyanate ester (CE) resins, silicone resins, and so on. Commonly reinforced fibers include glass fibers, quartz fibers, aramid fibers, and UHMWPE fibers. Glass fibers contain non-alkali glass fibers (E glass fibers), high-strength glass fibers (S glass fibers), high modulus glass fibers (M glass fibers), low dielectric glass fibers (D glass fibers) and high silica glass fibers [22, 23]. Among them, D glass fibers are suitable for airborne radomes with high requirements for wave-transparent performance and low requirements for mechanical properties; high silica glass fibers present the lowest ε and tan δ as well as excellent heat resistances. Quartz fibers are the most commonly reinforced fibers for high-performance airborne radomes due to the low ε and tan δ [24–26]. The first and second generations of radomes for F-15 fighter jets (USA) both adopted quartz fibers/cyanate resins wave-transparent composites as the skin of the sandwich structures. Aramid fibers are the ideal raw materials for airborne radomes due to low ε and tan δ and excellent impact strength [27–29]. UHMWPE fibers present excellent mechanical properties, low ε (3.0) and tan δ (10^{-4}), and have been applied to high-performance airborne radomes.

7.2.2 Shipboard Radomes

Shipboard radomes (Figure 7.3a) not only protect the radar antennas from the erosion of the natural environment, such as ocean waves, salt spray, rainstorms, and hail but also ensure the efficient transmission of the radar antenna signal to protect the ship's information security. Therefore, the shipboard radomes should possess high wave-transparent performance (low ε and $\tan\delta$), excellent mechanical properties, and heat and weather resistances [31].

At present, the cover wall structure of the large ship radomes is the C sandwich structure, composed of three layers of outer, middle, and inner skin and two core layers, as shown in Figure 7.3b [30]. The skin materials are generally made of polymer matrix wave-transparent composites, including glass fibers/epoxy resin composites and quartz fibers/cyanate resin composites. The core layer materials are usually made of low-density foams or honeycomb, and the interlayer is covered with frequency selective surface (FSS) layers. The C sandwich structures present the advantages of high strength, lightweight, and high wave-transparent performance. The type 120 shipboard radome (Figure 7.4a) fabricated by Harbin Habo Top Composites Co., Ltd. and the civilian shipboard radome designed by Leling Jingbei Rongye Composites Co., Ltd. (Figure 7.4b) were widely applied in the shipboard radomes.

7.2.3 Ground Radomes

Ground radomes (Figure 7.5) are not only applied to protect civil communication base station radar, satellite communication ground station radar, and radio

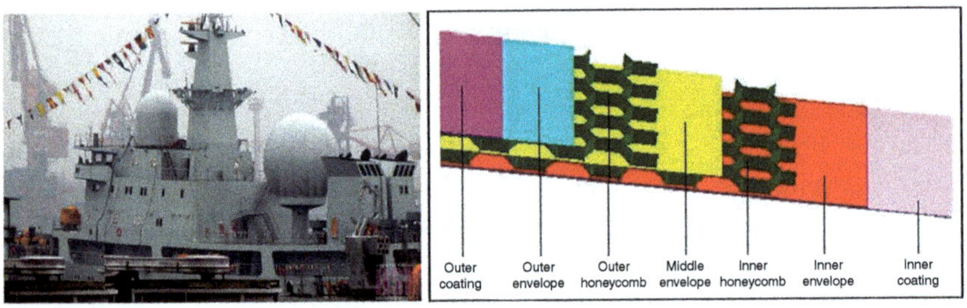

(a) (b)

Figure 7.3 Shipboard radomes (a) Credit: Sohu/https://www.sohu.com/a/193128064_114911/last accessed under Aug 22, 2023 and the wall structure (C sandwich structure, b) [30].

(a) (b)

Figure 7.4 120 shipboard radome (a) credit: Mingtang Group, and civil shipboard radome (b).

Figure 7.5 Ground radomes applied in different natural environments.

Table 7.2 Nominal radar frequency codes and bands.

Wave band	Nominal frequency range (MHz)	Wavelength range (cm)	Wave band	Nominal frequency range (MHz)	Wavelength range (cm)
P	230–1000	30.0–130.4	**X**	8000–12 000	2.5–3.75
L	1000–2000	15.0–30.0	**Ku**	12 000–18 000	1.67–2.5
S	2000–4000	7.5–15.0	**K**	18 000–26 500	1.11–1.67
C	4000–8000	3.75–7.5	**Ka**	26 500–40 000	0.75–1.11

astronomical telescope from harsh natural environments [32, 33] but also to ensure the reliability of the transmission for ground radar antenna signals transmission. Therefore, the ground radomes should present low ε and tan δ, excellent mechanical properties, and heat and weather resistances [34, 35].

Raw materials of the ground radomes generally consist of polymer matrix wave-transparent composites and interlayer materials (honeycomb or solid foam) [36]. The epoxy resins and UP resins with low cost are performed as polymer matrix, and the reinforced fibers mainly include various types of glass fibers, quartz, and aramid fibers. The ground radomes developed by Jinfa Technology Co., Ltd. (China), Lotte (Korea), and DuPon (USA) were widely applied in satellite ground stations, a meteorological radar station, and air traffic control radar station, which were suitable for the bands for L, S, C, X, Ku, Ka, and so on (Table 7.2) [37, 38].

7.2.4 Vehicle-Mounted Radomes

The vehicle radar antenna is usually placed in the open-air environment on the roof of the vehicle, which is susceptible to the interference of the external environment. Therefore, the vehicle-mounted radomes (Figure 7.6) usually present high wave-transparent performance, excellent mechanical properties, and environmental resistance [18, 39]. The vehicle radar radomes generally adopt a streamlined shape similar to that of the vehicle body, and the raw materials usually adopt glass fibers/epoxy or glass fibers/PTFE wave-transparent composites as the main structures and foams or honeycombs as the core materials.

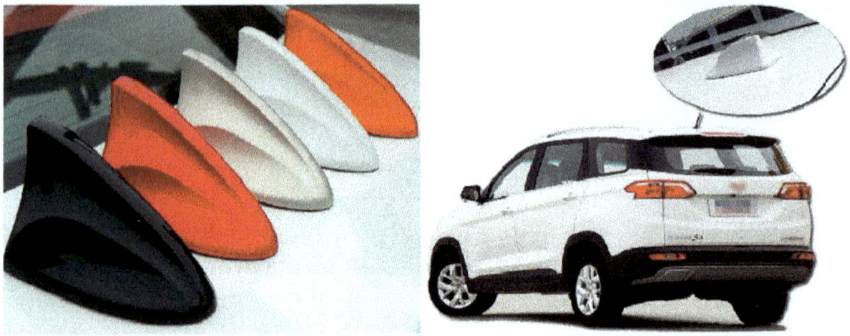

Figure 7.6 Typical images of the vehicle-mounted radomes. credit: 1688.com.

7.3 5G Communication Radomes

Currently, two 5G frequency bands delineated by the 3rd Generation Partnership Project (3GPP) are 450 to 6000 MHz and 24 250 to 52 600 MHz, respectively. The frequency bands of 5G electromagnetic waves allocated by China Telecom, China Mobile, and China Unicom are 3400–3500 MHz, 2515–2675 MHz, 4800–4900 MHz, and 3500–3600 MHz, respectively. With the rapid development of 5G communication technology, 5G electromagnetic waves present short wavelengths and transmission distances. Therefore, more 5G base stations need to be built to realize the comprehensive coverage of 5G signal, and the demand for 5G communication radomes has also increased substantially [40–42].

The role of the 5G communication radomes (Figure 7.7) is to protect the 5G antenna from external environmental interference and ensure the efficient transmission of the 5G signal. Therefore, the 5G communication radomes should present low ε and $\tan\delta$, excellent mechanical properties, and weather resistance [43, 44] (such as resistance to thermal oxygen aging, moisture, salt spray, and mold).

UP resins, epoxy resins, and PTFE resins are commonly used as polymer matrix in the 5G communication radomes, and glass fibers or quartz fibers as reinforcement. With the rapid development of the 5G antenna system, higher requirements are also

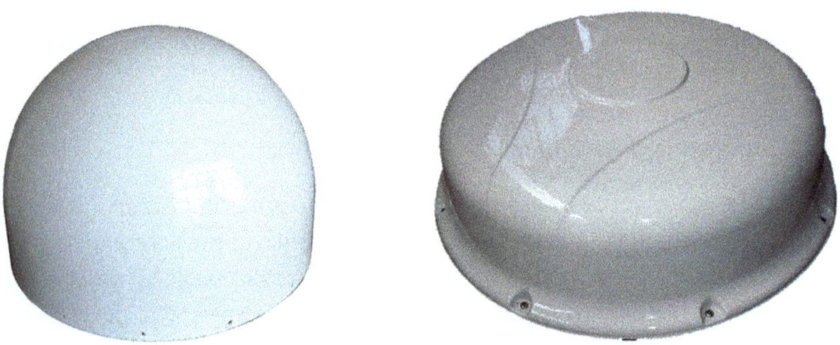

(a) (b)

Figure 7.7 The images of 5G communication radomes: hemispherical (a) and cylindrical (b).

put forward for the wave-transparent performance of radar radomes. Polypropylene (PP), polycarbonate (PC), and polydicyclic pentadiene (PDBCD) with low ε and $\tan\delta$ are gradually applied to 5G communication radomes [45].

Glass fibers/PP resins wave-transparent composites produced by Jinfa Technology and the ultra-strong glass fibers/PP resins wave-transparent composites produced by Huawei were both used for 5G communication radomes, about 40% lighter than that of traditional glass fibers reinforced plastic (FRP). PC/ABS composites developed by Tongyi Shares presented low ε and $\tan\delta$, excellent thermal conductivity, and weather and low-temperature resistances, which could be applied in harsh environments.

At present, the electromagnetic waves of the 5G communication are mostly millimeter waves with poor penetration [46]. Therefore, the 5G signals fade more severely with distance than the 4G signals. In order to achieve full coverage of 5G signals in the region, it is necessary to build far more radar antenna base stations to transmit the signals. At the same time, the demand for 5G communication radomes also increases tremendously. Developing polymer matrix wave-transparent composites with high specific strength/modulus, environmentally friendly, lightweight, low cost, low ε and $\tan\delta$ for 5G communication radomes is the common goal and development direction for domestic and foreign companies and researchers [47].

7.4 Printed Circuit Board

Print circuit board (PCB), as the support body of electronic components and the carrier for transmitting signals of the electronic components (Figure 7.8), which is widely used in the fields of aerospace/aerospace, medical equipment, communication services, computers, and civil industries (Figure 7.9).

As a key component of the integrated circuit, the PCB plays a vital role in supporting the interconnection of electronic components and ensuring their excellent connection and data transmission. PCB should present low ε and $\tan\delta$ to obtain high

(a) (b)

Figure 7.8 PCB (a) and its applications in electronic components credit: Shenzhen Dayu Industrial Co., Ltd. (b). Source: Jakkapan/Adobe Stock.

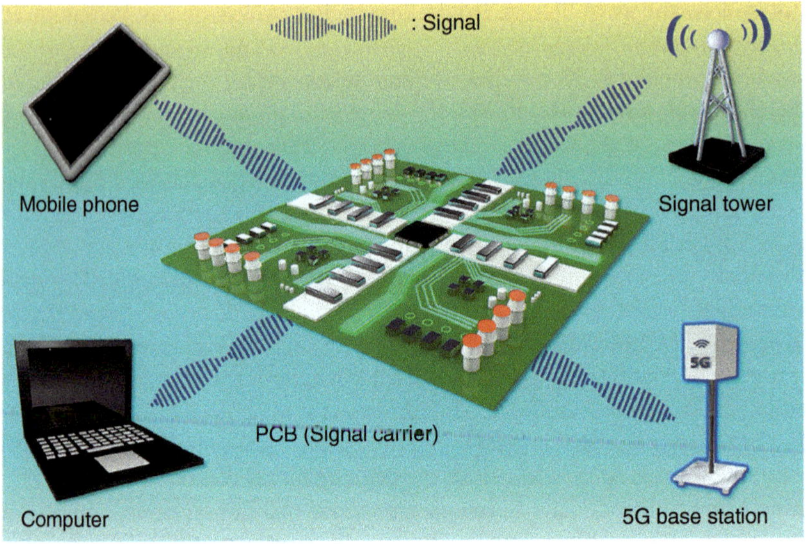

Figure 7.9 Schematic diagram of signal transmission of PCB in electronic equipment.

characteristic impedance and reduce signal interference and crosstalk during transmission [48, 49]. Moreover, PCB is also required to present excellent heat resistance, mechanical properties, bonding properties, and dimensional stabilities. At present, epoxy resins, PTFE resins, polyphenyl ether (PPO) resins, PI resins, and CE resins are commonly performed as the polymer matrix for PCB [50–52], the reinforced fibers including glass fibers, quartz fibers, arylon fibers, and UHMWPE fibers.

The glass fibers/FR-4 epoxy composites fabricated by the National Electronics Manufacturers Association (NEMA) were used as the standard for PCB in the industrial field. In the 1960s, Japan began to adopt glass fibers/epoxy composites as PCBs for industrial and civilian electronic equipment. In addition, Qisheng and Daikin Industrial Co., Ltd. of Japan, TTM Technology Corporation of the United States, and Kingboard Group Co., Ltd. of China adopted CE resins to prepare PCB. With the advent of the 5G, the performances and operating speed of electronic devices are greatly improved, which puts higher requirements on the comprehensive properties of the PCB [53, 54]. Zigong Tianlong Chemical Co., Ltd. and Nanya Circuit Board Co., Ltd. developed a series of fluorine-based resins, such as PTFE and fluorine-containing PI resins, which were widely applied in the field of PCB with high integration, high speed, and high frequency.

7.5 Summary

This chapter discusses the application and development trend of the polymer matrix wave-transparent composites. Polymer matrix wave-transparent composites with high-temperature resistance, high strength, and high wave-transparent are suitable for aircraft and airborne radar radomes. For ground, shipborne, and vehicular mounted radomes and 5G communication radomes, the polymer matrix wave-transparent composites must present excellent environmental resistance and high wave-transparent performance to apply in different environments. Polymer matrix wave-transparent composites for PCB should present the advantages of low ε and $\tan\delta$ as well as excellent processability. In addition, polymer matrix wave-transparent composites are also developed in the direction of multifunctionality, in addition to high wave-transparent performance, presenting high thermal conductivity or super-hydrophobicity.

References

1 Karch, C., Heidler, F., and Paul, C. (2021). Protection of aircraft radomes against direct lightning strikes—an overview. *Atmosphere* 12 (9): 1141.

2 Lysov, N., Temnikov, A., Chernensky, L. et al. (2021). A laboratory investigation of the probable mechanisms of the action of an artificial thunderstorm cell on model aircraft radomes. *Atmosphere* 12 (12): 1637.

3 Temnikov, A.G., Chernenskii, L.L., Orlov, A.V., and Antonenko, S.S. (2011). Breakdown of model aircraft radome dielectric shell in artificial charged aerosol clouds. *Technical Physics Letters* 37 (10): 907–910.

4 Temnikov, A., Chernensky, L., Belova, O. et al. (2022). Influence of kind of lightning stripe models on spectral characteristics of discharge phenomena inside aircraft nose radome model. *Journal of Electrostatics* 115: 103661.

5 Temnikov, A.G., Gilyazov, M.Z., Matveev, D.A. et al. (2011). Studying electromagnetic interference spectrum in antenna under aircraft radome using models with artificial charged aerosol clouds. *Technical Physics Letters* 37 (9): 845–848.

6 Shin, H.-J., Park, J.-Y., Hong, S.-C., and Lee, J.-R. (2017). In situ non-destructive evaluation of an aircraft UHF antenna radome based on pulse-echo ultrasonic propagation imaging. *Composite Structures* 160: 16–22.

7 Li, J., Wang, Y., Zhao, W. et al. (2022). High-performance quartz fiber/polysilazane and epoxy-modified cyanate ester microwave-transparent composites. *Advanced Composites and Hybrid Materials* 5 (3): 1830–1840.

8 Kim, D.-C., Choi, J.S., Shin, H.-S. et al. (2022). Effect of the manufacturing process on the equivalency qualification of glass fiber reinforced polymer. *Science and Engineering of Composite Materials* 29 (1): 403–411.

9 Elmahdy, A. and Verleysen, P. (2020). Mechanical behavior of basalt and glass textile composites at high strain rates: A comparison. *Polymer Testing* 81: 106224.

10 Gul, N. and Genc, O.B. (2022). Investigation of electrical and electromagnetic properties of quartz fiber reinforced polymer composite material by using modified paints with carbon nanoparticles (graphene/double-walled carbon nanotube). *Journal of Composite Materials* 56 (19): 3013–3027.

11 Kemkemian, S., Nouvel-Fiani, M., and Chamouard, E. (2011). Radar and electronic warfare cooperation: how to improve the system efficiency. *IEEE Aerospace and Electronic Systems Magazine* 26 (8): 32–38.

12 Kim, J., Park, C., and Hahn, J.W. (2022). Metal–semiconductor–metal metasurface for multiband infrared stealth technology using camouflage color pattern in visible range. *Advanced Optical Materials* 10 (6): 2101930.

13 Hu, J., Hu, Y., Ye, Y., and Shen, R. (2023). Unique applications of carbon materials in infrared stealth: A review. *Chemical Engineering Journal* 452: 139147.

14 Ahmad, H., Tariq, A., Shehzad, A. et al. (2019). Stealth technology: Methods and composite materials—A review. *Polymer Composites* 40 (12): 4457–4472.

15 Nair, R.U. and Jha, R.M. (2014). Electromagnetic design and performance analysis of airborne radomes: trends and perspectives [antenna applications corner]. *IEEE Antennas and Propagation Magazine* 56 (4): 276–298.

16 Yazeen, P.S.M., Vinisha, C.V., Vandana, S. et al. (2017). Electromagnetic performance analysis of graded dielectric inhomogeneous streamlined airborne radome. *IEEE Transactions on Antennas and Propagation* 65 (5): 2718–2723.

17 Xu, W., Duan, B.Y., Li, P., and Qiu, Y. (2017). Study on the electromagnetic performance of inhomogeneous radomes for airborne applications—Part II: the overall comparison with variable thickness radomes. *IEEE Transactions on Antennas and Propagation* 65 (6): 3175–3183.

18 Parameswaran, A., Sonalikar, H.S., and Kundu, D. (2021). Temperature-dependent electromagnetic design of inhomogeneous planar layer variable thickness radome for power transmission enhancement. *IEEE Antennas and Wireless Propagation Letters* 20 (8): 1572–1576.

19 Zhou, L., Wang, P., Pei, Y. et al. (2017). Design and characterization for dual-band and multi-band A-sandwich composite radome walls. *Composites Science and Technology* 149: 28–33.

20 Xu, W., Li, P., and Qiu, Y. (2019). Electromagnetic performance analysis of inhomogeneous airborne radomes for circular polarization applications. *IEEE Antennas and Wireless Propagation Letters* 18 (1): 74–78.

21 Vinisha, C.V., Mohammed Yazeen, P.S., Mahima, P. et al. (2017). Multi-layered graded porous radome design for dual-band airborne radar applications. *Electronics Letters* 53 (3): 189–191.

22 Choi, I., Kim, J.G., Seo, I.S., and Lee, D.G. (2012). Design of the hybrid composite face with electromagnetic wave transmission characteristics of low-observable radomes. *Composite Structures* 94 (11): 3394–3400.

23 Waqas H M, Shi D, Imran M, Khan S Z, Tong L, Ahad F E, Zaidi A A, Iqbal J, Ahmed W. Conceptual design of composite sandwich structure submarine radome. Materials2019;12(12):1966.

24 Liu, Z., Zhao, W., Yu, G., and Wu, L. (2021). Fabrication and mechanical behaviors of quartz fiber composite honeycomb with extremely low permittivity. *Composite Structures* 271: 114129.

25 Grace, L.R. (2015). The effect of moisture contamination on the relative permittivity of polymeric composite radar-protecting structures at X-band. *Composite Structures* 128: 305–312.

26 Lin, B., Wang, H., Wei, J., and Sui, T. (2021). Diamond wheel grinding characteristics of 3D orthogonal quartz fiber reinforced silica ceramic matrix composite. *Chinese Journal of Aeronautics* 34 (5): 404–414.

27 Choi, I., Kim, J.G., Lee, D.G., and Seo, I.S. (2011). Aramid/epoxy composites sandwich structures for low-observable radomes. *Composites Science and Technology* 71 (14): 1632–1638.

28 Xu, X., Zhang, B., Xing, L. et al. (2018). Characterization and analysis of the thermal damages of aramid/epoxy composite laminates induced by the dielectric heating effect of microwaves. *Composite Structures* 200: 371–379.

29 Nag, A., Rao, R.R., and Panda, P.K. (2021). High temperature ceramic radomes (HTCR) – A review. *Ceramics International* 47 (15): 20793–20806.

30 Shu, Y. (2020). Status and development trend of shipborne stealth radome. *Modern Radar* 42 (11): 10–14.

31 Lee, J.-E., Kwak, Y.-C., Jeong, E.-B. et al. (2022). Dynamic response analysis and verification of shipboard structure using composite materials and resilient mounts. *International Journal of Precision Engineering and Manufacturing* 23 (3): 333–345.

32 Liu, J., Li, L., Zuo, Y. et al. (2019). Analysis of performance degradation introduced by radome for high-precision GNSS antenna. *International Journal of Antennas and Propagation* 2019: 1529656.

33 Zhang, F.-s., Xu, J.-z., and Zu, L. (2017). The buckling behavior of radome with different braided angles under wind loading based on CFRP. *Composite Structures* 180: 904–913.

34 Marchetti, M., Arenare, D., Concaro, F., and Pasian, M. (2023). On de-pointing effects caused by snow accumulation on radomes for ground stations working at K-band: an analysis of the SNOWBEAR project. *IEEE Antennas and Propagation Magazine* 65 (2): 76–91.

35 Li, P., Xu, W.Y., and Song, L.W. (2016). A novel compensation strategy for the radiation characteristics of large dielectric radomes based on phase modification. *IEEE Antennas and Wireless Propagation Letters* 15: 1044–1047.

36 Chia, C.C., Lee, J.-R., and Park, C.Y. (2012). Radome health management based on synthesized impact detection, laser ultrasonic spectral imaging, and wavelet-transformed ultrasonic propagation imaging methods. *Composites Part B: Engineering* 43 (8): 2898–2906.

37 Oyan, M.J., Hamran, S.E., Hanssen, L. et al. (2012). Ultrawideband gated step frequency ground-penetrating radar. *IEEE Transactions on Geoscience and Remote Sensing* 50 (1): 212–220.

38 Yan, J.B., Alvestegui, D.G.-G., McDaniel, J.W. et al. (2017). Ultrawideband FMCW radar for airborne measurements of snow over sea ice and land. *IEEE Transactions on Geoscience and Remote Sensing* 55 (2): 834–843.

39 Poyatos-Martínez, D., Escot-Bocanegra, D., De Diego-Custodio, E. et al. (2011). Application of a spherical multi-purpose facility to the selection of the appropriate radome for an on-board POD antenna. *Journal of Electromagnetic Waves and Applications* 25 (8-9): 1243–1252.

40 Tan, H., Li, W., Wang, T. et al. (2015). The analysis on the candidate frequency bands of future mobile communication systems. *China Communications* 12 (Supplement): 140–149.

41 Zhang, J., Tang, P., Tian, L. et al. (2017). 6–100 GHz research progress and challenges from a channel perspective for fifth generation (5G) and future wireless communication. *Science China Information Sciences* 60 (8): 080301.

42 Karthikeya, G.S., Koul, S.K., Poddar, A.K., and Rohde, U. (2020). Path loss compensated millimeter wave antenna module integrated with 3D-printed radome. *International Journal of RF and Microwave Computer-Aided Engineering* 30 (10): e22347.

43 Huang, J., Cao, Y., Raimundo, X. et al. (2019). Rain statistics investigation and rain attenuation modeling for millimeter wave short-range fixed links. *IEEE Access* 7: 156110–156120.

44 Ha, T.D., Zhu, L., Alsaab, N. et al. (2023). Optically transparent metasurface radome for RCS reduction and gain enhancement of multifunctional antennas. *IEEE Transactions on Antennas and Propagation* 71 (1): 67–77.

45 Wang, J., Li, Y., Wang, J. et al. (2021). A low-profile vertically polarized magneto-electric monopole antenna with a 60% bandwidth for millimeter-wave applications. *IEEE Transactions on Antennas and Propagation* 69 (1): 3–13.

46 Nakamura, K., Iwasawa, N., Kawasaki, K. et al. (2020). The attenuation characteristics of millimeter-wave by snow accretion. *IEICE Communications Express* 9 (12): 674–678.

47 Wei, J., Zhang, J., Cao, X. et al. (2023). Durable superhydrophobic coatings for prevention of rain attenuation of 5G/weather radomes. *Nature Communications* 14 (1): 2862.

48 Zhang, B., Ye, F., Gao, Y. et al. (2016). Fabrication and dielectric properties of BADCy/$Ni_{0.5}Ti_{0.5}NbO_4$ composites for ultra-low-loss printed circuit board application. *Ceramics International* 42 (1): 234–241.

49 Spencer, T.J., Saha, R., Chen, J. et al. (2012). Air cavity transmission lines for off-chip interconnects characterized to 40 GHz. *IEEE Transactions on Components, Packaging and Manufacturing Technology* 2 (3): 367–374.

50 Deutsch, A., Surovic, C.W., Lanzetta, A.P. et al. (1996). Broadband characterization of low dielectric constant and low dielectric loss CYTUF cyanate ester printed circuit board material. *IEEE Transactions on Components, Packaging, and Manufacturing Technology: Part B* 19 (2): 331–337.

51 Imai, H., Sugimura, M., Kawasaki, M. et al. (2009). High-frequency propagation on printed circuit board using a material with a low dielectric constant, a low dielectric loss, and a flat surface. *IEEE Transactions on Components and Packaging Technologies* 32 (2): 415–423.

52 Abdelmalik, A.A., Borge, M.D., Nysveen, A. et al. (2016). Statistical analysis of dielectric breakdown of liquid insulated printed circuit boards. *IEEE Transactions on Dielectrics and Electrical Insulation* 23 (4): 2303–2310.

53 Lee, P., Jung, H., Yoo, C.-S., and Lee, H.H. (2023). Low dielectric constant characteristics of styrene and maleimide anhydride copolymer with modification for high frequency application of printed circuit board. *Polymers* 15 (9): 2078.

54 Wu, X., Xu, C.-a., Lu, M. et al. (2022). Preparation and characterization of high temperature resistant thermosetting polyphenylene ether resin. *Journal of Applied Polymer Science* 139 (39): e52858.

Index

Polymer Matrix Wave-Transparent Composites: Materials, Properties, and Applications, First Edition.
Junwei Gu, Yusheng Tang, Jie Kong, and Jing Dang.